AF745607

Secondary Metabolites - Trends and Reviews

Edited by Ramasamy Vijayakumar and Suresh Selvapuram Sudalaimuthu Raja

Published in London, United Kingdom

Secondary Metabolites - Trends and Reviews
http://dx.doi.org/10.5772/intechopen.98129
Edited by Ramasamy Vijayakumar and Suresh Selvapuram Sudalaimuthu Raja

Contributors
Ngoc-Van Thi Nguyen, Muhammad Zeeshan Bhatti, Hammad Ismail, Waqas Khan Kayani, Hudu Garba Mikail, Mamman Mohammed, Habib Danmalam Umar, Mohammed Musa Suleiman, Stella Omokhefe Bruce, Grazia Maria Scarpa, Vanda Prota, Nicola Schianchi, Federica Manunta, Ahmed M. Mohammed Shuikan, Ibrahim A. Arif, Wael N. Nabil Hozzein, Rakan M. Alshuwaykan, Ziming Yue, Te Ming Tseng, Varsha Singh, Josiane Argenta, Worlanyo Segbefia, Alyssa Miller, Julia Pérez-Ramos, Ana Laura Esquivel-Campos, Salud Pérez-Gutiérrez, Leonor Sánchez-Pérez, Nimsi Campos-Xolalpa, Ayush Mandwal, Kamlesh Prasad, Ravneet Kaur, Shubhra Shekhar, Ananth Anbu, Umadevi Ananth, Fernando Ramos, Frank L. Romero-Orejon, Ana María Muñoz, Luciana de la Fuente-Carmelino, Diana Jimenez-Champi, Eliana Contreras-López, Ivan Best, Luís Aguilar, Galina Satchanska, Ramasamy Vijayakumar, Suresh Selvapuram Sudalaimuthu Raja, Girish Nair

© The Editor(s) and the Author(s) 2022
The rights of the editor(s) and the author(s) have been asserted in accordance with the Copyright, Designs and Patents Act 1988. All rights to the book as a whole are reserved by INTECHOPEN LIMITED. The book as a whole (compilation) cannot be reproduced, distributed or used for commercial or non-commercial purposes without INTECHOPEN LIMITED's written permission. Enquiries concerning the use of the book should be directed to INTECHOPEN LIMITED rights and permissions department (permissions@intechopen.com).
Violations are liable to prosecution under the governing Copyright Law.

Individual chapters of this publication are distributed under the terms of the Creative Commons Attribution 3.0 Unported License which permits commercial use, distribution and reproduction of the individual chapters, provided the original author(s) and source publication are appropriately acknowledged. If so indicated, certain images may not be included under the Creative Commons license. In such cases users will need to obtain permission from the license holder to reproduce the material. More details and guidelines concerning content reuse and adaptation can be found at http://www.intechopen.com/copyright-policy.html.

Notice
Statements and opinions expressed in the chapters are these of the individual contributors and not necessarily those of the editors or publisher. No responsibility is accepted for the accuracy of information contained in the published chapters. The publisher assumes no responsibility for any damage or injury to persons or property arising out of the use of any materials, instructions, methods or ideas contained in the book.

First published in London, United Kingdom, 2022 by IntechOpen
IntechOpen is the global imprint of INTECHOPEN LIMITED, registered in England and Wales, registration number: 11086078, 5 Princes Gate Court, London, SW7 2QJ, United Kingdom

British Library Cataloguing-in-Publication Data
A catalogue record for this book is available from the British Library

Additional hard and PDF copies can be obtained from orders@intechopen.com

Secondary Metabolites - Trends and Reviews
Edited by Ramasamy Vijayakumar and Suresh Selvapuram Sudalaimuthu Raja
p. cm.
Print ISBN 978-1-80355-207-1
Online ISBN 978-1-80355-208-8
eBook (PDF) ISBN 978-1-80355-209-5

We are IntechOpen, the world's leading publisher of Open Access books

Built by scientists, for scientists

5,900+
Open access books available

145,000+
International authors and editors

180M+
Downloads

156
Countries delivered to

Our authors are among the
Top 1%
most cited scientists

12.2%
Contributors from top 500 universities

Clarivate Analytics
BOOK CITATION INDEX
INDEXED

WEB OF SCIENCE™

Selection of our books indexed in the Book Citation Index in Web of Science™ Core Collection (BKCI)

Interested in publishing with us?
Contact book.department@intechopen.com

Numbers displayed above are based on latest data collected.
For more information visit www.intechopen.com

For EU product safety concerns:
IN TECH d.o.o., Prolaz Marije Krucifikse Kozulić 3, 51000 Rijeka, Croatia,
info@intechopen.com or visit our website at intechopen.com.

Meet the editors

Dr. Ramasamy Vijayakumar is an assistant professor in the Department of Microbiology, Government Arts and Science College (formerly Bharathidasan University Constituent College), affiliated with Bharathidasan University, Perambalur, India. He has sixteen years of teaching experience as well as twenty years of research experience in the field of actinobacteriology. He has deposited around twenty-one nucleotide sequences in GenBank; published seventy-eight research articles, twelve book chapters, and two books; guided six Ph.D. candidates; and organized several seminars and workshops. He has received research grants from the DBT, TNSCST, and TANSCHE, Government of India. He filed one Indian patent. He is a life member of the National Academy of Biological Sciences, India, Mycological Society of India, and Microbiologist's Society, India. He is a member of the doctoral committee for Ph.D. scholars and a member of the board of studies for microbiology in various institutions. He received the 2013 Shri P.K. Das Memorial Best Faculty Award in Microbiology from the Nehru Arts and Science College, Coimbatore – India. As per the Indian Journal of Experimental Biology (Vol. 51, 2013), Dr. Vijayakumar was ranked second in the top five institutions in the field of actinobacterial research in India.

Dr. Suresh S.S. Raja is an assistant professor and head of the Department of Microbiology, Government Arts and Science College (formerly Bharathidasan University Constituent College), affiliated with Bharathidasan University, Perambalur, India. He has been teaching Microbiology both at undergraduate and post-graduate levels for the last twenty-two years. He has published twenty-five research articles in national and international journals and has attended numerous seminars, symposia, and workshops. He has documented seroprevalence of leptospirosis among farming communities in three districts of Tamil Nadu, South India. Medical microbiology remains his field of specialization. He has supervised three student projects funded by the state science academy. He has organized several seminars and training workshops. He is a member of the board of studies of various colleges and one university. He had been a member of a university inspection committee for approving Microbiology courses at colleges. He has delivered special lectures at academic institutions and extension programmes and is a member of professional bodies and voluntary organizations.

Contents

Preface

Metabolism is the set of life-sustaining biochemical reactions in organisms. Metabolites are the intermediary products of metabolism. Metabolic products have various functions, including fuel, structure, signaling, stimulatory and inhibitory effects on enzymes, the catalytic activity of their own, defense, and interactions with other organisms. While a primary metabolite is far more essential and directly involved in normal growth, development, and reproduction, a secondary metabolite, though not directly involved in those processes, usually has an important ecological function and provides survival benefit to the organism. All life forms, namely, microorganisms, plants, and animals, are involved in the production of secondary metabolites. In-depth exploration of herbal plants, animals, and microorganisms such as bacteria, actinobacteria, cyanobacteria, fungi, and algae has led to the discovery of novel secondary metabolites. Conventional procedures with the required biotechnological intervention will introduce novel secondary metabolic products with high pharmaceutical, agricultural, industrial, and environmental values.

This book consists of an introductory overview of secondary metabolites, followed by two main sections: "Secondary Metabolites: General Reviews and Biotechnological Interventions" and "Plant Secondary Metabolites." It includes thirteen chapters, six of which discuss biotechnological interventions in the production and research of secondary metabolites, and seven of which provide a comprehensive account of the secondary metabolites of plants. Chapters are contributed by authors from countries around the world, including Bulgaria, Canada, India, Italy, México, Nigeria, Pakistan, Perú, Saudi Arabia, Vietnam, and the United States.

The book is a useful resource for microbiologists, biotechnologists, biochemists, pharmacologists, and botanists. Students at all levels, scholars, scientists, and faculty members of various science disciplines will also find this book a valuable tool. We are thankful to all the contributors for the submission of their valuable work. We offer our special thanks and appreciation to Author Service Manager Ms. Marina Dusevic at IntechOpen for her encouragement and help throughout the publication process. We are also indebted to our colleagues and the management at Government Arts and Science College, Bharathidasan University, India.

Ramasamy Vijayakumar and Suresh Selvapuram Sudalaimuthu Raja
Department of Microbiology,
Government Arts and Science College,
(Formerly Bharathidasan University Constituent College),
Affiliated to Bharathidasan University,
Perambalur, Tamilnadu, India

Section 1

Introduction

Chapter 1

Introductory Chapter: Secondary Metabolites - An Overview

Girish Nair, Suresh Selvapuram Sudalaimuthu Raja and Ramasamy Vijayakumar

1. Introduction

The metabolism can be defined as the collection of all the biochemical reactions held in an organism. Metabolites are the intermediate products of metabolism. Metabolites have various functions, including fuel, structure, and signaling, stimulatory and inhibitory effects on enzymes, catalytic activity of their own, defense, and interactions with other organisms. A primary metabolite is directly involved in normal growth, development, and reproduction of the host cell. A secondary metabolite is not directly involved in those processes, but usually has an important ecological function [1]. Secondary metabolites are biochemical compounds with varied and sophisticated chemical structures, produced by strains of certain microbial, animal and by plant species. Products of secondary metabolism are that the metabolites are usually not produced during the phase of rapid growth (trophophase), but are synthesized during a subsequent production stage (idiophase). Herbal plants, animals, and microorganisms such as bacteria, actinobacteria, cyanobacteria, fungi, and algae attracted more attention in research that led to the discovery of secondary metabolites. The exploration of secondary metabolites from various resources subsequently led to the development of drugs for the treatment of human diseases of microbial origin. Routine screening of natural resources will introduce novel secondary metabolic products with high pharmaceutical value [2].

Secondary metabolites, complex group of natural metabolic products, serve as defense chemicals, quorum sensing metabolites in environmental interactions, symbiosis, transport of metals and solutes. And in doing so, they confer selective advantage of survival over competitors, though their absence does not compromise their vegetative growth [3]. Twenty-five percent of about 1 million natural secondary metabolites are biologically active. Plants contribute 60% of these metabolites, and microorganisms form the rest, among which fungi remain major (42%) producers of bioactive compounds [4, 5].

2. Taxonomy of secondary metabolites

There are five main classes of secondary metabolites such as terpenoids and steroides, fatty-acid-derived substances and polyketides, alkaloids, nonribosomal polypeptides, and enzyme cofactors [6].

IntechOpen

2.1 Terpenoids and steroides

Terpenes are the polymers of five carbon isoprene units and considered as the biggest class of secondary metabolites. When terpenes get modified by different functional groups and oxidized methyl groups at various positions, they form terpenoids. Depending on the carbon units, terpenoids can be divided into monoterpenes, sesquiterpenes, diterpenes, sesterpenes, and triterpenes. They find use in the anti-cancer treatment, as fragrance agent in cosmetics, and as food flavoring agent [7]. Steroids are diverse class of secondary metabolites and play an important physiological and biochemical function in the living organisms in which they are found. They are lipophilic, low molecular weight and are derived from cholesterol, the family of steroids includes sterols, bile acids, a number of hormones (both gonadal and adrenal cortex hormones), and some hydrocarbons. A number of synthetic steroids are being extensively used as anti-hormones, contraceptive drugs, anticancer agents, cardiovascular agents, osteoporosis drugs, antibiotics, anesthetics, anti-inflammatories, and anti-asthmatics. Many plant-derived sterols known as phytosterols are also used as dietary supplement as they are able to lower cholesterol in human body and prevent cancer [8].

2.2 Fatty-acid-derived substances and polyketides

A fatty acid is the carboxylic acid with aliphatic chain and is a form of energy reserve in the body called fats. Derivatives of fatty acid have a wide variety of industrial application such as plastics, lubricants, and fuels; they include hydroxy fatty acids, fatty alcohols, fatty acid methyl/ethyl esters, and fatty alkanes [9]. Polyketides (PKs) are produced by the action of polyketide synthases (PKSs) in animals, plants, fungi, and bacteria. These biologically active secondary metabolites display a high structural diversity and find many applications in treatment of various acute and chronic diseases. Examples include antibacterial (erythromycin and tetracycline), antitumor (anthracycline and doxorubicin), antifungal (amphotericin and griseofulvin), antiparasitic (avermectin), and anti-cholesterol (lovastatin) drugs. The acetyl tranferease, ketosynthase, thioesterase, and other such domains constitute polyketides. Linkage of acyl-coenzyme A (CoA) on the acyl carrier protein (ACP) facilitates biosynthesis of polyketides with catalytic support from AT domain [10].

2.3 Alkaloids

Plants are regarded as the oldest source of this natural occurring structurally diverse bioactive secondary metabolite. Some of the most widely recognized alkaloids, such as morphine, quinine, strychnine, and cocaine, are derived from plants. Alkaloids are small organic molecules containing nitrogen usually in a ring. In plants, they are mainly involved in defense against herbivores and pathogens. Rapid advances in molecular biology and metabolic engineering have led to discovery and synthesis of alkaloids also from microbes. Alkaloids can be classified according to their molecular weight, such as the indole alkaloids and isoquinoline alkaloids (each more than 4000 compounds). Other important groups include tropane alkaloids (~300 compounds), steroidal alkaloids (~450 compounds), and pyridine and pyrrolizidine alkaloids (respectively, ~250 and 570 compounds) [11].

2.4 Nonribosomal polypeptides

They come under the class of peptide secondary metabolites produced by microorganisms such as bacteria, cyanobacteria, fungi, and symbionts of higher eukaryotes. The nonribosomal peptides (NRPs) are synthesized by multidomain mega-enzymes named nonribosomal peptide synthetases (NRPSs), without the need for the cell ribosomal machinery and messenger RNAs. Their bioactivity and pharmaceutical properties can be evidenced by antibiotics (e.g., actinomycin, penicillin, cephalosporin, vancomycin), cytotoxics (e.g., bleomycin), and immunosuppressants (e.g., cyclosporines), which have found immense importance in the clinical industry [12].

2.5 Enzyme cofactors

The analysis of the cofactors is imperative in order to gain the understanding of the enzyme catalyzed reactions. Enzymes are proteins that catalyze vast repertoire of reactions found in nature. Generally the enzymes are composed 20 amino acid residues, but some may also require additional small molecules in the active site for the catalysis reaction to occur, these small molecules are known as cofactors. The cofactor can be a metal ion (e.g., $Fe+$) or small organic molecule [13].

3. Functions of secondary metabolites

3.1 Secondary metabolites as competitive weapons

The mechanism of natural defense has been evolved in microorganisms, and they achieve this by secretion of secondary metabolites. The best example could be the antibiotics, which can kill or inhibit the growth of competing microorganisms. Studies confirm that antibiotics are also involved in germination by stimulating spore formation, which can inhibit or stimulate spore formation. Formation of secondary metabolites and spores is regulated by similar factors. Thus the secondary metabolite slows down germination of spores until a less competitive environment and more favorable conditions for growth exist. It protects the dormant or initiated spore from consumption by amoebae and cleans the immediate environment of competing microorganisms during germination [1].

3.2 Secondary metabolites as metal transporting agents

Secondary metabolites act as metal precipitating or chelating agents in plants as the high bioaccumulation of the toxic trace metals can lead to abiotic stress that can cause oxidative damage to plant cells. Metal precipitation is achieved by low-molecular-weight compounds such as phenolics, amino acids, organic acids, and sugars as well as high-molecular-weight compounds such as mucilage and proteins in plants. In the rhizosphere or apoplastic space, the metals are excluded through chelation so as to avoid their entry into symplast. An example could be of the siderophores, which have high affinity for iron (Fe) and could solubilize ferric iron [14].

3.3 Secondary metabolites as agents for symbiotic relation with other organisms

In symbiotic relationship both the organisms are benefited from each other. The symbiotic association between soil fungi and roots is known as mycorrhizae.

Mycorrhizal roots can absorb much more phosphate than roots that have no symbiotic relationship with fungi. The fungi in turn protect the plant from damage by pathogens such as nematodes, *Fusarium*, *Pythium*, and *Phytophthera* by often using secondary metabolites such as antibiotics. Another example where the secondary metabolites mediate the symbiotic relationship is bacteria *Pseudomonas*, which act as plant growth-promoting bacteria, by colonizing the roots and producing antibiotics that limit the growth of other pathogenic bacteria as well as fungi [3].

3.4 Secondary metabolites as reproductive agent

Well-known sex hormones produced by fungi are trisporic acids, which are secondary metabolites of *Mucorales*. The trisporic acid was found in 1964 that caused enhanced carotene production in *Blakeslea trispora*. This was later shown to be the hormone that brought about zygophore production in *Mucor mucedo*. Zygophores (sexual hypae) are produced when vegetative hyphae of the two mating types of these heterothallic organisms approach one another. Trisporic acids are produced from mevalonic acid in a secondary metabolic pathway of which the early steps are present in both (+) and (−) sexes. Since distinct late steps are absent in these sexes, both strains must meet and come in contact in order to complete the pathway that forms trisporic acid [15]. Similarly, a secondary metabolite, sirenin, is also involved in sexual reproduction in *Allomyces*, a phycomycete by working as a chemotaxic hormone that brings together the female and male gametes [16].

3.5 Secondary metabolites as differentiation effectors

Differentiation occurs during the development of an organism, which can be a morphological change or chemical change. Secondary metabolites released during this time bring about differentiation. Sporulation, which is the process of formation of spores from vegetative phase, is connected with production of antibiotics. This is supported by several evidences such as antibiotic production by all sporulating microorganisms, sporulation and antibiotic synthesis are induced by depletion of some essential nutrient, genetic links between the synthesis of antibiotics and the formation of spores and antibiotics are frequently inhibitory to vegetative growth of their producers at concentrations produced during sporulation [3].

3.6 Secondary metabolites as agents of communication between organisms

Cell-to-cell communication has been hypothesized to evolve first in the unicellular organisms long before the appearance of specialized (glands, neurons, immune cells, blood cells) cells. In microorganisms small secondary metabolites act as informational cues to regulate gene expression. Homoserine lactones (HLs) are synthesized from S-adenosylmethionine by many Gram-negative bacteria that diffuse and regulate their population density. HLs function in *Psuedomonas aeruginosa*, an opportunistic pathogen responsible for many hospital acquired infection. It uses two HL signaling systems, which combined to regulate over 300 genes. HL signaling could bring about drastic changes in gene expression affecting secondary metabolism, the elaboration of virulence factors, sporulation, and biofilm formation. Similar to this is *Vibrio cholerae* that uses autoinducers such as CAI-1 to terminate host colonization, halting biofilm formation and virulence factor expression. The signaling is also seen as a mechanism of pathogenesis in Gram-positive bacteria such as *Staphylococcus aureus*. During infection when it enters the human body, it shows complex adaptive behavior that leads to changes in population density, time, and environment-specific. Part of this behavior is controlled by at least four two-component systems, one of which is the *agr* system, which uses a modified octapeptide in signaling [17].

4. Research on plants secondary metabolites

In the plant kingdom, more than 50,000 secondary metabolites have been discovered, and they exert a wide range of effects on the plant as well as other living organisms. Their functions involve induction of flowering, fruit set and abscission, maintenance of perennial growth or signal deciduous behavior. Many plant secondary metabolites act as antimicrobials and perform the role of attractants or, conversely, as repellents. They are used as herbs in the traditional medicine in many ancient communities as plant secondary metabolites have shown to possess various biological effects. Plant secondary metabolites are classified according to their chemical structures into several classes. The classes of secondary plant metabolites include phenolics, alkaloids, saponins, terpenes, lipids, and carbohydrates. They act as antibiotic, antifungal, and antiviral and therefore are able to protect plants from pathogens as well as serious leaf damage from the light because they contain important UV-absorbing compounds [18]. Further, secondary metabolites of plants also have ecological importance whereby they improve soil quality by influencing soil decomposition. Tannins and terpenes affect cycling of C and N by increasing N immobilization in the soil. They also defend plants from pathogens and diseases, attract pollinators, aid in seed dispersal, and help plants recover from injury [19].

5. Conclusion

Though numerically plants are largest contributors of secondary metabolites, unfathomed microbial metabolites of different ecological origin are treasure in store. The microbial biomolecules have several advantages over the metabolites of plant or animal origin, which finds many applications as briefed earlier. Microbial sources can be genetically modified to enhance the production of desired natural product by fermentation. The metabolic versatility makes microbes interesting objects for a range of economically important biotechnological applications. This book provides reviews and research articles on secondary metabolites of microbial, animal, and plant origin, which should benefit scientific community.

Author details

Girish Nair[1], Suresh Selvapuram Sudalaimuthu Raja[2] and Ramasamy Vijayakumar[2]*

1 Faculty of Natural and Agricultural Sciences, Department of Biochemistry, Genetics and Microbiology, University of Pretoria, Hatfield, Pretoria, South Africa

2 Department of Microbiology, Government Arts and Science College (Formerly Bharathidasan University Constituent College), Perambalur, India

*Address all correspondence to: rvijayakumar1979@gmail.com

IntechOpen

© 2022 The Author(s). Licensee IntechOpen. This chapter is distributed under the terms of the Creative Commons Attribution License (http://creativecommons.org/licenses/by/3.0), which permits unrestricted use, distribution, and reproduction in any medium, provided the original work is properly cited.

References

[1] Demain AL. Microbial production of primary metabolites. Die Naturwissenschaften. 1980;**67**:582-587. DOI: 10.1007/BF00396537

[2] Thirumurugan D, Cholarajan A, Suresh Raja SS, Vijayakumar R. An introductory chapter: Secondary metabolites. In: Vijayakumar R, Raja SSS, editors. Secondary Metabolites: Sources and Applications. Crotia: InTech – Open Science; 2018. p. 138. DOI: 10.5772/intechopen.79766

[3] Demain AL, Fang A. The natural functions of secondary metabolites. Advances in Biochemical Engineering/Biotechnology. 2000;**69**:1-39. DOI: 10.1007/3-540-44964-7_1

[4] Demain AL. Importance of microbial natural products and the need to revitalize their discovery. Journal of Industrial Microbiology & Biotechnology. 2014;**41**:185-201. DOI: 10.1007/s10295-013-1325-z

[5] Newman DJ, Cragg GM. Natural products as sources of new drugs over the 30 years from 1981 to 2010. Journal of Natural Products. 2012;**75**:311-335. DOI: 10.1021/np200906s

[6] McMurry JE. Organic chemistry with biological applications. In: Secondary Metabolites: An Introduction to Natural Products Chemistry. Stamford, USA: Cengage Learning Ltd; 2015. pp. 1016-1046

[7] Perveen S. Introductory chapter: Terpenes and Terpenoids. In: Perveen S, Al-Taweel A, editors. Terpenes and Terpenoids. Rijeka, Crotia: IntechOpen; 2018. pp. 1-13. DOI: 10.5772/intechopen. 79683

[8] Sultan A, Rauf Raza A. Steroids: A diverse class of secondary metabolites. Medicinal Chemistry. 2015;**5**:310-317. DOI: 10.4172/2161-0444.1000279

[9] Yu AQ, Pratomo Juwono NK, Leong SS, Chang MW. Production of fatty acid-derived valuable chemicals in synthetic microbes. Frontiers in Bioengineering and Biotechnology. 2014;**2**:78. DOI: 10.3389/fbioe.2014.00078

[10] Wang J, Zhang R, Chen X, Sun X, Yan Y, Shen X, et al. Biosynthesis of aromatic polyketides in microorganisms using type II polyketide synthases. Microbial Cell Factories. 2020;**19**:110. DOI: 10.1186/s12934-020-01367-4

[11] Ul-Ain Q, Khan H, Mubarak MS, Pervaiz A. Plant alkaloids as antiplatelet agent: Drugs of the future in the light of recent developments. Frontiers in Pharmacology. 2016;**7**:1-9. DOI: 10.3389/fphar.2016.00292

[12] Martínez-Núñez MA, López VEL. Nonribosomal peptides synthetases and their applications in industry. Sustainable Chemical Processes. 2016;**4**:13. DOI: 10.1186/s40508-016-0057-6

[13] Fischer JD, Holliday GL, Thornton JM. The CoFactor database: Organic cofactors in enzyme catalysis. Bioinformatics. 2010;**26**:2496-2497. DOI: 10.1093/bioinformatics/btq442

[14] Anjitha KS, Sameena PP, Puthur JT. Functional aspects of plant secondary metabolites in metal stress tolerance and their importance in pharmacology. Plant Stress. 2021;**2**:100038. DOI: org/10.1016/j.stress.2021.100038

[15] Wolf JC, Mirocha CJ. Control of sexual reproduction in *Gibberella zeae* (*Fusarium roseum* "Graminearum").

Applied and Environmental Microbiology. 1977;**33**:546-550

[16] Nutting WH, Rapoport H, Machlis L. The structure of sirenin. Journal of the American Chemical Society. 1968;**90**:6434-6438

[17] Dufour N, Rao RP. Secondary metabolites and other small molecules as intercellular pathogenic signals. FEMS Microbiology Letters. 2011;**314**:10-17. DOI: 10.1111/j.1574-6968.2010.02154

[18] Pagare S, Bhatia M, Tripathi N, Pagare S, Bansal YK. Secondary metabolites of plants and their role: Overview. Current Trends in Biotechnology and Pharmacy. 2015;**3**:293-304

[19] Hussein RA, El-Anssary AA. Plants secondary metabolites: The key drivers of the pharmacological actions of medicinal plants. Herbal Medicine. 2019;**1**:13. DOI: 10.5772/intechopen.76139

Section 2

Secondary Metabolites: General Reviews and Biotechnological Interventions

Chapter 2

Diverse Survival Functions of Secondary Metabolites in Nature

Ayush Mandwal

Abstract

Secondary metabolites are low molecular mass products of secondary metabolism which are usually produced by microorganisms experiencing stringent conditions. These metabolites are not essential for growth but serve diverse survival functions in nature. Besides offering survival advance to the producing organisms, they have several medicinal uses such as antibiotics, chemotherapeutic drugs, immune suppressants, and other medicines which benefited human society immensely for more than a century. This chapter provides an overview of various functions these secondary metabolites offer in nature from single-cell organisms to multicellular organisms. Furthermore, this chapter also discusses the underlying mechanisms behind their diverse functions and how these are regulated and synthesized under non-viable environmental conditions.

Keywords: secondary metabolites, antibiotics, *Streptomyces*, resistance/tolerence, cluster-situated regulators

1. Introduction

Secondary metabolites are biologically active small molecules that are not required for growth and development but which provide a competitive advantage to the producing organism [1]. These are small organic molecules that consist of unusual chemical structures which include β-lactam rings, cyclic peptides, depsipeptides containing unnatural and non-protein amino acids, unusual sugars and nucleosides, unsaturated bonds of polyacetylenes and polyenes, covalently bound chlorine and bromine; nitro-, hydroxamic acids, and so on. Their enormous diversity includes 22,000 terpenoids as well [2]. These complex molecules are commonly obtained from molds which make 17% of all antibiotics and actinomycetes make 74% [3]. The bacterial secondary metabolites are a source of many antibiotics, chemotherapeutic drugs, immune suppressants, and other medicines. Some species are relatively more prolific in secondary metabolism, for example, strains of *Streptomyces hygroscopicus* produce more than 180 different secondary metabolites [4]. It is estimated that the total number of microbial secondary metabolites so far discovered vary from 8000 up to 50,000 [5–7].

As microorganisms are rarely found in isolation, the presence of secondary metabolites in a microbial community exerts evolutionary pressure on both secondary metabolite-producing and non-producing members to develop means to withstand them. They either use secondary metabolites to have competitive advances or support the intra/interspecies communities against stressful environmental conditions. As these molecules play a vital role in the survival of various microbes, biosynthesis of these molecules is tightly regulated via various transcriptional factors which got triggered under stringent conditions.

This chapter is dedicated to the role of secondary metabolites as antibiotics by various species. The first section of the chapter begins with a discussion on how these molecules are utilized by the various organisms to ensure their survival in the environment. The second section of the chapter goes into the details of how secondary metabolites affect the metabolism of the organism and alter their own or other's organism susceptibility against antibiotics. And the last section discusses how these secondary metabolites as antibiotics are regulated and synthesized with a special focus on *Streptomyces* as they are a rich source of natural antibiotics due to their prolific secondary metabolism.

2. Secondary metabolites in nature

In recent years, the view that secondary metabolites facilitate the survival of the producer in competition with other living species has been expressed more widely [8, 9]. Some common arguments behind such a view are as follows: (1) Organisms that lack an immune system are prolific producers of these compounds which act as an alternative defense mechanism. (2) The compounds have sophisticated structures, mechanisms of action, and complex and energetically expensive pathways which can only exist if they provide survival advantage to the organism [10]. (3) They are produced in nature and used in competition between microorganisms, plants, and animals [11, 12]. (4) Biosynthetic genes of secondary metabolites are clustered, which would only be selected for if the product conferred a selective advantage, and the absence of non-functional genes in these clusters. (5) The presence of resistance and regulatory genes in these clusters, and lastly by not least the non-producers have clustering of resistance genes.

Besides providing a survival advantage to microbes, secondary metabolites with antibacterial and antifungal properties can cause public health problems if found in soil, straw, and agricultural products. These are usually considered to be mycotoxins, but they are nevertheless antibiotics. And the natural production of such toxic metabolites is one of our major public health problems in the field and during the storage of crops. Natural soil and wheat straw contain patulin [13] and aflatoxin is known to be produced on corn, cottonseed, peanuts, and tree nuts in the field [14]. These toxins can cause hepatotoxicity, teratogenicity, immunotoxicity, mutation, cancer, and death [15].

Below list provide some of the functions of secondary metabolites with examples found across various species from single cell microbes to multicellular organisms.

1. Agents of chemical warfare in nature

    - According to Cavalier-Smith [16], secondary metabolites are most useful to the organisms producing them as competitive weapons. Antibiotics are more effective than macromolecular toxins such as animal venoms because of their higher diffusibility into cells and broader modes of action and diverse molecular structure varieties possible.

- Microbe vs. microbe:
  - In nature, competition between various fungi has been demonstrated in virtually every type of fungal ecosystem including coprophilous, carbinocolous, lignicolous, fungicolous, phylloplane, rhizosphere, marine, and aquatic [17].
  - Bacteria produce antibiotics when they need them for survival. For instance, myxobacteria can grow on *E. coli* only if the cell density is more than 10^7 myxobacteria/ml [18]. Such high cell density in the local environment produces high concentrations of lytic enzymes and antibiotics needed to grow on *E. coli*.
- Bacteria vs. amoebae:
  - As eukaryotes cells such as amoebae, a protozoans cell, feed on bacteria [19], bacteria found their ways to protect themselves against amoebae and other protozoans in general. Both *Serratia marcescens* and *Chromobacterium violaceum* bacteria produce antibiotically-active pigments namely prodigiosin and violacein, respectively to protect themselves from amoebae. These molecules can either encyst the protozoa or kill them.
- Microorganisms vs. higher plants:
  - Over 150 microbial compounds called phytotoxins have been reported that are active against plants [20]. Several such phytotoxins (e.g., phaseolotoxin, rhizobitoxine, and syringomycin) show typical antibiotic activity against other microorganisms and are thus belong to a class of antibiotics.
  - Plants produce antibiotics called phytoalexins after being exposed to plant pathogenic microorganisms in order to protect themselves [21]. They are of low molecular weight, weakly active, and indiscriminate, that is, they inhibit both prokaryotes and eucaryotes including higher plant cells and mammalian cells.
- Microorganisms vs. insects:
  - Certain fungi produce secondary metabolites for entomopathogenic activity: infecting and killing insects. *Beauveria bassiaria* fungus produces one such compound called bassianolide, a cyclodepsipeptide, which elicits atonic symptoms in silkworm larvae [22]. Another pathogen, *Metarrhizium anisophae*, produces the peptidolactone toxins known as destruxins [23].
  - Similarly, to fight back against bacterial infections, insects produce antibacterial proteins [24]. Some of these proteins are lysozyme, sarcotoxins, cecropins, and defensins. These proteins are either bactericidal or bacteriostatic by nature.

- Microorganisms vs. higher animals:

    - It is beneficial for microbes to make fresh food as objectionable as possible to large organisms as quickly as possible [25]. They produce secondary metabolites such as antibiotics and toxins which are toxic to large animals such as livestock. Thus, large animals will refuse to consume moldy feed which ensures the availability of food sources for various microbes.

    - If in case, animals and plants do get infected from microbes, they produce various peptides which kill microbes by permeabilizing their cell membranes as a way to defend against microbial infection [26].

2. Metal transport agents

- Certain secondary metabolites can act as metal transport agents. Siderophores (also known as sideramines) containing molecules function in the uptake, transport, and solubilization of iron. Siderophores are complex molecules that solubilize ferric ion which has a solubility of only 10^{-18} mol/L at pH 7.4 and have an extremely high affinity for iron ($K_d = 10^{-20}$–10^{-50}). Another group of molecules includes the ionophoric antibiotics, for example, macrotetrolide antibiotics, which function in the transport of certain alkali-metal ions such as potassium and affect its permeability through the cell membranes.

- Iron-transport factors in many cases are antibiotics by nature. They are on the borderline between primary and secondary metabolites since they are usually not required for growth but do stimulate growth under iron-deficient conditions. Antibiotic activity is due to the ability of these compounds to starve other species of iron when the latter lack the ability to take up the Fe-sideramine complex. Such antibiotics include nocardamin [27] and desferritriacetylfusigen [28].

3. Effects of microbial secondary metabolites on antibiotic tolerance and resistance

Secondary metabolites can bring profound variation in microbial physiology, metabolism, and stress responses [29]. Several evidence suggest that these molecules can modulate microbial susceptibility to commonly used antibiotics. This section explores which types of secondary metabolites alter antimicrobial susceptibility, and how and why this phenomenon occurs.

In a given environment, microorganisms are rarely found in isolation. Due to such reasons, the presence of secondary metabolites in a microbial community exerts evolutionary pressure on both secondary metabolite-producing and non-producing members to develop means to withstand them. Generally, secondary metabolites alter the state of metabolism which directly has an impact on an organism's ability to withstand antibiotics assault by either increasing its tolerance or making itself more resistant.

Although antibiotics tolerance and resistance are often treated in a similar manner, they offer different abilities to the microorganism. The phenomenon of antibiotic tolerance is the ability of organism to survive transient antibiotic exposure while resistance is the ability of organism to grow in the presence of antibiotics at a given concentration [30–32]. Another generic term commonly used is antibiotic resilience which refers to the ability of a bacterial population to be refractory to antibiotic treatment, which can arise from an increase in tolerance and/or resistance.

There are common modes of action through which secondary metabolites molecules can alter antibiotic efficacy in both single-species and polymicrobial communities. Specifically, secondary metabolites can regulate multidrug resistance efflux systems, can modulate the toxicity of antibiotics through interactions with reactive oxygen species (ROS) and can induce antibiotic tolerance to provide an overlooked route for the evolution of antibiotic resistance.

The knowledge of such interactions between secondary metabolites and antibiotic efficacy is beneficial as this could be applied to optimize the use of existing antimicrobial drugs and generate targets for novel therapeutic strategies.

3.1 Induction of efflux systems

One common mechanism bacteria use to tolerate clinical antibiotic treatment is by activation of efflux pumps that export toxins out of the cell [33, 34]. However, efflux pumps exist long before human use of synthetic antibiotics and therefore are presumed to have originally evolved to transport other, naturally occurring, substrates, such as secondary metabolites [35, 36]. Importantly, efflux pumps vary in their specificity, with regard to both their regulation and their substrate affinity [37]. Therefore, it is essential to understand how the secondary metabolite interacts with the transcriptional regulation of the efflux system, as well as which classes of drug the efflux system can transport before one predict whether a secondary metabolite will increase antibiotic resilience in its producer through the induction of a particular efflux system or not. A well-known example of a secondary metabolite that affects the efflux system is indole. Indole is a signaling molecule self-produced by *E. coli* that triggers the expression of certain multidrug efflux pumps in enteric bacteria [38, 39]. It is also found that subpopulation of mutants in *E. coli* which is more resilient towards norfloxacin and gentamicin antibiotics treatment produces high levels of indole which give population-level resistance [40]. This behavior is characterized as a "charity" as the subpopulation responds in favor to support the rest of the community. It was also found that indole achieve such a role by upregulating the MdtEF-TolC efflux system [40].

As mentioned earlier, efflux pumps are evolved to transport in general various molecules including secondary metabolites, efflux pumps can also provide protection against secondary metabolites that are toxic to their producers. For instance, phenazines are redox-active and toxic secondary metabolites produced by the *Pseudomonas aeruginosa*, which activates expression of the MexGHI-OpmD efflux system via the redox-sensing transcription factor SoxR [41–43]. In addition, phenazines also serve other important roles both for the microbes and other species as it can increase *P. aeruginosa* virulence in the lungs of patients with cystic fibrosis, or help in protecting plants against fungal pathogens [44, 45].

A related phenomenon has been observed in the Gram-positive bacterium *Streptomyces coelicolor*. This bacterium produces a natural antibiotic, actinorhodin, that stimulates the expression of a transporter similar to those that export tetracycline [46, 47].

3.2 Modulation of oxidative stress

The idea of oxidative stress become well known when it was proposed that regardless of the specific cellular targets of various antibiotic classes, bactericidal antibiotics exert their lethal effects in part by inducing oxidative stress [48]. Although this hypothesis resulted in major controversies [49, 50], evidence reviewed elsewhere [51] suggests that bactericidal antibiotics do impact cellular redox states and that the resulting increase in ROS and oxidative stress can contribute to cell death. Such phenomena relate well with secondary metabolites as they also interface with cellular redox homeostasis and oxidative stress responses. Below three different modes of action are discussed by which secondary metabolites can potentially antagonize or potentiate the toxicity of antibiotics: upregulation of oxidative stress response genes; direct detoxification of ROS; and increased endogenous ROS generation.

3.2.1 Upregulation of defenses against oxidative stress

Secondary metabolites that upregulate the expression of oxidative stress responses can prime bacterial cells for tolerance and/or resistance to antibiotics, which is similar to the protective effects of exposure to sublethal concentrations of oxidants such as H_2O_2 [52]. Among this class of metabolites, indole is a non-toxic molecule produced by *E. coli*, which activates various genes such as thioredoxin reductase, DNA-binding protein (Dps), and alkyl hydroperoxide reductases that activate only during oxidative stress response [53]. Furthermore, the frequency of *E. coli* persisters can increase by at least an order of magnitude if it is exposed to indole to three different classes (fluoroquinolones, aminoglycosides, and β-lactams) antibiotics. In addition, deletion of oxyR substantially diminishes this effect, which demonstrates that upregulation of oxidative stress responses by a secondary metabolite can contribute to bacterial persistence [53].

Another example of the secondary metabolite is pyocyanin (PYO) produced by *P. aeruginosa*. PYO increases superoxide dismutase activity [54] and upregulates the transcription of several other oxidative stress response genes, including those encoding alkyl hydroperoxide reductases, thioredoxin reductase, catalase, and iron-sulfur cluster biogenesis machinery [43]. Interestingly, PYO has been shown to increase the frequency of gentamicin-resistant mutants in *P. aeruginosa* cultures that is independent of drug efflux, as PYO does not upregulate aminoglycoside-transporting efflux pumps [55]. A plausible explanation for this phenomenon is that PYO-induced oxidative stress responses counteract ROS-related gentamicin toxicity. The reasoning behind the explanation is as follows: (1) Gentamicin is known to promote increased intracellular ROS levels [56, 57]. (2) Pretreating cells with oxidants can prime them to tolerate antibiotics [52]. This consequently could decrease the rate at which spontaneous mutants are lost from the population [58], thus the frequency of resistant mutants increases as observed experimentally. PYO has been demonstrated to promote the growth of *P. aeruginosa* in the presence of other antibiotics such a β-lactam antibiotic carbenicillin and other aminoglycosides such as kanamycin, streptomycin, and tobramycin [59]. These antibiotics have been shown to perturb cellular redox states [60, 61] rather than as substrates for PYO-regulated efflux systems [55, 62]. This suggests that the observed decreases in antibiotic efficacy could be related to PYO-induced oxidative stress responses.

3.2.2 Detoxification of ROS

Another way to protect against antibiotic assaults of oxidative stress is by directly detoxifying antibiotic-induced ROS by upregulating the antioxidant activity. Ergothioneine, is one of two major sulfur-containing redox buffers in mycobacteria, along with mycothiol which help in detoxifying ROS during the stress response. This molecule is so important for the *Mycobacterium tuberculosis* that loss of ergothioneine biosynthesis genes decreases minimum inhibitory concentrations (MICs) for various clinical antibiotics such as rifampicin, bedaquiline, clofazimine, and isoniazid. Additionally, loss of gene also decreases the survival rate by at least 30–60% under treatment at the MICs compared to wild type [63].

3.2.3 Synergistic interactions between secondary metabolites and antibiotics

So far secondary metabolites have been demonstrated to decrease antibiotic efficacy by attenuating oxidative stress. However, they can also amplify the toxicity of antibiotics by increasing ROS generation. 2-heptyl-3-hydroxy-4-quinolone is one example, also known as the *Pseudomonas* quinolone signal (PQS) produced by *P. aeruginosa*. PQS is a redox-active molecule that possesses both antioxidant properties and pro-oxidant activity as it can reduce not only free radicals but also metal ions. Reduction in metal ions concentration such as iron is lethal for the cell as it facilitates ROS formation through the Fenton reaction [64].

3.3 Interspecies antibiotic resilience

So far, we have discussed examples of how secondary metabolites affect their producers susceptibility to antibiotics. However, secondary metabolites can also modulate interspecies antibiotic resilience. Such study is beneficial as it will potentially show how interactions among members of a polymicrobial infection might affect antibiotic treatment outcomes [65, 66]. For example, one study has demonstrated that indole, which is produced by *E. coli*, can increase antibiotic tolerance of *Salmonella enterica subsp. enterica serovar Typhimurium* [38, 67]. *S. Typhimurium* does not produce indole, yet it becomes more than threefold tolerant against ciprofloxacin in the presence of exogenously added indole, as well as in the case where it is co-cultured with indole-producing *E. coli* [67]. Indole induces the same OxyR-regulated oxidative stress responses in *S. Typhimurium*, as in *E. coli*, and the deletion of oxyR decreases tolerance against ciprofloxacin in *S. Typhimurium* mediated by it [67]. Thus, indole acts like an interspecies modulators of antibiotic resilience. Besides indole, putrescine is another secondary metabolite that acts similar to indole. For example, *Burkholderia cenocepacia* produces putrescine to protect itself from polymyxin B but it also protects its neighboring species in the co-cultures, including *E. coli* and *P. aeruginosa* [68]. There are definitely more such secondary metabolites that exist which have the potential of being interspecies modulators of antibiotic resilience.

Importantly, although the above examples suggest that certain secreted secondary metabolites have the potential to raise the community-wide level of antibiotic resilience in polymicrobial communities, it may not be this case always. As mentioned earlier that secondary metabolites can be toxic or non-toxic and can increase or decrease resilience against antibiotics, it is important to consider whether the stress caused by the secondary metabolite is tolerable to the non-producing species. If the molecule-caused toxicity outweighs the benefits it provides against antibiotics, the non-producing species would not gain a benefit. In such a case, the secondary metabolite might even act synergistically with the clinical antibiotic [69].

4. The regulation of the secondary metabolism of *Streptomyces*

So far, we have discussed what different kinds of secondary metabolites exist, how they affect cellular metabolism and help in survival in competitive and stressful environments. This last section focuses on the factors on which the biosynthesis of secondary metabolites is regulated. This will show how different conditions lead to the secretion of secondary metabolites which leads to phenomena discussed in the previous two sections. As the exploration of regulation of secondary metabolites among various species is still in its early stages except for Streptomycetes, this section will focus on its regulatory mechanisms behind biosynthesis of secondary metabolites.

Streptomycetes and other actinobacteria are renowned as a rich source of natural products of clinical, agricultural, and biotechnological value. Sequencing genomes of numerous streptomycetes has revealed that they all possess the capacity to produce multiple secondary metabolites [70, 71] implies that it can repel a large number of competitors using either individual or combination of molecules using their synergistic characteristics [72]. The genes of enzymes responsible for the production of individual secondary metabolites are found clustered. Furthermore, these clustered genes are commonly associated with regulatory gene/s that regulate their transcription or resistance genes.

This section discusses some of the factors which regulate the production of secondary metabolites in *Streptomyces*.

4.1 Cluster-situated regulators (CSRs)

Generally, a single regulatory gene regulates several gene clusters associated with secondary metabolites production. This way multiple chemical signals which can trigger activation of the regulatory gene can activate specific or multiple genes clustered corresponding to secondary metabolite production. Some of the gene clustered include the clusters for streptomycin in *S. griseus* and actinorhodin (ACT) in *S. coelicolor*. Their corresponding transcription factors are called StrR and ActII-ORF430 respectively [73]. Both StrR and ActII-ORF430 transcription factors directly activate the transcription of genes of the corresponding clusters that encode biosynthetic enzymes. Moreover, evidence suggests that the cellular level of a CSR is the principal factor that determines the level of transcription of the biosynthetic genes it targets. This correlates closely with the level of secondary metabolite produced [74, 75]. Thus, factors that control the production of ActII-ORF4 and StrR will ultimately regulate the production of ACT and streptomycin respectively. Both these transcription factors are under regulation with many activators, repressors, and inducers which are explained well in detail here [76].

4.2 The stringent response and nutrient deprivation

Stringent response is an stress response of bacteria in reaction to amino-acid starvation, fatty acid limitation, iron limitation, heat shock, and other stress conditions. During stringent response, accumulation of (p)ppGpp enables bacteria to survive sustained periods of nutrient deprivation. For several *Streptomyces* species, mutations that block the synthesis of (p)ppGpp (guanosine tetra- and penta-phosphate) have been found to alter antibiotic production and hinder morphological development [77]. In general, the stringent response enhances transcription of numerous genes associated with the stationary phase of batch culture and stress responses. Stimulation of (p)ppGpp synthesis, either by subjecting growing cultures to amino acid starvation [78] or inducing expression of a truncated version of relA that confers ribosome independent (p)ppGpp synthetase activity [79, 80] increases the level of actII-ORF4 transcription and production of the corresponding antibiotics.

4.2.1 Regulation of secondary metabolism by carbon

The availability and source of carbon have a substantial effect on the production of antibiotics and morphological development [81]; for example, glucose blocks production of ACT by *S. coelicolor* [82]. Lack of carbon source triggers a stringent response which as explained above leads to accumulation of (p)ppGpp which increases the level of actII-ORF4 transcription and production of the corresponding antibiotics.

4.2.2 Regulation of secondary metabolism by nitrogen

Numerous studies have shown that the source of nitrogen can influence the production of antibiotics. In the presence of sources of nitrogen that are favorable for growth, production of many, but not all, secondary metabolites is reduced [83–85]. One interpretation of this tendency is that by supplying a good source of nitrogen, the available carbon can be used for growth and generating biomass. Thus cell does not experience or experience of lesser extent of the stringent condition in the presence of suitable nitrogen source.

4.3 Upsetting of zinc and iron homeostasis

Zinc is an essential trace element and cofactor required for the structure and function of many proteins. Being important, it is under tight regulation by Znr, a zinc-responsive transcriptional repressor that regulates genes encoding a high-affinity uptake system for zinc, as well as zinc-free paralogues of ribosomal proteins in many bacteria, including streptomycetes [86, 87]. Znr also directly represses a promoter within the cluster of coelibactin, a non-ribosomally synthesized peptide predicted to have siderophore (metal-chelating) activity in *S. coelicolor* [88, 89]. AbsC, a pleiotropic regulator which is required for the production of ACT and RED (chromosomally-encoded antibiotics, the prodiginine complex RED, which is red in color), represses promoters of the coelibactin cluster under the specific condition of low zinc [88]. Although the underlying regulatory mechanism is still unknown, under low zinc concentration if upregulation of genes encoding a high-affinity uptake system for zinc via Znr does not work, AbsC potentially becomes active which increases the production of ACT and RED for antibiotics production.

Iron is another essential metal that is under tight regulation [90]. Members of the DmdR (divalent metal-dependent) family i.e., DmdR1 and DmdR2 are the key regulatory components of iron homeostasis in *S. coelicolor* [91, 92]. The dmdR1 gene overlaps with adm gene on the opposite strand and disruption of the overlapping gene increases the production of RED and ACT which leads to antibiotics production [91]. Although the details of the DmdR1/Adm system remain to be uncovered, it is likely that physiological cellular stress indirectly affects antibiotic production.

4.4 Extracellular signaling molecules

One extracellular signaling molecule g-butyrolactones has been shown to regulate antibiotic production in many streptomycetes cultures [93–95]. Such a mechanism is beneficial for the community survival as extracellular signaling molecules are diffusible in solid media. This way even a single actinomycete can stimulate antibiotic production in another when grown next to each other on an agar plate [96], thus protecting the entire community against competitive microbes.

5. Conclusions

Although secondary metabolites are not considered essential for the growth and development of microorganisms, they serve diverse survival functions in nature. These molecules allowed both antagonistic and symbiotic relationships between various species from single-cell organisms to multicellular organisms. Without such relationships, the natural ecosystem whether it is soil, or lake, or forest would not be filled with rich and diverse life forms that we find on this planet. Thus, it can be said that secondary metabolites are as essential as primary metabolites in a environment with multi-species communities coexisting together.

Author details

Ayush Mandwal
University of Calgary, Calgary, Canada

*Address all correspondence to: ayushmandwal5@gmail.com

IntechOpen

© 2022 The Author(s). Licensee IntechOpen. This chapter is distributed under the terms of the Creative Commons Attribution License (http://creativecommons.org/licenses/by/3.0), which permits unrestricted use, distribution, and reproduction in any medium, provided the original work is properly cited.

References

[1] Demain AL, Fang A. The natural functions of secondary metabolites. In: Fiechter A, editor. History of Modern Biotechnology I. Advances in Biochemical Engineering/Biotechnology. Vol. 69. Berlin, Heidelberg: Springer; 2001. pp. 1-39. DOI: 10.1007/3-540-44964-7_1

[2] Connolly JD, Hill RA, et al. Dictionary of terpenoids. London,UK: Chapman & Hall; 1991

[3] Miyadoh S. Research on antibiotic screening in Japan over the last decade: A producing microorganism approach. Actinomycetologica. 1993;7(2):100-106. DOI: 10.3209/saj.7_100

[4] Zedan H. The economic value of microbial diversity. SIM News. 1993;**43**: 178-185

[5] Lechevalier HA, Lechevalier MP. Biology of actinomycetes. Annual Reviews in Microbiology. 1967;**21**(1):71-100

[6] Omura S. Trends in the search for bioactive microbial metabolites. Journal of Industrial Microbiology. 1992;**10** (3–4):135-156. DOI: 10.1007/bf01569759

[7] Reichenbach H, Höfle G. Biologically active secondary metabolites from myxobacteria. Biotechnology Advances. 1993;**11**(2):219-277. DOI: 10.1016/0734-9750(93)90042-l

[8] Stone MJ, Williams DH. On the evolution of functional secondary metabolites (natural products). Molecular Microbiology. 1992;**6**(1): 29-34. DOI: 10.1111/j.1365-2958.1992.tb00834.x

[9] Williams DH, Stone MJ, Hauck PR, Rahman SK. Why are secondary metabolites (natural products) biosynthesized? Journal of Natural Products. 1989;**52**(6):1189-1208. DOI: 10.1021/np50066a001

[10] Katz E, Demain AL. The peptide antibiotics of *Bacillus*: Chemistry, biogenesis, and possible functions. Bacteriological Reviews. 1977;**41**(2): 449-474

[11] Demain A. Do antibiotics function in nature. Search. 1980;**11**(5):148-151

[12] Demain AL, Solomon NA. Biology of industrial microorganisms. vol. 6. Oxford,UK: Butterworth-Heinemann; 1985

[13] Norstadt FA, McCalla TM. Microbial populations in stubble-mulched soil. Soil Science. 1969;**107**(3):188-193. DOI: 10.1097/00010694-196903000-00006

[14] Hesseltine CW, Rogers RF, Shotwell OL. Aflatoxin and mold flora in North Carolina in 1977 corn crop. Mycologia. 1981;**73**(2):216. DOI: 10.2307/3759642

[15] Trail F, Mahanti N, Linz J. Molecular biology of aflatoxin biosynthesis. Microbiology. 1995;**141**(4):755-765

[16] Cavalier-Smith T. Origins of secondary metabolism. In: Secondary Metabolites: Their Function and Evolution. Chichester, UK: John Wiley & Sons; 1992. pp. 64-87

[17] Gloer JB. The chemistry of fungal antagonism and defense. Canadian Journal of Botany. 1995;**73**(S1): 1265-1274. DOI: 10.1139/b95-387

[18] Rosenberg E. Myxobacteria: development and cell interactions. Berlin/Heidelberg, Germany: Springer Science & Business Media; 2012

[19] Habte M, Alexander M. Further evidence for the regulation of bacterial populations in soil by protozoa. Archives of Microbiology. 1977;**113**(3):181-183. DOI: 10.1007/bf00492022

[20] Fischer H, Bellus D. Phytotoxicants from microorganisms and related compounds. Pesticide Science. 1983; **14**(3):334-346. DOI: 10.1002/ps.2780140316

[21] Darvill AG, Albersheim P. Phytoalexins and their Elicitors-A Defense Against Microbial Infection in Plants. Annual Review of Plant Physiology and Plant Molecular Biology. 1984;35(1):243-275. DOI: 10.1146/annurev.pp.35.060184.001331

[22] Kanaoka M, Isogai A, Suzuki A. Syntheses of bassianolide and its two homologs, enniatin C and decabassianolide. Agricultural and Biological Chemistry. 1979;**43**(5):1079-1083

[23] Lee S, izumiya N, Suzuki A, Tamura S. Synthesis of a cyclohexadepsipeptide, protodestruxin. Tetrahedron Letters. 1975;**16**(11):883-886. DOI: 10.1016/S0040-4039(00)72010-7

[24] Kimbrell DA. Insect antibacterial proteins: Not just for insects and against bacteria. BioEssays. 1991;**13**(12):657-663. DOI: 10.1002/bies.950131207

[25] Janzen DH. Why fruits rot, seeds mold, and meat spoils. The American Naturalist. 1977;**111**(980):691-713. DOI: 10.1086/283200

[26] Nissen-Meyer J, Nes IF. Ribosomally synthesized antimicrobial peptides: Their function, structure, biogenesis, and mechanism of action. Archives of Microbiology. 1997;**167**(2–3):67-77. DOI: 10.1007/s002030050418

[27] Keller-Schierlein W, Prelog V. Stoffwechselprodukte von Actinomyceten. 30. Mitteilung. Über das Ferrioxamin E; ein Beitrag zur Konstitution des Nocardamins. Helvetica Chimica Acta. 1961;**44**(7):1981-1985. DOI: 10.1002/hlca.19610440721

[28] Anke H. Metabolic products of microorganism. 163 Desferritriaceetylfusigen, an antibiotic from *Aspergillus deflectus*. The Journal of Antibiotics. 1977;**30**(2):125-128

[29] Perry EK, Meirelles LA, Newman DK. From the soil to the clinic: The impact of microbial secondary metabolites on antibiotic tolerance and resistance. Nature Reviews Microbiology. 2021;19:1-14. DOI: 10.1038/s41579-021-00620-w

[30] Kester JC, Fortune SM. Persisters and beyond: Mechanisms of phenotypic drug resistance and drug tolerance in bacteria. Critical Reviews in Biochemistry and Molecular Biology. 2013;**49**(2):91-101. DOI: 10.3109/10409238.2013.869543

[31] Brauner A, Fridman O, Gefen O, Balaban NQ. Distinguishing between resistance, tolerance and persistence to antibiotic treatment. Nature Reviews Microbiology. 2016;**14**(5):320-330. DOI: 10.1038/nrmicro.2016.34

[32] Balaban NQ, Helaine S, Lewis K, Ackermann M, Aldridge B, Andersson DI, et al. Definitions and guidelines for research on antibiotic persistence. Nature Reviews Microbiology. 2019;**17**(7):441-448. DOI: 10.1038/s41579-019-0196-3

[33] Li XZ, Plésiat P, Nikaido H. The challenge of efflux-mediated antibiotic resistance in gram-negative bacteria. Clinical Microbiology Reviews. 2015; **28**(2):337-418. DOI: 10.1128/cmr.00117-14

[34] Piddock LJV. Clinically relevant chromosomally encoded multidrug resistance efflux pumps in bacteria. Clinical Microbiology Reviews. 2006; **19**(2):382-402. DOI: 10.1128/cmr.19.2.382-402.2006

[35] Piddock LJV. Multidrug-resistance efflux pumps? Not just for resistance. Nature Reviews Microbiology. 2006; **4**(8):629-636. DOI: 10.1038/nrmicro1464

[36] Martinez JL, Sánchez MB, Martínez-Solano L, Hernandez A, Garmendia L, Fajardo A, et al. Functional role of bacterial multidrug efflux pumps in microbial natural ecosystems. FEMS Microbiology Reviews. 2009;**33**(2): 430-449. DOI: 10.1111/j.1574-6976.2008.00157.x

[37] Du D, Wang-Kan X, Neuberger A, van Veen HW, Pos KM, Piddock LJV, et al. Multidrug efflux pumps: Structure, function and regulation. Nature Reviews Microbiology. 2018;**16**(9):523-539. DOI: 10.1038/s41579-018-0048-6

[38] Nikaido E, Giraud E, Baucheron S, Yamasaki S, Wiedemann A, Okamoto K, et al. Effects of indole on drug resistance and virulence of *Salmonella enterica* serovar Typhimurium revealed by genome-wide analyses. Gut Pathogens. 2012;**4**(1):5-5. DOI: 10.1186/1757-4749-4-5

[39] Nishino K, Nikaido E, Yamaguchi A. Regulation of multidrug efflux systems involved in multidrug and metal resistance of *Salmonella enterica* serovar Typhimurium. Journal of Bacteriology. 2007;**189**(24):9066-9075. DOI: 10.1128/jb.01045-07

[40] Lee HH, Molla MN, Cantor CR, Collins JJ. Bacterial charity work leads to population-wide resistance. Nature. 2010;**467**(7311):82-85. DOI: 10.1038/nature09354

[41] Dietrich LEP, PriceWhelan A, Petersen A, Whiteley M, Newman DK. The phenazine pyocyanin is a terminal signalling factor in the quorum sensing network of *Pseudomonas aeruginosa*. Molecular Microbiology. 2006;**61**(5): 1308-1321. DOI: 10.1111/j.1365-2958.2006.05306.x

[42] Sakhtah H, Koyama L, Zhang Y, Morales DK, Fields BL, Price-Whelan A, et al. The *Pseudomonas aeruginosa* efflux pump MexGHI-OpmD transports a natural phenazine that controls gene expression and biofilm development. Proceedings of the National Academy of Sciences. 2016;**113**(25):E3538-E3547. DOI: 10.1073/pnas.1600424113

[43] Meirelles LA, Newman DK. Both toxic and beneficial effects of pyocyanin contribute to the lifecycle of *Pseudomonas aeruginosa*. Molecular Microbiology. 2018;**110**(6):995-1010. DOI: 10.1111/mmi.14132

[44] Lau GW, Hassett DJ, Ran H, Kong F. The role of pyocyanin in *Pseudomonas aeruginosa* infection. Trends in Molecular Medicine. 2004;**10**(12):599-606. DOI: 10.1016/j.molmed.2004.10.002

[45] Mavrodi DV, Blankenfeldt W, Thomashow LS. Phenazine compounds in fluorescent *Pseudomonas* spp. biosynthesis and regulation*. Annual Review of Phytopathology. 2006;**44**(1): 417-445. DOI: 10.1146/annurev.phyto.44.013106.145710

[46] Tahlan K, Ahn SK, Sing A, Bodnaruk TD, Willems AR, Davidson AR, et al. Initiation of actinorhodin export in *Streptomyces coelicolor*. Molecular Microbiology. 2007; **63**(4):951-961. DOI: 10.1111/j.1365-2958.2006.05559.x

[47] Willems AR, Tahlan K, Taguchi T, Zhang K, Lee ZZ, Ichinose K, et al.

Crystal structures of the *Streptomyces coelicolor* TetR-like protein ActR alone and in complex with actinorhodin or the actinorhodin biosynthetic precursor (S)-DNPA. Journal of Molecular Biology. 2008;**376**(5):1377-1387. DOI: 10.1016/j.jmb.2007.12.061

[48] Kohanski MA, Dwyer DJ, Hayete B, Lawrence CA, Collins JJ. A common mechanism of cellular death induced by bactericidal antibiotics. Cell. 2007;**130**(5):797-810. DOI: 10.1016/j.cell.2007.06.049

[49] Liu Y, Imlay JA. Cell death from antibiotics without the involvement of reactive oxygen species. Science. 2013;**339**(6124):1210-1213. DOI: 10.1126/science.1232751

[50] Keren I, Wu Y, Inocencio J, Mulcahy LR, Lewis K. Killing by bactericidal antibiotics does not depend on reactive oxygen species. Science. 2013;**339**(6124):1213-1216. DOI: 10.1126/science.1232688

[51] Dwyer DJ, Collins JJ, Walker GC. Unraveling the physiological complexities of antibiotic lethality. Annual Review of Pharmacology and Toxicology. 2015;**55**(1):313-332

[52] Mosel M, Li L, Drlica K, Zhao X. Superoxide-mediated protection of *Escherichia coli* from antimicrobials. Antimicrobial Agents and Chemotherapy. 2013;**57**(11):5755-5759. DOI: 10.1128/aac.00754-13

[53] Vega NM, Allison KR, Khalil AS, Collins JJ. Signaling-mediated bacterial persister formation. Nature Chemical Biology. 2012;**8**(5):431-433. DOI: 10.1038/nchembio.915

[54] Hassett DJ, Charniga L, Bean K, Ohman DE, Cohen MS. Response of *Pseudomonas aeruginosa* to pyocyanin: Mechanisms of resistance, antioxidant defenses, and demonstration of a manganese-cofactored superoxide dismutase. Infection and Immunity. 1992;**60**(2):328-336. DOI: 10.1128/iai.60.2.328-336.1992

[55] Meirelles LA, Perry EK, Bergkessel M, Newman DK. Bacterial defenses against a natural antibiotic promote collateral resilience to clinical antibiotics. PLoS Biology. 2021;**19**(3):e3001093. DOI: 10.1371/journal.pbio.3001093

[56] Priuska EM, Schacht J. Formation of free radicals by gentamicin and iron and evidence for an iron/gentamicin complex. Biochemical Pharmacology. 1995;**50**(11):1749-1752. DOI: 10.1016/0006-2952(95)02160-4

[57] Prayle A, Watson A, Fortnum H, Smyth A. Side effects of aminoglycosides on the kidney, ear and balance in cystic fibrosis. Thorax. 2010;**65**(7):654-658. DOI: 10.1136/thx.2009.131532

[58] Alexander HK, MacLean RC. Stochastic bacterial population dynamics restrict the establishment of antibiotic resistance from single cells. Proceedings of the National Academy of Sciences of the United States of America. 2020;**117**(32):19455-19464. DOI: 10.1073/pnas.1919672117

[59] Zhu K, Chen S, Sysoeva TA, You L. Universal antibiotic tolerance arising from antibiotic-triggered accumulation of pyocyanin in *Pseudomonas aeruginosa*. PLoS Biology. 2019;**17**(12):e3000573. DOI: 10.1371/journal.pbio.3000573

[60] Dwyer DJ, Belenky PA, Yang JH, MacDonald IC, Martell JD, Takahashi N, et al. Antibiotics induce redox-related physiological alterations as part of their lethality. Proceedings of the National Academy of Sciences. 2014;**111**(20):E2100-E2109

[61] Belenky P, Ye J, Porter CM, Cohen N, Lobritz M, Ferrante T, et al. Bactericidal antibiotics induce toxic metabolic perturbations that lead to cellular damage. Cell Reports. 2015;**13**(5):968-980 DOI: 10.1016/j.celrep.2015.09.059

[62] Lister PD, Wolter DJ, Hanson ND. Antibacterial-resistant *Pseudomonas aeruginosa*: Clinical impact and complex regulation of chromosomally encoded resistance mechanisms. Clinical Microbiology Reviews. 2009;**22**(4):582-610. DOI: 10.1128/cmr.00040-09

[63] Saini V, Cumming B, Guidry L, Lamprecht DA, Adamson J, Reddy V, et al. Ergothioneine maintains redox and bioenergetic homeostasis essential for drug susceptibility and virulence of *Mycobacterium tuberculosis*. Cell Reports. 2016;**14**(3):572-585. DOI: 10.1016/j.celrep.2015.12.056

[64] Häussler S, Becker T. The *Pseudomonas* quinolone signal (PQS) balances life and death in *Pseudomonas aeruginosa* populations. PLoS Pathogens. 2008;**4**(9):e1000166. DOI: 10.1371/journal.ppat.1000166

[65] Bottery MJ, Pitchford JW, Friman VP. Ecology and evolution of antimicrobial resistance in bacterial communities. The ISME Journal. 2021;**15**(4):939-948. DOI: 10.1038/s41396-020-00832-7

[66] Welp AL, Bomberger JM. Bacterial community interactions during chronic respiratory disease. Frontiers in Cellular and Infection Microbiology. 2020;**10**:213. DOI: 10.3389/fcimb.2020.00213

[67] Vega NM, Allison KR, Samuels AN, Klempner MS, Collins JJ. *Salmonella typhimurium* intercepts *Escherichia coli* signaling to enhance antibiotic tolerance. Proceedings of the National Academy of Sciences. 2013;**110**(35):14420-14425. DOI: 10.1073/pnas.1308085110

[68] El-Halfawy OM, Valvano MA. Chemical communication of antibiotic resistance by a highly resistant subpopulation of bacterial cells. PLoS One. 2013;**8**(7):e68874. DOI: 10.1371/journal.pone.0068874

[69] Wood KB, Cluzel P. Trade-offs between drug toxicity and benefit in the multi-antibiotic resistance system underlie optimal growth of *E. coli*. BMC Systems Biology. 2012;**6**(1):48. DOI: 10.1186/1752-0509-6-48

[70] Corre C, Challis GL. New natural product biosynthetic chemistry discovered by genome mining. Natural Product Reports. 2009;**26**(8):977-986. DOI: 10.1039/b713024b

[71] Nett M, Ikeda H, Moore BS. Genomic basis for natural product biosynthetic diversity in the actinomycetes. Natural Product Reports. 2009;**26**(11):1362-1384. DOI: 10.1039/b817069j

[72] Challis GL, Hopwood DA. Synergy and contingency as driving forces for the evolution of multiple secondary metabolite production by *Streptomyces* species. Proceedings of the National Academy of Sciences. 2003;**100**(Suppl. 2):14555-14561. DOI: 10.1073/pnas.1934677100

[73] Vujaklija D, Horinouchi S, Beppu T. Detection of an A-factor-responsive protein that binds to the upstream activation sequence of strR, a regulatory gene for streptomycin biosynthesis in *Streptomyces griseus*. Journal of Bacteriology. 1993;**175**(9):2652-2661

[74] Gramajo HC, Takano E, Bibb MJ. Stationaryphase production of the antibiotic actinorhodin in *Streptomyces coelicolor* A3(2) is transcriptionally

regulated. Molecular Microbiology. 1993; 7(6):837-845. DOI: 10.1111/j.1365-2958.1993.tb01174.x

[75] Tomono A, Tsai Y, Yamazaki H, Ohnishi Y, Horinouchi S. Transcriptional control by A-factor of strR, the pathway-specific transcriptional activator for streptomycin biosynthesis in *Streptomyces griseus*. Journal of Bacteriology. 2005;**187**(16):5595-5604. DOI: 10.1128/jb.187.16.5595-5604.2005

[76] van Wezel GP, McDowall KJ. The regulation of the secondary metabolism of *Streptomyces*: New links and experimental advances. Natural Product Reports. 2011;**28**(7):1311-1333. DOI: 10.1039/c1np00003a

[77] Hopwood DA. Forty years of genetics with *Streptomyces*: From in vivo through in vitro to in silico. Microbiology. 1999;**145**(9):2183-2202

[78] Strauch E, Takano E, Baylts HA, Bibb MJ. The stringent response in *Streptomyces coelicolor* A3(2). Molecular Microbiology. 1991;**5**(2):289-298. DOI: 10.1111/j.1365-2958.1991.tb02109.x

[79] Hesketh A, Sun J, Bibb M. Induction of ppGpp synthesis in *Streptomyces coelicolor* A3(2) grown under conditions of nutritional sufficiency elicits actIIORF4 transcription and actinorhodin biosynthesis. Molecular Microbiology. 2001;**39**(1):136-144. DOI: 10.1046/j.1365-2958.2001.02221.x

[80] Sun J, Hesketh A, Bibb M. Functional analysis of relA and rshA, two relA/spoT homologues of *Streptomyces coelicolor* A3(2). Journal of Bacteriology. 2001;**183**(11):3488-3498. DOI: 10.1128/jb.183.11.3488-3498.2001

[81] Sánchez S, Chávez A, Forero A, García-Huante Y, Romero A, Sánchez M, et al. Carbon source regulation of antibiotic production. The Journal of Antibiotics. 2010;**63**(8):442-459. DOI: 10.1038/ja.2010.78

[82] Lee HN, Im JH, Lee MJ, Lee SY, Kim ES. A putative secreted solute binding protein, SCO6569 is a possible AfsR2-dependent down-regulator of actinorhodin biosynthesis in *Streptomyces coelicolor*. Process Biochemistry. 2009; **44**(3):373-377. DOI: 10.1016/j.procbio.2008.12.002

[83] Merrick M, Edwards R. Nitrogen control in bacteria. Microbiological Reviews. 1995;**59**(4):604-622

[84] Shapiro S. Regulation of Secondary Metabolism in Actinomycetes. CRC Press; 1989

[85] Sanchez S, Demain AL. Metabolic regulation of fermentation processes. Enzyme and Microbial Technology. 2002;**31**(7):895-906. DOI: 10.1016/s0141-0229(02)00172-2

[86] Shin JH, Oh SY, Kim SJ, Roe JH. The zinc-responsive regulator Zur controls a zinc uptake system and some ribosomal proteins in *Streptomyces coelicolor* A3(2) â–¿. Journal of Bacteriology. 2007; **189**(11):4070-4077. DOI: 10.1128/jb.01851-06

[87] Panina EM, Mironov AA, Gelfand MS. Comparative genomics of bacterial zinc regulons: Enhanced ion transport, pathogenesis, and rearrangement of ribosomal proteins. Proceedings of the National Academy of Sciences. 2003;**100**(17):9912-9917. DOI: 10.1073/pnas.1733691100

[88] Kallifidas D, Pascoe B, Owen GA, Strain-Damerell CM, Hong HJ, Paget MSB. The zinc-responsive regulator Zur controls expression of the coelibactin gene cluster in *Streptomyces coelicolor*. Journal of Bacteriology. 2009;

192(2):608-611. DOI: 10.1128/jb.01022-09

[89] Hesketh A, Kock H, Mootien S, Bibb M. The role of absC, a novel regulatory gene for secondary metabolism, in zincdependent antibiotic production in *Streptomyces coelicolor* A3 (2). Molecular Microbiology. 2009; **74**(6):1427-1444. DOI: 10.1111/j.1365-2958.2009.06941.x

[90] Hantke K. Iron and metal regulation in bacteria. Current Opinion in Microbiology. 2001;**4**(2): 172-177. DOI: 10.1016/s1369-5274(00)00184-3

[91] Tunca S, Barreiro C, Coque JJR, Martin JF. Two overlapping antiparallel genes encoding the iron regulator DmdR1 and the Adm proteins control sidephore and antibiotic biosynthesis in *Streptomyces coelicolor* A3 (2). The FEBS Journal. 2009;**276**(17): 4814-4827

[92] Tunca S, Barreiro C, Sola-Landa A, Coque JJR, Martn JF. Transcriptional regulation of the desferrioxamine gene cluster of *Streptomyces coelicolor* is mediated by binding of DmdR1 to an iron box in the promoter of the desA gene. The FEBS Journal. 2007;**274**(4): 1110-1122

[93] Willey JM, Gaskell AA. Morphogenetic signaling molecules of the streptomycetes. Chemical Reviews. 2011; **111**(1):174-187. DOI: 10.1021/cr1000404

[94] Nishida H, Ohnishi Y, Beppu T, Horinouchi S. Evolution of butyrolactone synthases and receptors in *Streptomyces*. Environmental Microbiology. 2007;**9**(8):1986-1994. DOI: 10.1111/j.1462-2920.2007.01314.x

[95] Hopwood DA. Complex enzymes in microbial natural product biosynthesis, Part A: overview articles and peptides. Cambridge, Massachusetts: Academic Press; 2009;458

[96] Ueda K, Kawai S, Ogawa HO, Kiyama A, Kubota T, Kawanobe H, et al. Wide distribution of interspecific stimulatory events on antibiotic production and sporulation among *Streptomyces* species. The Journal of Antibiotics. 2000;**53**(9):979-982

Chapter 3

Secondary Metabolites: The Natural Remedies

Hudu Garba Mikail, Mamman Mohammed, Habib Danmalam Umar and Mohammed Musa Suleiman

Abstract

The chapter discusses the meaning and origin of some important classes of secondary metabolites such as alkaloids, terpenoids, tannins, flavonoids, saponins, glycosides, and phenolic compounds, etc., produced by some bacteria, fungi, or plants. Very important drugs that are used clinically are derived from these secondary metabolites. Several reports obtained in scientific journals and books written by different scientists working or who have worked in the fields of natural products medicine were reviewed. These different classes of secondary metabolites have shown activity against varied diseases, and compounds that are of novel structure and activity have been isolated and characterized from them. The chapter highlights the economic impacts of these chemical compounds including their role in improving human and animal health and well-being by serving as sources of some antibiotics, anticancer, anti-inflammatory, antifertility, antidiabetics, analgesics, growth promoters, etc. Secondary metabolites are also used to enhance agricultural productivity, they find uses as pesticides, insecticides, and preservatives. Some folkloric uses of secondary metabolites chemical compounds based on reliable sources of information and genuine scientific investigations are highlighted.

Keywords: secondary metabolites, natural remedies, phytochemical constituents, bioactive compounds

1. Introduction

Metabolomics is the study of metabolites within biofluids, cells, tissues, or organisms [1]. Whereas collectively, metabolites and their interactions are known as metabolome [2].

Metabolites are small molecules produced by metabolic reactions; these molecules are intermediate or end products of metabolic reactions. The metabolic reactions are catalyzed by naturally occurring enzymes within the organisms' cells [3]. Compounds derived from primary and secondary metabolism are known as primary and secondary metabolites, respectively.

Primary metabolites are indispensable compounds used by organisms for their growth, development, and reproduction; these compounds are synthesized by the cells as a result of metabolism during the growth phase. Primary metabolites are referred to as central metabolites due to their key role in maintaining normal physiological processes. Primary metabolites include vitamins (B2 and B12), lactic acid, amino acids, polyols, alcohols such as ethanol, nucleotides, organic acids, etc. [3, 4].

The current chapter discusses the meaning and origin or sources of some important classes of secondary metabolites such as alkaloids, terpenoids, tannins, flavonoids, saponins, cardiac glycosides, phenolic compounds, etc., the economic impacts of secondary metabolite compounds including their role in improving human and animal health and well-being (as antibiotics, anticancer, anti-inflammatory, antifertility, antidiabetics, pain relievers, growth promoters, etc.). The chapter addresses the role of secondary metabolites in enhancing agricultural productivity (as pesticides, insecticides, preservatives, etc.); it also discusses the important present-day drugs derived from secondary metabolites, as well as some important biological/pharmacological effects or activities of different classes of the secondary metabolites and their folkloric usage based on reliable sources of information and genuine scientific investigations.

1.1 The meaning and origin of important classes of secondary metabolites

Secondary metabolites also known as phytochemical constituents, bioactive compounds, specialized metabolites, secondary products, or toxins are organic compounds produced by organisms such as plants, fungi, or bacteria as a result of secondary metabolic processes that lead to production and accumulation of diverse chemical compounds known as secondary metabolites. These compounds are not required for primary metabolic processes by the organisms [3–5]. Secondary metabolites are formed toward the end of the growth phase; thus, they are not directly involved in the normal physiologic processes of the organism such growth and development as well as reproductive processes. Instead, they increase the organism's survivability through mediation of ecological interaction, to the organism, this serves as a selective advantage [4, 5]. Interspecies defenses such as defense against herbivory by plants are part of the important roles of secondary metabolites. However, humans use secondary metabolites as medicines, recreational drugs, flavorings, pigments, etc. [6].

Secondary metabolites are classified commonly based on their vast structural diversity, biosynthesis, and function. According to the literature, over 2140, 000 secondary metabolites are known; however, the main classes of secondary metabolites are five, which include alkaloids, terpenoids and steroids, nonribosomal polypeptides, polyketides and fatty-acid-derived substances, and enzyme cofactors [7].

1.2 The origin and sources of some important classes of secondary metabolites

Secondary metabolite is a term coined in 1910 by a Medicine and Physiology Nobel Prize laureate, Albert Kossel [8]. Friedrich Czapek, a Polish botanist, 30 years later described them as metabolic nitrogenous end products [9].

Secondary metabolites are produced by plants, fungi, or bacteria as well as many marine organisms such as snails, corals, tunicates, and sponges [10]. There are 150, 000–200, 000 bioactive compounds derived from the plant kingdom, 50, 000–100, 000 from animal kingdom, and 22,000–23, 000 from microbes [11].

1.3 Plant secondary metabolites

Plants are the major sources of secondary metabolites; they produced 80% of the known secondary metabolites occurring in nature [10]. Secondary metabolites are used by carnivorous plants to attract, capture, digest, and assimilate the prey [12]. One of the early known plant secondary metabolites is morphine, isolated in 1804 [11].

1.4 Fungal secondary metabolites

In 1928, Alexander Fleming while working at St Mary's Hospital in London discovered the most known secondary metabolite, the penicillin. Penicillin was discovered experimentally from a mold, the *Penicillium notatum* [13, 14].

1.5 Bacterial secondary metabolites

Oligosaccharide, b-lactam, polyketide, non-ribosomal pathways, and shikimate are the main secondary metabolite production pathways in bacteria [15]. Although bacterial secondary metabolites have some beneficial effects, many are toxic to mammals through secretion of exotoxin, botulinum toxin secreted by *Clostridium botulinum* bacteria is a very good example [15].

1.6 The alkaloids

The name alkaloid was introduced by Carl Friedrich Wilhelm Meißner, in the year 1819. The name was derived from Latin root alkali, rooted from Arabic word al-qalwi meaning plants ashes. The wide usage of the word alkaloid came after J. Oscar's publication in the year 1880 in Albert Ladenburg, the chemical dictionary [16].

A large variety of organisms produced alkaloids; these chemical compounds are derived from plants, bacteria, fungi, and animals [17]. Morphine was the first individual alkaloid isolated in 1804 from the opium poppy plant (*Papaver somniferum*) [18].

1.7 The cardiac glycosides

The ancient Romans, Syrians, and Egyptians used cardiac glycosides contained in plant extracts for medicinal purposes, the plant extracts from *Urginea maritima* (Scilla), squill, or sea onion were used as emetics and heart tonics. African warriors in the medieval age used Strophanthus species as arrows head poison against their targets. Cardiac glycosides were established in the twentieth century as agent for the treatment of heart failure [19].

Early writings of 1250 BC mentioned *Digitalis purpurea*; digitalis was included in herbal collections used in prescription by the Welsh family physicians. The origin of digitalis was from the foxglove plant. A botanist and physician of English origin, William withering in the eighteenth century described the foxglove plant's clinical effects in a published monograph. He was the first investigator of the systemic bioactivity of digitalis. "An account of the Foxglove and some of its medical uses with practical remarks on dropsy, and other diseases" is a book authored by William Withering in 1785 reporting the toxicity and indications of digitalis [19].

Plant is main source of cardiac glycosides; however, bufadienolide was isolated from frogs and mammalian tissues that are rich sources of endogenous digitalis; this show that animal species are also good sources of cardiac glycosides [20].

1.8 The flavonoids

Flavonoids or bioflavonoids are yellow compounds derived from the Latin word Flavus, meaning yellow, their natural coloration [21]. Albert Szent-Györgyi and some group of scientists in the 1930s discovered that crude yellow extracts from lemons, oranges, etc., were more effective at preventing scurvy than vitamin C. They referred to these compounds as citrin or vitamin P, which were later discovered to be hesperidin, neohesperidin, etc., belonging to flavonoids rather than the vitamins [22].

Flavonoids are compounds belonging to polyphenolic structural class of secondary metabolites. They are widely found in vegetables, fruits, flowers, wine, tea, grains, roots, bark, and stem [23, 24]. Flavonoid compounds are found in several parts of plants, they are products extracted from plants using various extraction techniques such as chromatography [25].

1.9 The phenolic compounds

Phenolic compounds are secondary metabolites produced by the secondary metabolic pathways of plants [26]. They are derived from pentose phosphate and shikimic acid of plants through metabolization of phenylpropanoid [27, 28]. The composition of phenolic substances or polyphenols includes tannins, flavonoids, lignans, coumarins, and phenolic acids [26], colored anthocyanins [29]; these compounds are naturally found in vegetables, fruits, leaves, and roots among other products of plant origin [26, 27].

1.10 The tannins

Tannins are group of astringent and complex polyphenolic compounds found in plants, which can bind and precipitate proteins; the word tannin was derived from the usage of this compound in tanning animal hides and skins to make leather [30]; the term was first introduced in 1796 [31]. Commonly, tannins are found in wood, buds, fruits, leaves, stems, roots, seeds, and in the bark of trees [32]. Condensed tannins are the most abundant polyphenols, which are virtually found in plant families [33].

1.11 The terpenoids

Terpenoids or isoprenoids are modified terpenes [34, 35]; terpenoids usually contain additional functional group and oxygen [35]. These chemical compounds are the largest class of secondary metabolites representing 60% of the natural products known [36].

2. The biological activities of secondary metabolites

Unique structural diversity is provided by natural products when compared with standard combinatorial chemistry; these give opportunities for discovering novel lead compounds with low molecular weight. The world's biodiversity evaluation of natural products for potential biological activity is less than 10%; thus, a lot of useful novel natural lead compounds await discovery [37].

Terrestrial plants are the major source of secondary metabolites; other sources include fungi, bacteria, as well as several marine organisms [10].

2.1 The pharmacological activity of plant-derived secondary metabolites

2.1.1 Antibacterial activity

Natural antibiotics are secondary metabolites produced by microbes that inhibit bacterial growth by targeting essential cellular processes such as the synthesis of the bacterial cell wall, DNA/RNA, and proteins. They are not essential for the growth of the organism (and usually produce at the end of the exponential phase of their growth). They have diverse roles, such as in cellular differentiation, nutrient sequestration, metal transport, ecological interactions, and defense [38, 39].

Between 1935 and 1968, 12 classes of antibiotics were launched and approved for use as drugs. However, between 1969 and 2000, the number dropped markedly, with only two classes introduced. Out of the 30 antibiotics launched between the year 2003 and 2015, 16 belong to natural products and their derivatives. They include three new classes of natural antibiotics—two actinomycete: the lipopeptide daptomycin in 2003 and fidaxomicin (of the tiacumicin family) in 2010. The third is a fungal product: retapamulin derived from pleuromutilin and approved in 2007 for topical use [38, 40].

Newman and Cragg reported the introduction of several natural secondary metabolites that have been reported to possess potent antibacterial activity including: anthrasil, omadacycline, dalbavacin, plazomicin, ceftaroline fasamil acetate, lefamulin, sarecycline, eravacycline imi-cilast-relebactam, etc. [41].

2.1.2 Anti-inflammatory activity

Inflammation is a normal biological process that occurs as a response to microbial infection, chemical irritation, or tissue injury. It is usually initiated by moving the immune cells from blood vessels and release of mediators to the damage site. It is then followed by reinforcement with inflammatory cells, release of reactive oxygen species (ROS), reactive nitrogen species (RNS), and proinflammatory cytokines to fight the foreign pathogens and repairing the injured tissues. In general, normal inflammation is rapid and self-limiting, but unresolved and prolonged inflammation causes various chronic disorders. As a pathologic condition, inflammation can include a wide range of diseases such as rheumatic and immune-mediated conditions, diabetes, cardiovascular accident, etc. [38, 42]. Aswad and coworkers reported the use of moupinamide, capsaicin, and hypaphorine—natural products—with high scores in their indexing of potential anti-inflammatory drug candidates [43]. Mona et al. also reported more than 15 herbs, where their anti-inflammatory effects have been evaluated in clinical and experimental studies including *Curcuma longa*, *Zingiber officinale*, *Rosmarinus officinalis*, *Borago officinalis* [42].

2.1.3 Anticancer activity

Cancer is one of the leading causes of death (second to cardiovascular diseases) in the world, despite the availability of wide range of anticancer drugs. The estimated cancer burden in the world as reported by the World Health Organization (WHO) is 18.1 million new cases and 9.6 million deaths as at 2018 [38, 44]. Presently, research efforts are directed toward the discovery of natural products with anticancer potential [45]. Several secondary metabolites have been reported to possess anticancer potential; some of these compounds have the capacity to prevent oxidative stress and inflammation that causes damage to DNA, which in turn leads to carcinogenesis [45]. Natural products such as irinotecan, vincristine, vinblastine, etoposide, and paclitaxel from plants, actinomycin D and mitomycin C from bacteria as well as marine-derived bleomycin are widely used in the treatment of various cancers [44].

Also, fruits and vegetables are plant sources that are known to contain vitamins, minerals, folate, plant sterols, carotenoids, and various phytochemicals such as flavonoid and polyphenols—natural product compounds that are associated with reduced cancer mortality and risk [46]. The critical relationship of fruit and vegetable intake and cancer prevention has been thoroughly documented. It has been suggested that major public health benefits could be achieved by substantially increasing consumption of these foods [38].

Herbs and spices such as ginger, capsicum, curcumin, clove, rosemary, sage, oregano, and cinnamon are very rich in antioxidants due to the high content of phenolic compounds and have been shown to counteract reactive oxygen species (ROS)-mediated damage in different human cancers [47]. Many cyclic peptides and their derivatives obtained from marine organisms have been shown to possess anticancer, antimicrobial, anti-inflammatory, antiproliferative, and antihypertensive properties [46]. Furthermore, lactoferrin, a multifunctional protein found in bovine and camel milk, has also been reported to possess anticancer effect [48].

2.1.4 Antiviral activity

Natural compounds are an important source for the discovery and the development of novel antiviral drugs because of their availability and expected low side effects. Naturally occurring compounds with antiviral activity have been recognized as early as 1940s. The search for effective drugs against human immunodeficiency virus (HIV) is the need of hour. Most of the work related with antiviral compounds revolves around inhibition of various enzymes associated with the life cycle of viruses. Structure-function relationship between secondary metabolites and the HIV enzyme inhibitory activity has been observed [38].

2.1.5 Hepatoprotective activity

Diseases of the liver have been classified as high priority areas of health care, as an estimate by the World Health Organization shows approximately 500 million people of the world are suffering from a severe form of liver disorders that may lead to chronic hepatitis. Hepatic disorders can be caused by exposure to agents such as drugs, viruses, parasites, and toxins. Such an exposure usually may result in degeneration and inflammation of the liver; furthermore, it results in fibrosis and cirrhosis [49]. In addition, different chronic diseases such as diabetes may lead to development of hepatic clinical manifestations.

Several flavonoids such as catechin, apigenin, quercetin, naringenin, rutin, and venoruton are reported for their hapatoprotective activities [38]. Muhammad and coworkers review studies conducted on the composition, pharmacology, and nature of some selected plants in the light of possible mechanism deduced from experimental trials [49]. Also, a comprehensive review by Meng et al. [50], listed several plants and products that have been used in the prevention and treatment of chemically induced liver damages [50].

2.1.6 Important present-day drugs derived from plants secondary metabolites

Many drugs with wide range of pharmacological activities were derived from alkaloids [51]. Some of the important drugs derived from alkaloids include:

1. Quinine—antimalaria [51]

2. Morphine—analgesic [52]

3. Codeine—analgesic, antitussive [53]

4. Ephedrine—antiasthma [51]

5. Galantamine—cholinomimetics [54]

6. Homoharringtonine—anticancer [50]

7. Quinidine—antiarrhythmic [52]

8. Vincamine—vasodilator [52]

9. Chelerythrine—antibacterial [55]

10. Piperine—antihyperglycemic [56]

11. Atropine—anticholinergic [57]

12. Pilocarpine—cholinergic agonist [58]

13. Paclitaxel—anticancer [59]

14. Ergotamine—anti-migraine [60]

15. Reserpine—antihypertensive [60]

16. Vinblastine and vincristine—anticancer [60]

17. Physostigmine—anti-mydriatic, etc. [60]

2.1.7 The pharmacological activities of cardiac glycosides

The effects of cardiac glycosides mainly for increasing heart muscle force of contraction and reducing heart rate are beneficial for treating cardiac arrhythmias and congestive heart failure; cardiac glycosides have long been used to manage these ailments. The commonest cardiac glycosides used clinically include digoxin, digitoxin, ouabain, and bufalin [61]. Other forms of cardiac glycosides are antiarin, thevetin A and B, peruvoside, neriifolin, thevetoxin, ruvoside, theveridoside, cerberin, convallarin, convallamarin, convallatoxin, glucoscillarene A, proscillaridine A, scillarene A, scilliglaucoside and scilliphaeoside, marinobufagenin, oleandrin, folineriin, adynerin, digitoxigenin, marinobufagenin, telocinobufagin [62]. Among these substances, literature has also reported the therapeutic uses of acetyldigoxin, digitoxin, digoxin, gitoformate, gitoxin, lanatoside C, metildigoxin (β-methyldigoxin), ouabain (strophanthin-g), peruvoside, proscillaridin, strophanthin-k [63], apart from digoxin, digitoxin, ouabain, and bufalin earlier mentioned [61].

2.1.8 The pharmacological activities of flavonoids and phenolic compounds

From plants, over 8000 phenolic compounds have been reported [64]. Interestingly, flavonoids make up half of these phenolic compounds [64]. Effectively, flavonoids and several other phenolic compounds have been reported to possess antibacterial, anti-inflammatory, antioxidants, anticancer, cardioprotective, immunomodulatory, and skin radioprotective effects from UV light. More so, these compounds are good pharmaceutical candidates for medical application [65]. Several flavonoids including apigenin, galangin, flavone and flavonol glycosides, isoflavones, flavanones, and chalcones have been shown to possess potent antibacterial activity [38].

2.1.9 The pharmacological activities of tannins

Certain carcinogenic incidences, such as esophageal cancer, have been related to tannins-rich foods consumption, especially the herbal tea and betel nuts. However, several reports showed that tannins' carcinogenic effects are not due to tannins themselves but likely due to components associated with the tannins [66]. Many literatures revealed negative association between cancer incidences and consumption of tannins components and tea polyphenols, suggesting their anticarcinogenic effects [66].

The antimutagenic and antimicrobial activities of tannins have been documented. Tannins inhibit the growth of viruses, bacteria, yeast, and many fungi. It has also been reported that propyl gallate and tannic acid inhibit aquatic bacteria and food-borne bacteria; this action is not reported for gallic acid. In food processing industry, catfish fillets' shelf-life can be enhanced using the tannic acid antimicrobial property. The antihypertensive, hypolipidemic, coagulative, and immunomodulatory effects of tannins have been reported [66].

2.1.10 The pharmacological properties of terpenoids

Terpenoids being the most abundant compounds in natural products have been reported to possess antibacterial, antimalarial, antiviral, hypoglycemic, neuroprotective, and anti-inflammatory activities. Furthermore, literatures have also documented the effects of terpenoids in treating and preventing cardiovascular diseases, antioxidation, immunoregulation, and promotion of transdermal absorption of substances [67].

2.2 The pharmacological activities of fungal-derived secondary metabolites

2.2.1 Some important drugs of fungal origin

2.2.1.1 Antibiotics

The beginning was the discovery of penicillin by Alexander Fleming from penicillium mold; penicillin is one of the most known antibiotics in use, and the beta lactam antibiotics penicillin and cephalosporin were all derived from fungus [68]. Other antibiotics derived from fungus include alamethicin, brefeldin A, aphidicolin, citromycin, fumagillin, cerulenin, eupenifeldin, fusidic acid, fusafungine, itaconic acid, usnic acid, helvolic acid, nigrosporin B, verrucarin A, vermiculine, etc. [68]. Tiamulin, retaparmulin, and valnemulin are antibiotics derived from pleuromutilin [68].

2.2.1.2 Antifungal agents

Antifungal griseofulvin is a derivative of penicillium species [69], azoxystrobin, echinocandins, strobilurin, micafungin, anidulafungin, and caspofungin are all antifungal agents originally derived from fungus [70].

2.2.1.3 Immunosuppressive agents

Bredinin, cyclosporin, mycophenolic acid, myriocin, endocrocin, and gliotoxin are all immunosuppressants isolated from fungus [71].

2.2.1.4 Potential antiviral agents

Compounds from several mushrooms such as *Ganoderma lucidum*, *Grifola frondose*, *Garnoderma colossus*, *Lentinus edodes*, *Hypsizygus marmoreus*, *Scleroderma citrinum*, *Cordyceps militaris*, *Trametes versicolor*, *Flammulina velutipes*, *Fomitopsis officinalis* are under research for potential antiviral activities validations [72, 73].

2.2.1.5 Potential antidiabetic and antimalarial agents

Ternatin and many other fungal isolates have potential hypoglycaemic effects [74]. Potential antimalarial agents of fungal origin under scientific elucidations include antiamoebin, codinaeopsin, zervamicins, and efrapeptins [75].

2.3 The pharmacological activities of bacterial-derived secondary metabolites

Pharmaceutical agents of bacterial origin include antibiotics, immunomodulators, nematicides, antitumor agents, coccidiostatic agents, enzyme inhibitors, and insecticides. Interestingly, *Escherichia coli* is used as a host in molecular biology for synthesis of recombinant proteins [76]. Selman Abraham Waksman, the father of antibiotics, discovered actinomycin; this effort was followed by the discovery of streptomycin in 1944 [77]. Other clinically important antibiotics derived from bacteria include bacitracin [78], polymyxin B [79], gentamicin [80], amphotericin b [81], tetracycline [82], erythromycin [83], rifamycin [84], vancomycin [82], neomycin [85], streptomycin [86], and chloramphenicol [87]. Etc.

3. The role of secondary metabolites in enhancing agricultural productivity

The resistance against herbivores and pathogens is a role played decisively by the chemical protection nature of plants, the secondary metabolites; they are plant features important especially for protection against a wide range of microorganisms such as bacteria, viruses, fungi, arthropods, herbivores, and vertebrates [88]. Soil decomposition is influenced by plant secondary metabolites by increasing nitrogen immobilization in the soil; cycling of carbon (C) and nitrogen (N) is affected by terpenes and tannins [89].

Exudates from plants roots contain secondary metabolites that can attract, kill, or deter underground microbes, herbivorous insects, and nematodes, competing plants and underground injuries are also inhibited [90]. Plants secondary metabolites contain potential toxic substances used for defense against insects; these chemical compounds can be utilized for design of future insecticides with multiple or specific targets [91]. A good example of an insecticidal compound of such nature is pyrethrin derived from the flowers of Pyrethrum cinerariaefolium plant; pyrethroids are the synthetic analogs of pyrethrin [92].

In terms of animals' productivity, animals that ingest forages containing different plants secondary metabolites get their meat and dairy products enhanced in terms of biochemical richness making them good for human consumption [93].

In today's food industries, plants secondary metabolites are used extensively as flavoring, coloring, and texturizing agents. Preservation and anti-browning are done with metabolites possessing antioxidative properties [94].

4. Folkloric usage of secondary metabolites based on reliable sources of information and genuine scientific investigations

The fact that animals and humans have been in existence before the advent of orthodox medicine is a proof that plants have been quite effective in treating diseases. The folkloric use of plant medicine has a long history [95]. From the earliest times, man acquired knowledge of the adverse and beneficial effects of plants from observations on animals. To distinguish edible from poisonous plants, grazing animals were observed and the plants not eaten were considered poisonous [96]. About 80% of the rural population today depends largely on medicinal plants for primary health care [97]. About 25% of all prescription drugs in developed countries are obtained directly or indirectly from plants [98].

Plants produce valuable organic compounds, some of which have potentials in treating ailments in both animals and humans [99]. Of the 252 drugs considered as basic and essential by the WHO, 11% are exclusively of plant origin and a significant number are synthetic drugs obtained from natural precursors [100]. In 1997, the world market for phytomedicinal products was estimated at US$10 billion [101]. This prompted the WHO to consider phytotherapy in its alternative or complementary health program. Locally produced plant medicines can be cheaper than imported synthetic drugs. One striking example is an herbal wound powder (Himax®) in Sri Lanka that was found to be as effective as an imported powder (Neomex®) and comparatively 80–90% cheaper [102].

The most easily accessible, affordable, and inexpensive sources of treatment in the primary healthcare system throughout the world are medicinal plants; there is a long history for the therapy of various disease conditions traditionally in various regions of the world [103].

Natural products' earliest records were depicted on clay tablets from Mesopotamia (2600 BC) in cuneiform; there are documented evidences of the folkloric of the use of oils derived from *Commiphora species* and *Cupressus sempervirens* that are still in use today to treat inflammation, coughs, and colds [104]. The Egyptian pharmaceutical record "the Ebers Papyrus" (2900 BC) documented over 700 drugs of plant origins; these agents include infusions, gargles, ointments, and pills. The Chinese folkloric record books such as the Materia Medica (1100 BC) with 52 prescriptions, the Tang Herbal (659 AD) with 850 drugs, and the Shennong Herbal (100 BC) with 365 drugs provide records of natural products' uses [104]. Theophrastus (300 BC), the Greek natural scientist and philosopher, is an expert in dealing with medicinal herbs, while Dioscorides (100 AD), the Greek physician, documented the uses and storage of medicinal herbs [104]. The monasteries in Germany, England, France, and Ireland preserved this knowledge during the Dark and Middle Western Ages. Preservation of the Greek and Roman knowledge was done by the Arabs. They also expanded of their own resources; this is done with the Indian and Chinese unfamiliar herbs to the Greek and Roman world [104]. In the eighth century, it was the Arabs who privately own pharmacies. Avicenna, a Persian physician, pharmacist, poet, and philosopher, contributed a lot to the science of medicine and pharmacy through his notable work such as the "Canon Medicine" [104].

4.1 Some reported medicinal uses of secondary metabolites

4.1.1 Alkaloids

Alkaloids have a wide range of pharmacological effects including antimalarial (quinine), antiasthma (ephedrine), anticancer (homoharringtonine), vasodilatory (vincamine), antiarrhythmic (quinidine), analgesic (morphine), antibacterial (chelerythrine), and antihyperglycemic activities (e.g., piperine) [37].

4.1.2 Anthraquinones

Huang et al. [105] and other teams clearly demonstrated that anthraquinones, such as emodin, aloe-emodin, and rhein, inhibit the growth and proliferation of various cancer cells, such as lung adenocarcinoma, myelogenous leukemia, neuroblastoma, hepatocellular carcinoma, bladder cancer, and others through cell death and survival's modulation. Several anthraquinones are able to inhibit the replication of viruses or even directly kill enveloped or unenveloped strains [106]. Senna, cascara, frangula, rhubarb, and aloe are commonly used for their laxative effects [107].

4.1.3 Flavonoids

Flavonoids have various health-promoting effects such as antioxidative, anti-inflammatory, anticarcinogenic, and antimutagenic. Flavonoids have antioxidant effects associated with various diseases such as Alzheimer's disease, cancer, atherosclerosis [108].

4.1.4 Cardiac glycosides

The most important use of the cardiac glycosides is its effects in treatment of cardiac failure. In cardiac failure, or congestive heart failure, heart cannot pump sufficient blood to maintain body needs. During each heart contraction, there is an influx of Na^+ and an outflow of K^+. Before the next contraction, Na^+, K^+-ATPase must reestablish the concentration gradient pumping Na^+ into the cell against a concentration gradient. This process requires energy, which is obtained from hydrolysis of ATP to ADP by Na^+, K^+-ATPase. Cardiac glycosides inhibit Na^+, K^+-ATPase, and consequently increase the force of myocardial contraction [109].

4.1.5 Saponins

Saponins exhibit a biological role and medicinal properties such as anti-inflammatory [110], antibacterial, antifungal, antiviral, insecticidal, anticancer, cytotoxic, and molluscicidal action [111].

4.1.6 Terpenes and steroids

Terpenes include substances such as floral fragrances, which serve as insect attractants, pine oil, growth inhibitors, plant hormones (gibberellic acid and abscisic acid), and some of which are insecticidal. About 30,000 terpenes have been identified; they all possess repeating five-carbon isoprene units (a five-carbon ring) [112].

Artemesinin is a sequiterpene, which originated from the Chinese medicinal plant Quinhao (*Artemisia annua*). It was used to treat fever medicine for over two millennia. It was mentioned in the 52 Remedies recovered from the Mawangdui Tomb dating from the Han Dynasty 206 BC – 221 BC located in Henan Province [113]. Placitaxol (a diterpene) is quite effective in treating against ovarian, breast, colon, non-small-cell lung cancer, and malignant melanoma [114]. Terpenoids (diterpenoids, sesquiterpenoids, triterpenoids) and lignoids also have antiviral activities. A number of them inhibit replication of inhibit coronaviruses, including SARS-Corona Virus. Betulinic acid and savinin are competitive inhibitors of a protease (an enzyme that breaks down proteins) produced by the SARS-CoV 3CL virus [114]. It will be worthwhile testing the effect of these terpenoids on SARs-CoV 2, the cause of recent Covid-19 pandemic.

4.1.7 Alkylresorcinols

Secondary metabolites are known for their angiogenic or wound healing activity, new compounds such as the new alkylresorcinols isolated from the lipophilic extract of Urginea indica L. bulbs have been reported to possess wound healing activity following experimental trauma [115].

Conflict of interest

Authors declare no conflict of interest.

Author details

Hudu Garba Mikail[1*], Mamman Mohammed[2], Habib Danmalam Umar[3] and Mohammed Musa Suleiman[2]

1 Faculty of Veterinary Medicine, Department of Veterinary Pharmacology and Toxicology, University of Abuja, Nigeria

2 Faculty of Veterinary Medicine, Department of Veterinary Pharmacology and Toxicology, Ahmadu Bello University Zaria, Nigeria

3 Faculty of Pharmaceutical Sciences, Department of Pharmacognosy, Ahmadu Bello University Zaria, Nigeria

*Address all correspondence to: mghudu@yahoo.com; hudu.mikail@uniabuja.edu.ng

IntechOpen

© 2022 The Author(s). Licensee IntechOpen. This chapter is distributed under the terms of the Creative Commons Attribution License (http://creativecommons.org/licenses/by/3.0), which permits unrestricted use, distribution, and reproduction in any medium, provided the original work is properly cited.

References

[1] Daviss B. Growing pains for metabolomics. The Scientist. 2005; **19**(8):25-28

[2] Jordan KW, Nordenstam J, Lauwers GY, Rothenberger DA, Alavi K, Garwood M, et al. Metabolomic characterization of human rectal adenocarcinoma with intact tissue magnetic resonance spectroscopy. Diseases of the Colon and Rectum. 2009;**52**(3):520-525. DOI: 10.1007/DCR.0b013e31819c9a2c

[3] Meštrović T. What are Metabolites?—News Medical [Internet]. 2018. Available from: https://www.news-medical.net/medical/authors/tomislav-mestrovic

[4] González-Mera IF, González-Falconí DE, Morera V, et al. Secondary metabolites in plants: Main classes, phytochemical analysis, and pharmacological activities. Revista Bionatura. 2019;**4**(4):1000-1009. DOI: 10.21931/RB/2019.04.04.11

[5] Návarová H, Bernsdorff F, Döring AC, Zeier J. Pipecolic acid, any endogenous mediator of defense amplification and priming, is a critical regulator of inducible plant immunity. The Plant Cell. 2012;**24**(12):512341. DOI: 10.1105/tpc.112.103564

[6] Secondary Metabolites-Knowledge Encyclopedia. [Internet]. 2016. Available from: www.biologyreference.com [Accessed: May 10, 2016]

[7] McMurry JE. Organic chemistry with biological applications. In: Secondary Metabolites: An Introduction to Natural Products Chemistry. Stamford, USA: Cengage Learning Ltd; 2015. pp. 1016-1046

[8] Jones ME. Albrecht Kossel, a biographical sketch. The Yale Journal of Biology and Medicine. 1953;**26**(1):80-97

[9] Bourgaud F, Gravot A, Milesi S, Gontier E. Production of plant secondary metabolites: A historical perspective. Plant Science. 2001;**161**(5):839-851. DOI: 10.1016/S0168-9452(01)00490-3

[10] Berdy J. Bioactive microbial metabolites. The Journal of Antibiotics. 2005;**58**(1):1-26

[11] Ramasamy V, Suresh SS. In: Vijayakumar R, Raja SSS, editors. Secondary metabolites sources and applications. London: Intech Open Access Publishers; 2018. DOI: 10.5772/intechopen.79766

[12] Hatcher CR, Ryves DB, Millett J. The function of secondary metabolites in plant carnivory. Annals of Botany. 2020;**125**(3):399-411. DOI: 10.1093/aob/mcz191

[13] Conniff R. Penicillin: Wonder Drug of World War II. History Net. [Internet]. 2017. Available from: https://www.historynet.com/penicillin-wonder-drug-world-war-ii.htm [Accessed: April 11, 2020]

[14] Macfarlane G. Alexander Fleming: The Man and The Myth. Cambridge, Mass.: Harvard University Press; 1984. ISBN 978-0-19-281884-3

[15] Gokulan K, Khare S, Cerniglia C. Metabolic pathways: Production of secondary metabolites of bacteria. In: Encyclopedia of Food Microbiology. Amsterdam, Netherland: Academic Press, Elsevier Ltd., 2014. pp. 561-569. ISBN 978-0-12-384733-1

[16] Oscar J. Alkaloide. In: Ladenburg, Handwörterbuch der Chemie. Vol. 1. Breslau, Germany: Eduard Trewendt; 1882. pp. 213-422

[17] Le Quesne PW. Alkaloids: Biochemistry, ecology, and medicinal applications. Edited by Margaret F. Roberts (University of London) and Michael Wink (University of Heidelberg). Plenum Press, New York, NY. ISBN 0-306-45465-3. Journal of Natural Products. 1999;**62**(4):664-664. DOI: 10.1021/np980259j

[18] Luch A. Molecular, Clinical, and Environmental Toxicology. New York City: Springer; 2009. p. 20. ISBN 978-3-7643-8335-0

[19] Norn S, Kruse P. Cardiac glycosides: From ancient history through Withering's foxglove to endogeneous cardiac glycosides. Dansk Medicinhistorisk Årbog. 2004:119-132. Danish

[20] El-Mallakh RS, Brar KS, Yeruva RR. Cardiac glycosides in human physiology and disease: Update for entomologists. Insects. 2019;**10**(4):102. DOI: 10.3390/insects10040102

[21] Delage B. Flavonoids. Corvallis, Oregon: Linus Pauling Institute, Oregon State University; 2015

[22] Clemetson AB. Vitamin C. Vol. I. Boca Raton, Florida, United States: CRC Press; 2018. ISBN 978-1-351-08601-1

[23] Burak M, Imen Y. Flavonoids, and their antioxidant properties. Turkiye Klin Tip Bil Derg. 1999;**19**:296-304

[24] Ovando C, Hernandez D, Hernandez E, et al. Chemical studies of anthocyanins: A review. Food Chemistry. 2009;**113**:859-871

[25] Havsteen B. The biochemistry and medical significance of the flavonoids. Pharmacology & Therapeutics. 2002;**2002**(96):67-202

[26] Luna Guevara ML, Ochoa Velasco CE, Carranza PH, Contreras Cortes LEU, Luna Guevara JJ. Composition, physico-chemical properties, and antioxidant capacity of Renealmia alpinia (Rottb.) Maas fruit. Nutritional and antioxidants attributes of Renealmia alpinia (Rottb.) Maas fruit. Revista de la Facultad de Ciencias Agrarias UNCuyo. 2018;**50**(2):377-385. ISSN impreso 0370-4661. ISSN (en línea) 1853-8665

[27] Randhir R, Lin YT, Shetty K. Phenolics, their antioxidant and antimicrobial activity in dark germinated fenugreek sprouts in response to peptide and phytochemical elicitors. Asia Pacific Journal of Clinical Nutrition. 2004;**v**(13):295-307

[28] Lin D, Xiao M, Zhao J, Li Z, Xing B, Li X, et al. An overview of plant phenolic compounds and their importance in human nutrition and management of type 2 diabetes. Molecules. 2016;**21**(10): 1374. DOI: 10.3390/molecules21101374

[29] Babbar N, Oberoi HS, Sandhu SK, Bhargav VK. Influence of different solvents in extraction of phenolic compounds from vegetable residues and their evaluation as natural sources of antioxidants. Journal of Food Science and Technology. 2014;**51**:2568-2575

[30] Sieniawska E, Baj T. Pharmacognosy: Fundamentals, Applications, and Strategies. Amsterdam, Netherland: Elsevier Science; 2017. DOI: 10.1016/C2014-0-01794-7. ISBN 978-0-12-802104-0

[31] Evans W.C. Trease and Evans Pharmacognosy. 13th ed. London: Baillière Tindall; 1989. pp. 386-393

[32] Serrano J, Puupponen-Pimiä R, Dauer A, Aura AM, Saura-Calixto F.

Tannins: Current knowledge of food sources, intake, bioavailability, and biological effects. Molecular Nutrition & Food Research. 2009;**53**(Suppl. 2): S310-S329. DOI: 10.1002/mnfr.200 900039

[33] Hättenschwiler S, Vitousek P. The role of polyphenols in terrestrial ecosystem nutrient cycling. Trends in Ecology and Evolution. 2000;**15**:238-243. DOI: 10.1016/S0169-5347(00)01861-9

[34] Houghton I. The Physiology of Cannabis Terpenes and Terpenoids—A Brief Overview. Elliot Barker. 2016

[35] International Union of Pure and Applied Chemistry. IUPAC Compendium of Chemical Terminology. North Carolina, United States: IUPAC; 2019 (2005-2021). DOI: 10.1351/goldbook. T06279

[36] Firn R. Nature's Chemicals: The Natural Products that Shaped our World. Oxford Scholarship Online. United Kingdom: Oxford University Press; 2010. DOI:10.1093/acprof: oso/9780199566839.001.0001. ISBN-13: 9780199566839

[37] Cragg GM, Newman DJ. Biodiversity: A continuing source of novel drug leads. Pure and Applied Chemistry. 2005;**77**:7-24

[38] Shashank K, Abhay KP. Chemistry and biological activities of flavonoids: An overview. The Scientific World Journal. 2013;**2013**:162750. 16 pages. DOI: 10.1155/2013/162750

[39] Lara R. Antibiotic discovery, Labome the World Laboratories. Materials and Methods. 2018;**8**:2671. DOI: 10.13070/ mm.en.8.2671

[40] Conly JM, Johnston BL. Where are all the new antibiotics? The new antibiotic paradox. Canadian Journal of Infectious Diseases and Medical Microbiology. 2005;**16**:892058. 2 pages. DOI: 10.1155/2005/892058

[41] Newman DJ, Cragg GM. Natural products as sources of new drugs over the nearly four decades from 01/1981 to 09/2019. Journal of Natural Products. 2020;**83**(3):770-803. DOI: 10.1021/acs. jnatprod.9b01285

[42] Mona G, Sina O, Mohammad BO. Review of anti-inflammatory herbal medicines. Advances in Pharmacological and Pharmaceutical Sciences. 2016; **2016**:9130979. 11 pages. DOI: 10.1155/ 2016/9130979

[43] Aswad M, Rayan M, Abu-Lafi S, et al. Nature is the best source of anti-inflammatory drugs: Indexing natural products for their anti-inflammatory bioactivity. Inflammation Research. 2018;**67**:67-75. DOI: 10.1007/ s00011-017-1096-5

[44] Huang M, Lu JJ, Ding J. Natural products in cancer therapy: Past, present and future. Natural Products Bioprospecting. 2021;**11**:5-13. DOI: 10.1007/s13659-020-00293-7

[45] Block KI, Gyllenhaal C, Lowe L, et al. Designing a broad-spectrum integrative approach for cancer prevention and treatment. Seminars in Cancer Biology. 2015;**35**(Suppl):S276-S304

[46] Dionysia T, Işıl T, Gökhan K, Gökay MB, Vaso Z, Athanasia P, et al. Mining natural products with anticancer biological activity through a systems biology approach. Oxidative Medicine and Cellular Longevity. 2021;**2021**:9993518. 17 pages. DOI: 10.1155/2021/9993518

[47] Sammar M, Abu-Farich B, Rayan I, Falah M, Rayan A. Correlation between cytotoxicity in cancer cells and free

radical-scavenging activity: In vitro evaluation of 57 medicinal and edible plant extracts. Oncology Letters. 2019;**18**(6):6563-6571

[48] de Mejia EG, Dia VP. The role of nutraceutical proteins and peptides in apoptosis, angiogenesis, and metastasis of cancer cells. Cancer Metastasis Reviews. 2010;**29**(3):511-528

[49] Muhammad A, Tariq K, Kaneez F, Qurat ul Ain A, Muhammad O, Ali TK, et al. Selected hepatoprotective herbal medicines: Evidence from ethnomedicinal applications, animal models, and possible mechanism of actions. Phytotherapy Research. 2018;**32**(2):199-215. DOI: 10.1002/ptr.5957

[50] Meng X, Li Y, Li S, Gan RY, Li HB. Natural products for prevention and treatment of chemical-induced liver injuries. Comprehensive Reviews in Food Science and Food Safety. 2018;**17**:472-495. DOI: 10.1111/1541-4337.12335

[51] Kittakoop P, Mahidol C, Ruchirawat S. Alkaloids as important scaffolds in therapeutic drugs for the treatments of cancer, tuberculosis, and smoking cessation. Current Topics in Medicinal Chemistry. 2014;**14**(2):239-252. DOI: 10.2174/1568026613666131216105049

[52] Raymond SS, Jonathan SJ, Watkins-Pitchford MJ. The Essence of Analgesia and Analgesics. Cambridge, Chambersburg, New York: Cambridge University Press; 2010. pp. 82-90. ISBN 978-1139491983

[53] Prommer E. Role of codeine in palliative care. Journal of Opioid Management. 2010;7(5):401-406. DOI: 10.5055/jom.2011.0081

[54] Russo P, Frustaci A, Del Bufalo A, Fini M, Cesario A. Multitarget drugs of plants origin acting on Alzheimer's disease. Current Medicinal Chemistry. 2013;**20**(13):1686-1693. DOI: 10.2174/0929867311320130008

[55] Cushnie TP, Cushnie B, Lamb AJ. Alkaloids: An overview of their antibacterial, antibiotic-enhancing and antivirulence activities. International Journal of Antimicrobial Agents. 2014;**44**(5):377-386. DOI: 10.1016/j.ijantimicag.2014.06.001

[56] Qiu S, Sun H, Zhang AH, Xu HY, Yan GL, Han Y, et al. Natural alkaloids: Basic aspects, biological roles, and future perspectives. Chinese Journal of Natural Medicines. 2014;**12**(6):401-406. DOI: 10.1016/S1875-5364(14)60063-7

[57] Hillier K. Atropine. In: xPharm: The Comprehensive Pharmacology Reference, S.J. Enna and David B. Bylund, editors. Amsterdam, Netherland: Elsevier; 2008. pp. 1-5. ISBN 978-0-08-055232-3. DOI: https://doi.org/10.1016/B978-008055232-3.61267-X.

[58] De Abreu IN, Sawaya AC, Eberlin MN, Mazzafera P. Production of Pilocarpine in Callus of Jaborandi (Pilocarpus microphyllus Stapf). In Vitro Cellular & Developmental Biology. 2005;**41**(6):806-811. DOI: 10.1079/IVP2005711

[59] Fischer J, Ganellin CR. Analogue-Based Drug Discovery. Hoboken, New Jersey, United States: John Wiley and Sons; 2006. p. 512. ISBN 9783527607495

[60] Evans WC. Trease and Evans Pharmacognosy. 16th ed. London, U.K.: W. B. Saunders Ltd; 2009. pp. 191-393

[61] Rengul CA, Irem D. In: Vinood BP, Rajkumar R, Victor RP, editors. Cardiac Glycosides and Oxidative Stress in Liver Cancer: The Liver Oxidative Stress and Dietary Antioxidants. Amsterdam, Netherland: Elsevier; 2018. pp. 55-61

[62] Roberts DM, Gallapatthy G, Dunuwille A, Chan AB. Pharmacological treatment of cardiac glycoside poisoning. British Journal of Clinical Pharmacology. 2015;**81**(13):488-495. DOI: 10.1111/bcp.12814

[63] Aronson JK. Meyler's Side Effects of Drugs: The International Encyclopedia of Adverse Drug Reactions and Interactions. 16th ed. Vol. 7. 2015. p. 7674. eBook ISBN: 9780444537164. Print Book ISBN: 9780444537171.

[64] Ahmed SI, Hayat MQ, Tahir M, Mansoor Q, Ismail M, Keck K, et al. Pharmacologically active flavonoids from the anticancer, antioxidant and antimicrobial extracts of *Cassia angustifolia* Vahl. BMC Complementary and Alternative Medicine. 2016;**16**:460

[65] Andreu L, Nuncio-Jáuregui N, Carbonell-Barrachin ÁA, Legua P, Hernández F. Antioxidant properties and chemical characterization of Spanish *Opuntia ficus-indica* Mill. cladodes and fruits. Journal of the Science of Food and Agriculture. 2018;**98**:1566-1573

[66] Chung KT, Wong TY, Wei CI, Huang YW, Lin Y. Tannins and human health: A review. Critical Reviews in Food Science and Nutrition. 1998;**38**(6):421-464. DOI: 10.1080/10408699891274273

[67] Yang W, Chen X, Li Y, Guo S, Wang Z, Yu X. Advances in pharmacological activities of terpenoids. Natural Product Communications. 2020;**15**(3):1-13. DOI: 10.1177/1934578X20903555

[68] Broadbent D. Antibiotics produced by fungi. The Botanical Review. 1966;**32**(3):219-242. DOI: 10.1007/BF02858660

[69] Block SS. Disinfection, Sterilization, and Preservation. Philadelphia, United States: Lippincott Williams and Wilkins; 2001. p. 631. ISBN 978-0683307405

[70] Richardson MD, Warnock DW. Fungal Infection Diagnosis and Management. Hoboken, New Jersey, United States: John Wiley and Sons; 2003. ISBN 978-1-4051-15780

[71] Kim H, Baker JB, Park Y, Park HB, DeArmond PD, Kim SH, et al. Total synthesis, assignment of the absolute stereochemistry, and structure-activity relationship studies of subglutinols A and B. Chemistry: An Asian Journal. 2010;**5**(8):1902-1910. DOI: 10.1002/asia.201000147

[72] Pradeep P, Manju V, Ahsan MF. In: Agrawal DC, Dhanasekaran M, editors. Antiviral Potency of Mushroom Constituents, Medicinal Mushrooms: Recent Progress in Research and Development. Singapore: Springer; 2019. pp. 275-297. DOI: 10.1007/978-981-13-6382-5_10. ISBN 9789811363825

[73] Friedman M. Mushroom polysaccharides: Chemistry and antiobesity, antidiabetes, anticancer, and antibiotic properties in cells, rodents, and humans. Food. 2016;**5**(4):80. DOI: 10.3390/foods5040080

[74] Lo HC, Wasser SP. Medicinal mushrooms for glycemic control in diabetes mellitus: History, current status, future perspectives, and unsolved problems (review). International Journal of Medicinal Mushrooms. 2011;**13**(5):401-426. DOI: 10.1615/intjmedmushr.v13.i5.10

[75] Nagaraj G, Uma MV, Shivayogi MS, Balaram H. Antimalarial activities of peptide antibiotics isolated from fungi. Antimicrobial Agents and Chemotherapy. 2001;**45**(1):145-149. DOI: 10.1128/aac.45.1.145-149.2001

[76] Lancini G, Demain AL. Bacterial Pharmaceutical Products. In: Rosenberg E, DeLong EF, Lory S, Stackebrandt E, Thompson F, editors. The Prokaryotes. Berlin, Heidelberg: Springer; 2013. pp. 257-280. DOI: 10.1007/978-3-642-31331-8_28

[77] Waksman SA. The farther of antibiotics. The Journal of Biological Chemistry. 2004;**279**(48):e7

[78] Johnson BA, Anker H, Meleney FL. Bacitracin: A new antibiotic produced by a member of the *B. subtilis* group. Science. 1945;**102**(2650):376-377. DOI: 10.1126/science.102.2650.376

[79] Bennett JE, Dolin R, Blaser MJ, Mandell GL. Mandell, Douglas, and Bennett's Principles and Practice of Infectious Diseases E-Book. Amsterdam, Netherlands: Elsevier Health Sciences; 2009. p. 469. ISBN 9781437720600

[80] Weinstein MJ, Luedemann GM, Oden EM, Wagman GH, Rosselet JP, Marquez JA, et al. Gentamicin, a new antibiotic complex from Micromonospora. Journal of Medicinal Chemistry. 1963;**6**(4):463-464. DOI: 10.1021/jm00340a034

[81] Walker SR. Trends and Changes in Drug Research and Development. Berlin, Heidelberg, Germany: Springer Science & Business Media; 2012. p. 109. ISBN 9789400926592

[82] Nelson ML, Levy SB. The history of the tetracyclines. Annals of the New York Academy of Sciences. 2011;**1241**(Antimicrobial Therapeutics Review):17-32. DOI: 10.1111/j.1749-6632.2011.06354.x. ISSN 0077 8923

[83] World Health Organization. World Health Organization Model List of Essential Medicines: 21st List 2019. Geneva: World Health Organization; 2019. hdl: 10665/325771. WHO/MVP/EMP/IAU/2019.06

[84] Sensi P. History of the development of rifampin. Reviews of Infectious Diseases. 1983;**5**(Suppl. 3):S402-S406. DOI: 10.1093/clinids/5.supplement_3.s402

[85] Butler MS, Hansford KA, Blaskovich MA, Halai R, Cooper MA. Glycopeptide antibiotics: Back to the future. The Journal of Antibiotics. 2014;**67**(9):631-644. DOI: 10.1038/ja.2014.111

[86] Woodruff HB, Selman A. Waksman, winner of the 1952 Nobel Prize for physiology or medicine. Applied and Environmental Microbiology. 2014;**80**(1):2-8. DOI: 10.1128/AEM.01143-13

[87] Pongs O. Chapter 3: Chloramphenicol. In: Hahn FE, editor. Mechanism of Action of Antibacterial Agents. Antibiotics Volume V Part 1. Berlin, Heidelberg: Springer Berlin Heidelberg. 1979. pp. 26-42. ISBN 978-3-642-46403-4

[88] Wink M. Plant breeding: Importance of plant secondary metabolites for protection against pathogens and herbivores. Theoretical and Applied Genetics. 1988;**75**:225-233. DOI: 10.1007/BF00303957

[89] Smolander A, Kanerva S, Adamczyk B, Kitunen V. Nitrogen transformations in boreal forest soils—does composition of plant secondary compounds give any explanations? Plant and Soil. 2012;**350**:1-26. DOI: 10.1007/s11104-011-0895-7

[90] Rasmann S, Köllner TG, Degenhardt J, Hiltpold I, Toepfer S, Kuhlmann U, et al. Recruitment of entomopathogenic nematodes by insect-damaged maize roots. Nature.

2005;**434**:732-737. DOI: 10.1038/nature03451

[91] Rattan RS. Mechanism of action of insecticidal secondary metabolites of plant origin. Crop Protection. 2010;**29**(9):913-920. DOI: 10.1016/j.cropro.2010.05.008

[92] Ensley SM. Pyrethrins and pyrethroids, In: Ramesh C. Gupta, editor. Veterinary Toxicology. 3rd ed. Cambridge, Massachusetts, United States: Academic Press; 2018. pp. 515-520. DOI: 10.1016/B978-0-12-811410-0.00039-8. ISBN 9780128114100

[93] Clemensen AK, Provenza FD, Hendrickson JR, Grusak MA. Ecological implications of plant secondary metabolites—phytochemical diversity can enhance agricultural sustainability. Frontiers in Sustainable Food Systems. 2020;**4**:233. DOI: 10.3389/fsufs.2020.547826

[94] Kallscheuer N, Classen T, Drepper T, Marienhagen J. Production of plant metabolites with applications in the food industry using engineered microorganisms. Current Opinion in Biotechnology. 2019;**56**:7-17. DOI: 10.1016/j.copbio.2018.07.008

[95] Shaw D. Risks or remedies? Safety aspects of herbal remedies. Journal of Royal Society of Medicine. 1998;**91**:294-296

[96] Nwude N, Ibrahim MA. Plants used in traditional veterinary medical practice in Nigeria. Journal of Veterinary Pharmacology and Therapeutics. 1980;**3**:261-273

[97] Akinyemi B. Recent concept in plaque formation. Journal of Clinical Pathology. 2000;**30**:13-16

[98] Newman DJ, Cragg G, Snader KM. The influence of natural products upon drug discovery. Natural Product Reports. 2000;**17**:175-285

[99] Ghani A. Introduction to Pharmacognosy. 1st ed. Zaria, Nigeria: Ahmadu Bello University Press; 1990. p. 58

[100] Rates SMK. Plants as source of drugs. Toxicon. 2001;**39**:603-613

[101] Soldati F. The registration of medicinal plant products, what quality of documentation should be required? The industrial point of view. In: World Congress on Medicinal and Aromatic Plants for Human Welfare 2, Abstracts. Mendoza: ICMPA/ISHS/SAIPOA; 1997. p. L-48

[102] Anjaria JV. Traditional (indigenous) veterinary medicine project. Final report, Livestock evelopment project. Sri Lanka Asian Development Bank. Gannoruwa, Peradeniya: Sri Lanka Veterinary Research Institute; 1986. pp. 15-24

[103] Umashankar DD. Plant secondary metabolites as potential usage in regenerative medicine. The Journal of Phytopharmacology. 2020;**9**:270-273

[104] Rauf A, Jehan N. The folkloric use of medicinal plants in public health care. In: Public Health. Madrid, Spain: SM group; 2015. pp. 1-12. Available from: www.smgebooks.com

[105] Huang Q, Lu G, Shen HM, Chung MCM, Ong CN. Anti-cancer pr operties of anthraquinones from rhubarb. Medical Research Review. 2007;**27**:609-630. DOI: 10.1002/med.20094

[106] Li S, Yang WTC, Lai CC, Huang SH, Liao JM, Wan L, et al. Antiviral activity of aloe-emodin against influenza A virus via galectin-3 up-regulation. European Journal of Pharmacology.

2014;**738**:125-132. DOI: 10.1016/j.ejphar.2014.05.028

[107] International Agency for Research on Cancer (IARC). Some Traditional Herbal Medicines, Some Mycotoxins, Naphthalene and Styrene. In: IARC Monographs on the Evaluation of Carcinogenic Risks to Humans. Vol. 82. Lyon, France: IARC Press; 2002

[108] Metodiewa D, Kochman A, Karolczak S. Evidence for antiradical and antioxidant properties of four biologically active N, N,diethylaminoethyl ethers of flavanone oximes: A comparison with natural polyphenolic flavonoid (rutin) action. Biochemistry and Molecular Biology International. 1997;**41**:1067-1075

[109] Farnsworth NF. Biological and phytochemical screening of plants. Journal of Pharmaceutical Sciences. 1966;**55**:225-276

[110] Just ML, Recio MC, Giner RM, Cuéllar MJ, Marĩez S, Bilia AR, et al. Anti-inflammatory activity of unusual lupane saponins from Bupleurum fruticescens. Planta Medica. 1998;**64**(5):404-407

[111] Abed El Aziz MMA, Ashour AS, Melad ASG. A review on saponins from medicinal plants: Chemistry, isolation, and determination. Journal of Nonomedicine Research. 2019;**7**(4):282-288

[112] Kaiser R. The Scent of Orchids: Olfactory and Chemical Investigations. Basel: Editiones Roche; 1993

[113] Harper DJ. Early Chinese Medical Literature: The Mawangdui Medical Manuscripts. London: Kegan Paul Intern; 1998

[114] Evangelista LF (managing editor). MIMS Annual. 1995. p. 1082

[115] Mikail HG, Karvouni H, Kotsiou A, Tesseromatis C, Magiatis P. New alkylresorcinols from a lipophilic extract of Urginea indica L. bulbs showing experimental trauma healing activity. Fitoterapia. 2015;**101**:41-45

Chapter 4

Secondary Metabolites from Natural Products

Stella Omokhefe Bruce

Abstract

Natural products are substances that are confined from living organisms, they are in the form of primary or secondary metabolites. Secondary metabolites are compounds with varied chemical structures, produced by some plants and strains of microbial species. Unlike primary metabolites (nucleotides, amino acids, carbohydrates, and lipids) that are essential for growth, secondary metabolites are not. Secondary metabolites are produced or synthesized during the stationary stage. In this chapter, we will discuss secondary metabolites from natural products synthesized mainly by plants, fungi, and bacteria. Plants synthesize a large diversity of secondary metabolites; plant secondary metabolites are split into four groups namely alkaloids, phenolic compounds, terpenoids, and glucosinolates. Several classes of fungal and bacterial secondary metabolites, their sources, and pharmacological uses associated with the secondary metabolites are also discussed. Therefore, several classes of secondary metabolites are responsible for the biological and pharmacological activities of plants and herbal medicines.

Keywords: secondary metabolites, natural products, alkaloids, phenolic compounds, terpenes

1. Introduction

Secondary metabolites are natural products synthesized mainly by plants, fungi and bacteria. Secondary metabolites are molecules with low molecular weight and various biological activities and chemical structures [1]. Secondary metabolites are also called specialized metabolites; they generally mediate ecological interactions by increasing their ability to survive [2]. Secondary metabolites function as a defense against herbivores and other interspecies in plants; and it was first established by A. kossel in 1910, and was discovered 20 years later as an end product of nitrogen metabolism by Friedrich Czapek a Botanist [3].

2. Plant secondary metabolites

Plants are capable of manufacturing diverse types of organic compounds which are grouped into primary and secondary metabolites [3]. Some secondary

IntechOpen

metabolites are phenylpropanoids or cinnamic acids, which protect plants from UV damage [4]. Since ancient times, the plant secondary metabolite's biological effects in humans have been known. The herb *Artemisia annua* contains Artemisinin, which is widely used in herbal or traditional medicine. Plant secondary metabolites can be divided into four major classes: alkaloids, phenolic compounds, terpenes, and glucosinolates [5, 6].

2.1 Alkaloids

Plants are natural products and the oldest source of alkaloids, examples of the most widely recognized alkaloids are morphine, quinine, strychnine, and cocaine [7]. Alkaloids are present as water-soluble salts of organic acids, esters, tannins (Cinchona bark) or in plant tissues [7, 8].

Most alkaloids are isolated in the form of crystalline, non-odorous, nonvolatile and amorphous compounds, low molecular weight alkaloids, such as arecoline and pilocarpine, non-oxygen atom alkaloids such as sparteine and nicotine occur in the liquid form, these are all from plant matrices. Majority of alkaloids are colorless with a bitter taste, apart from colchicine and berberine. Alkaloids are derived from plant sources and a diverse group of nitrogen-containing basic compounds, which contain one or more nitrogen atoms. Chemically they are heterogeneous. Based on chemical structures, they are classified into two broad categories [9]:

Examples of plants with alkaloids include, *Datura stramonium*, *Atropa belladonna*, *Erythroxylum coca*, *Solanaceae* (nightshade) plant family, *Papaver somniferum*, and *Catharanthus roseus* [9].

Alkaloids (about 20,000) are isolated from plants, but it have also been found in microorganisms, marine organisms such as algae, dinoflagellates, and pufferfish, and terrestrial animals such as insects, salamanders, and toads [10].

Classification based on the botanical origin of the alkaloids, their Sources and pharmacological properties are listed below (**Table 1**). For example., *Papaver* (opium) alkaloids, *Cinchona* alkaloids, *Rauvolfia* alkaloids, *Catharanthus* alkaloids, *Strychnos* alkaloids, Ergot alkaloids, cactus alkaloids, and *Solanum* alkaloids [10], while the structures of some alkaloids are shown in **Figure 1**.

2.2 Phenolic compounds

Plant secondary metabolism produces phenolic compounds with chemical structures of one hydroxyl aromatic ring. These phenolic compounds are classified based on their carbon chain [11]. Phenolic compounds are found in plant tissues, fruits and vegetables and are also ubiquitously distributed phytochemicals. Phenolic compounds are synthesized through phenylpropanoid and shikimic acid pathways [12]. Phenolic compounds possess numerous bioactive properties and health-protective effects, although they are not nutrients, therefore postharvest treatments have been used to enhance or preserve the phenolic compounds in fruits and vegetables [12]. Phenolic compounds possess an aromatic ring with one or more hydroxyl substituents that can be divided into several classes, which are common chemical structures essential for health benefits [13].

Plant materials like (Tropical Root and Crops) contain two classes of phenolic compounds as hydroxybenzoic acids and hydroxycinnamic acids. Phenolic compounds are present in Nigerian ***Centaurea perrottetii*** **DC. [family COMPOSITAE]** and other related genera (*Cheirolophus, Rhaponticoides,* and *Volutaria*) [14].

Alkaloid	Source	Properties
Ajmaline	*Rauvolfia serpentina*	Antiarrhythmic, antihypertensive
Caffeine	*Coffea arabica*	Stimulant, insecticide
Camptothecin	*Camptotheca acuminata*	Antineoplastic
Cocaine	*Erythroxylon coca*	Analgesic, narcotic, local anesthetic
Codeine	*Papaver somniferum*	Analgesic, antitussive
Emetine	*Uragoga ipecacuanha*	Antiamoebic, expectorant, emetic
Hyoscyamine	*Atropa belladonna* and others	Anticholinergic
Morphine	*P. somniferum*	Analgesic, narcotic
Nicotine	*Nicotiana tabacum*	Stimulant
Pilocarpine	*Pilocarpus jaborandi*	Cholinergic
Quinidine	*Cinchona* spp.	Antiarrhythmic
Quinine	*Cinchona* spp.	Antimalarial
Reserpine	*R. serpentina*	Tranquilizer
Scopolamine	*Hyoscyamus niger* and others	Sedative, anticholinergic
Strychnine	*Strychnos nux-vomica*	Stimulant, poison
Taxol	*Taxus brevifolia*	Antineoplastic
Vinblastine and vincristine	*Catharanthus roseus*	Antineoplastic

Table 1.
Spurces and pharmacological uses of selected plant-derived alkaloids.

The phenolic compounds found in plants are represented in **Table 2**, while the categories of phenolic compounds and their representative compounds are shown in **Figure 2**. Phenolic compounds survive in plant material, in either a soluble or a bound form [15, 16].

2.3 Terpenoids

Terpenes are a unique group of hydrocarbon-based natural products whose structures are derived from isoprene. Terpenoid secondary metabolites occur in plant tissue types often secured in secretory structures [17]. Over 30,000 members of terpenes are in an enormous class of natural products, they have been used for a broad variety of purposes including medicine, flavoring and perfume [18]. Terpenes as a broad group with ecological roles, that exhibit a range of deadly to entirely edible toxicity, which include antimicrobial properties and other properties [19, 20].

Plants and flowering plants (angiosperms) subdivisions have colonized the majority of the terrestrial surface, courtesy of rich levels of specialization and the relationships with other organisms [21].

Terpenes are important plant metabolites that include substances like floral fragrances that serve as plant hormones (gibberellic and abscisic acid), growth inhibitors, insect attractants, pine oil, and insecticides [22].

Terpenoids or isoprenoids are high in plants where many can be considered secondary metabolites and have fundamental roles in the metabolism of all organisms [23]. Terpenoid secondary metabolism in plants began with the recruitment of genes

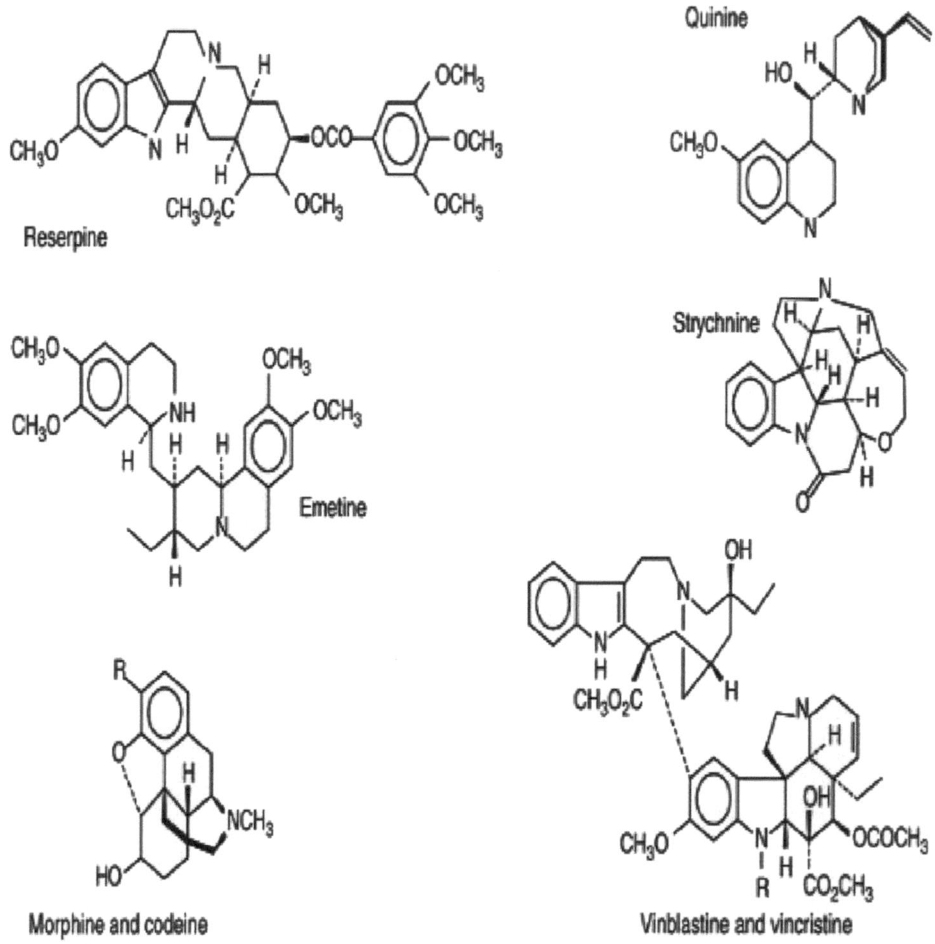

Figure 1.
Structures of some alkaloids. Note that the structures of morphine and codeine are based on the same skeleton, but are decorated with different functional groups in the position represented by 'R'. In morphine, this group is –OH, while in codeine it is CH_2O. Similarly, vinblastine and vincristine are based on the same skeleton, but differ in the nature of the R-group, which for vinblastine is $–CH_3$ and for vincristine is –CHO.

from primary metabolism [24] and accelerated due to the proliferation of cytochrome P450 and terpene synthase gene families in the genomes of plants [25].

Terpenoids play various physiological and ecological functions in plant life and human through direct and indirect plant defenses, because of their enormous applications in the pharmaceutical, food and cosmetics industries [26]. Examples of terpenoids from plant species are 1). Artemisinin, present in *A. annua*, Chinese wormwood. 2). Tetrahydrocannabinol, present in *Cannabis sativa,* cannabis. 3). Azadirachtin, present in *Azadirachta indica,* the (Neem tree). 4). Saponins, glycosylated triterpenes present in *Chenopodium quinoa*, quinoa [27, 28].

2.4 Glucosinolates

The pungent smell of plants (mustard, cabbage, and horseradish) is due to mustard oils produced from glucosinolates [29]. Glucosinolates are biosynthesized from amino acids, which consists of three glucosinolate subtypes (aliphatic, indole

Polyphenolic Compounds	Example	Fruit Source
Phenolic acids	*Hydroxycinnamic acids* Caffeic acid Chlorogenic acid Ferulic acid Sinapic acid Cattaric acids Neochlorogenic acid *p*-Coumaric acid	Blackberry, raspberry, strawberry, blackcurrant, blueberry, cranberry, pear, sweet cherry, apple, orange, grapefruit, lemon, and peach
	Hydroxybenzoic acids Ellagic acid Gallic acid	Strawberry, raspberry, grapes, longan seed, and pomegranate
Flavonoids	*Flavonols* Myricetin Quercetin Kaempferol Isorhamnetin	Apples, apricots, grapes, plums, bilberries, cranberries, olive, elderberries, currants, cherries, blackberries, and blueberries
	Flavanones Naringenin Hesperetin	Lemon, orange, grapefruit, and tangerine
	Flavones Apigenin Luteolin Tangeretin Nobiletin	Citrus fruits and pear
	Flavan-3-ols (+)-Catechin (−)-Epicatechin (−)-Epicatechin 3-gallate (−)-Epigallocatechin-3-gallate (+)-Gallocatechin (−)-Epigallocatechin Procyanidins Prodelphinidins	Apples, apricots, grapes, peaches, nectarines, raspberries, cherries, blackberries, blueberries, cranberries, pears, plums, and raisins
	Anthocyanins Cyanidin 3-galactoside Cyanidin 3-glucoside Cyanidin 3-arabinoside Cyanidin 3-xyloside Malvidin Delphinidin Pelargonidin	Blackberries, blackcurrant, blueberries, black grape, elderberries, strawberries, cherries, plums, cranberry, pomegranate, and raspberry
	Dihydrochalcones Phloretin Phloridzin	Apple
Stilbenes	Resveratrol *trans*-Resveratrol	Grapes
Tannins	Catechin polymers Epicatechin polymers Ellagitannins Proanthocyanidins Tannic acids	Grape seed/skin, apple juice, strawberries, raspberries, pomegranate, walnuts, peach, blackberry, and plum
Lignans	Secoisolariciresinol Matairesinol	Pear

Table 2.
Selected phenolic compounds found in plants.

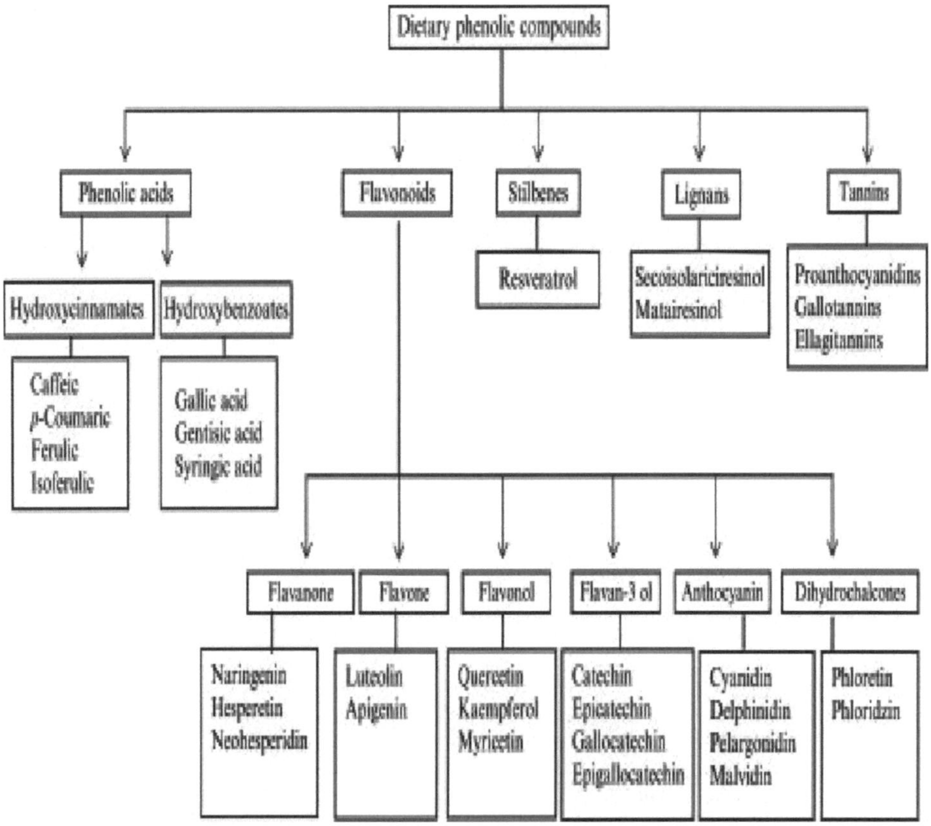

Figure 2.
Categories of phenolic compounds.

and aromatic glucosinolates) that have their corresponding precursors. Aliphatic glucosinolates are derived from isoleucine, alanine, valine, methionine, and leucine. Indole and aromatic glucosinolates are obtained from phenylalanine or tyrosine and tryptophan. Examples of the three classes of glucosinolates represented by 3methyl-sulfinylpropyl glucosinolate; indol3ylmethyl glucosinolate; and benzyl glucosinolate in **Figure 3**.

Glucosinolates are responsible for the pungent properties present in mustard, rucola, horseradish, cruciferous vegetables, and nasturtium and they are sulfur and nitrogen-containing glycosides, which protect against carcinogenesis [30].

The glucosinolates of sulforaphane (Glucoraphanin) present in broccoli, cabbage, and cauliflower (cruciferous vegetables) are responsible for protection against carcinogenesis. The Brown (*Brassica juncea*), white (*Brassica alba*) and black (*Brassica nigra*) mustards are examples of mustard seed with the family Brassicaceae [31, 32].

Secondary metabolites in plants (glucosinolates, isothiocyanates, S-methyl cysteine, allyl sulfurs, phytates, phytoestrogens) likely to protect against cancers, and antioxidant properties (phenolic compounds, flavonoids) [32].

Isothiocyanates are present in cruciferous vegetables, which is the product of the degradation of glucosinolates. *S*-methyl cysteine is a sulfur-containing phytochemicals found in all brassica vegetables [33, 34].

DOI: http://dx.doi.org/10.5772/intechopen.102222

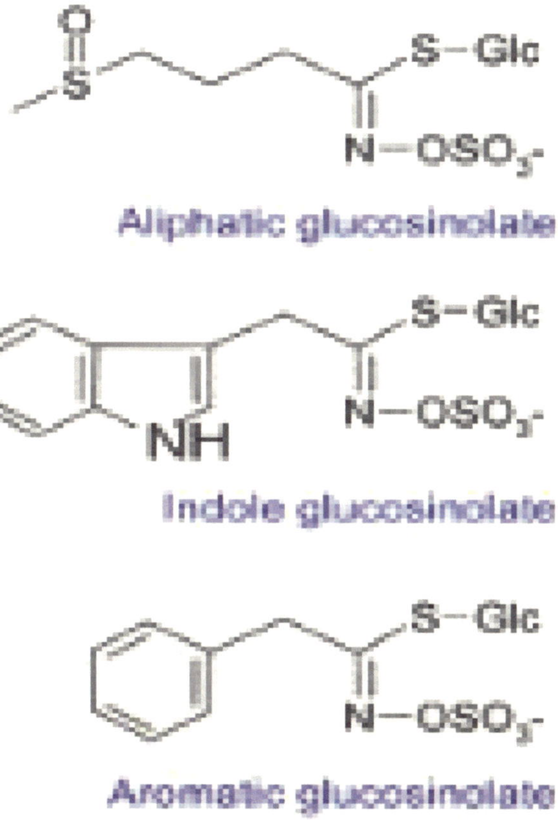

Figure 3.
Glucosinolates.

Glucosinolates contain metabolites found in the plant *Arabidopsis thaliana*. The strong taste of foods (horseradish, wasabi, and mustard) is as a result of glucosinolates [35, 36].

Over 130 glucosinolate compounds have been identified in plants, and one way that they vary is by the amino acid precursor that is incorporated during glucosinolates biosynthesis [37].

3. Fungal secondary metabolites

Fungi are eukaryotic organisms that can utilize various solid substrates of their biochemical and biological evolution and are also known to inhabit almost all ecological niches of the Earth. Some of the solid substrates utilized by fungi are decaying and dead material, such as live plants (endophytic, parasitic, and mycorrhizal fungi), lichens (lichenicolous and endolichenic fungi), insects (entomopathogenic fungi) and herbivore dung (saprophytic and coprophilous fungi). A characteristic feature of many of these fungi (filamentous growth and complex morphology), is their ability to produce secondary metabolites which are useful in pharmaceutical, agrochemical industries and food with different biological activities [38, 39].

In the production of secondary metabolites which occurs after fungal growth has stopped because of nutrient limitations but an abundant carbon source available, it is then possible to manipulate their formation. Some endophytic fungi can produce secondary metabolites known from plants. Examples include production paclitaxel (Taxol®) and camptothecin, by *Taxomyces andreanae* and *Nothapodytes foetida*, respectively, and a synthetic precursor of an anticancer drug, podophyllotoxin, by *Phialocephala fortinii* [39].

The several classes of fungal secondary metabolites are polyketides (aflatoxin and fumonisins), nonribosomal peptides (sirodesmin, peramine, siderophores) and terpenes (T-2 toxin, deoxynivalenol (DON)), indole terpenes (paxiline and lolitrems) as represented in **Figure 4**. Polyketides are building blocks of natural products and are the largest group of metabolites occurring in their greatest number. They are the most sought-after molecules because of their wide spectrum of activities (clinical, industrial and economical activities). Non-ribosomal peptides are catalyzed without mRNA template by a complex enzyme called Nonribosomal peptide-synthetase (*NRPS*) enzymes. The peptide is modified by accessory enzymes similar to polyketides and often includes noncanonical amino acids. Nonribosomal peptide-synthetase (*NRPS*) enzymes include B-lactam antibiotics, cyclosporine A and echinocandin [40, 41].

The first FDA-approved secondary metabolite was Lovastatin, to lower cholesterol levels. In oyster mushrooms [42], red yeast rice [43], and Pu-erh [44], Lovastatin occurs naturally in low concentrations. Their mode of action is inhibition of HMG-CoA reductase, and it is the enzyme responsible for converting HMG-CoA to mevalonate.

Fungal secondary metabolites are dangerous to humans. The fungi *Claviceps purpurea*, a member of the ergot group, typically growing on rye, when ingested results in the death of humans. In *C. purpurea*, a build-up of poisonous alkaloids lead to spasms and seizures, Itching, diarrhea, psychosis or gangrene and paresthesias [45].

Fungi are organisms that produce a wide range of natural products often called secondary metabolites; many natural products are of agricultural, medical, and

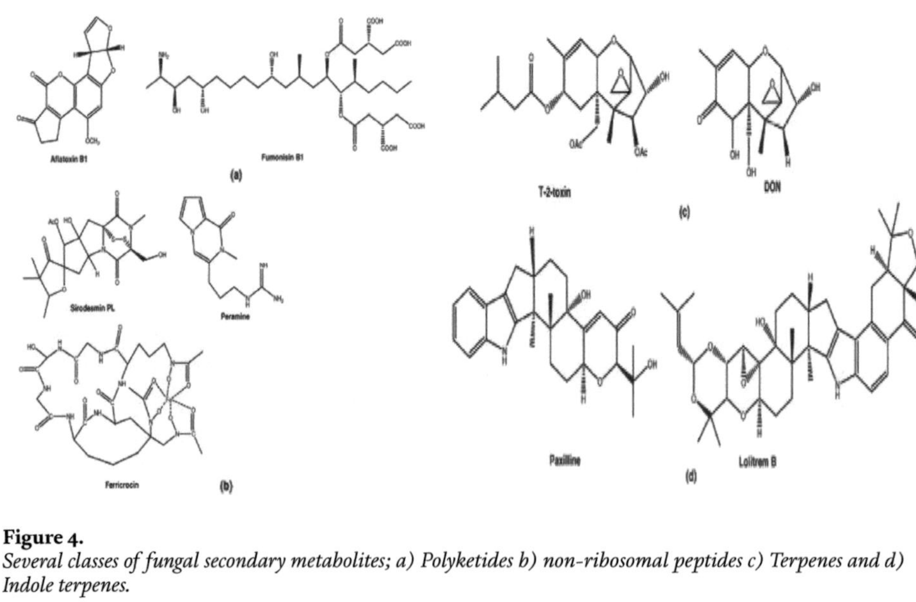

Figure 4.
Several classes of fungal secondary metabolites; a) Polyketides b) non-ribosomal peptides c) Terpenes and d) Indole terpenes.

industrial importance. Examples of natural products causing harm (mycotoxins), while others are advantageous (antibiotics) to humans [46, 47]. The biosynthesis of natural products is usually associated with cell differentiation or development, the establishment of a G-protein-mediated growth pathway in *Aspergillus nidulans* regulates both asexual sporulation and natural product biosynthesis [48].

Secondary metabolism is connected with sporulation processes in microorganisms [49, 50], including fungi [51, 52]. Secondary metabolites connected with sporulation can be classified into three groups: (i) Sporulation activated by metabolites (*A. nidulans* [53–56]), (ii) Sporulation structures from pigments (melanins [57, 58]), and (iii) toxic metabolites secreted at the time of sporulation by growing colonies (the biosynthesis of some deleterious natural products, such as mycotoxins [48, 59]). These examples of fungal secondary metabolites are shown in **Table 3**.

Natural products are essential for sporulation, examples of fungal strains that are sporulated and deficient in secondary metabolite production are *Penicillium urticae* patulin mutants [52] and *A. nidulans* sterigmatocystin mutants [67]. Secondary metabolites such as brevianamides A and B produced by *Penicillium brevicompactum* [60], some natural products have subtle effects on sporulation, as recent studies of *A. nidulans* sterigmatocystin mutants suggest that they display a decrease in asexual spore production [61, 62].

Secondary metabolites have easily visible effects on morphological differentiation in fungi, mycelium excretes compounds that can prompt sexual and asexual sporulation in other fungi [63–65], these compounds have not been identified but are assumed to be natural products produced as the mycelia ages. Other natural product such as *Fusarium graminearum* enhances perithecial production in *F. graminearum* and produces an estrogenic mycotoxin called zearalenone, an inhibitor of zearalenone synthesis, which inhibits the sexual development of this fungus [66].

Butyrolactone I, produced by the fungus *Aspergillus terreus,* is an inhibitor of eukaryotic cyclin-dependent kinases, which increases sporulation [68]. Some secondary metabolites trigger sporulation and influence the development of the

Secondary metabolite	Producing fungus	Association with development	References
Linoleic-acid derived psi factor	*Aspergillus nidulans*	Induces sporulation; affects ratio of asexual to sexual spore development	[54–57]
Zearalenone	*Fusarium graminearum*	Induces sporulation; enhances perithecial formation	[60]
Butyrolactone I	*Aspergillus terreus*	Induces sporulation and lovastatin production	[61]
Melanin	*Colletotrichum lagenarium*	Associated with appressorial formation	[62]
Melanin	*Alternaria alternata*	UV protection of spore	[62]
Melanin	*Cochliobolus heterotrophus*	Required for spore survival	[63]
Spore pigment	*Aspergillus fumigatus*	Required for virulence	[64]
Mycotoxins	*Aspergillus* spp.	Produced after sporulation	[65, 66]
Patulin	*Penicillium urticae*	Antibiotic; produced after sporulation	[53]

Table 3.
Fungal secondary metabolites.

producing organism and neighboring members of the same species. Natural product biosynthetic gene clusters can be conserved between organisms, for example, the sterigmatocystin-aflatoxin biosynthetic gene cluster in several *Aspergillus* spp. [69].

4. Bacterial secondary metabolites

The bacterial secondary metabolites are natural products source of anticholesterol agents, immune suppressants, antibiotics, antitumor agents, and other medicines; secondary metabolite-producing microorganisms synthesize these bioactive and complex molecules at the late phase and stationary phase of their growth [70–72] as shown in **Figure 5a**. In bacteria, the actinomycetes (streptomycetes) produce a significant number of chemically distinct secondary metabolites [73–76]. Other major sources include soil pseudomonas, bacilli, and myxococci [77–80]. An example of a bacterial secondary metabolite is botulinum toxin synthesized by *Clostridium botulinum*, with a positive and negative effect on humans. However, botulinum toxin has multiple medical uses for the treatment of muscle spasticity, migraine and cosmetics use [81].

Bacterial production of secondary metabolites starts in the stationary phase in response to environmental stress and lack of nutrients. Secondary metabolite synthesis in bacteria, allow them to better interact with their ecological niche and it is not essential for their growth. The b-lactam, shikimate, polyketide and non-ribosomal are the synthetic pathways for secondary metabolite production [82] as shown in **Figure 5b**. B-lactam family of cephalosporins antibiotics have been used to treat bacterial infections for 40 years and above. Gram-positive bacteria, Gram-negative bacteria, and fungi are the major sources of b-lactam antibiotics. The shikimate pathway contributes to the basic building blocks for aromatic metabolites and amino acids, which can serve as antibacterial agents. In the bacterial secondary metabolite, two enzymes can transfer a complete enolpyruvoyl moiety to a metabolic pathway, 5-enolpyruvoyl shikimate 3-phosphate synthase and chorismate synthase that require a reduced cofactor, flavin mononucleotide, for its activation. When secreted those found in the prokaryotic cell wall are endotoxins, while those poisonous compounds are known as exotoxins. Other examples of bacterial secondary metabolites are phenazine, polyketides, nonribosomal peptides, ribosomal peptides, glucosides, and alkaloids.

4.1 Phenazine

Bacteria are natural phenazines, phenazines are heterocyclic, nitrogenous compounds that differ in their physical and chemical properties. Phenazines are significant for their potential impact on bacterial interactions and biotechnological processes. It exhibits a wide range of biological activities, Pyocyanin, from *Pseudomonas aeruginosa*. Other phenazines from *Pseudomonas* sp. and *Streptomyces* sp. (Natural Products of Actinobacteria Derived from Marine Organisms) [83].

Phenazines produced by various bacteria species and excrete them in high quantities in the environment in a visible form to the naked eye, they are nitrogen-containing colored aromatic secondary metabolites. The main use of phenazines is to

DOI: http://dx.doi.org/10.5772/intechopen.102222

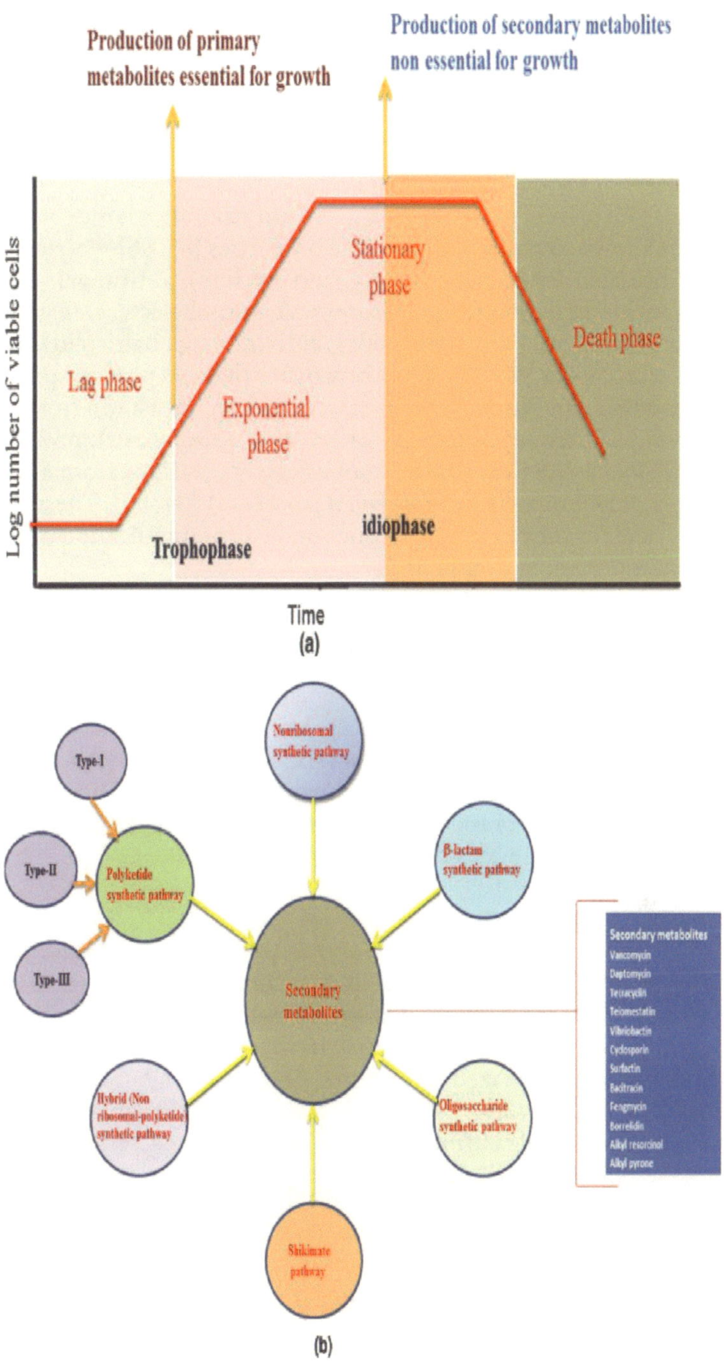

Figure 5.

a) the secondary metabolite-producing microorganisms synthesize these bioactive and complex molecules at the late phase and stationary phase of their growth. b) Secondary metabolic pathway reactions are conducted by an individual enzyme or multienzyme complexes. Intermediate or end-products of primary metabolic pathways are channeled from their systematic metabolic pathways that lead to the synthesis of secondary metabolites.

protect plants (biocontrol field), because of their antimicrobial properties. Examples of bacteria species able to produce phenazines are *Pseudomonas* spp. (including *P. aeruginosa*, *P. fluorescens*, and *Pseudomonas chlororaphis*) [84].

4.2 Polyketides

Polyketides from plants, bacteria, fungi, and animals, are a large group of secondary metabolites known to possess remarkable properties [85, 86]. Polyketides possess some bioactivities such as antibacterial (e.g., tetracycline), antifungal (e.g., amphotericin B), immune-suppressing (e.g., rapamycin), anti-cholesterol (e.g., lovastatin), anti-inflammatory activity (e.g., flavonoids), antiviral (e.g., balticolid), and anti-cancer (e.g., doxorubicin) [87–93]. Some organisms that can produce polyketides are plants (e.g., emodin from *Rheum palmatum*), fungi (e.g., lovastatin from *Phomopsis vexans*), bacteria (e.g., tetracycline from *Streptomyces aureofaciens*), protists (e.g., maitotoxin-1 from *Gambierdiscus australes*), mollusks (e.g., elysione from *Elysia viridis*), and insects (e.g., stegobinone from *Stegobium paniceum*) [94–99]. These organisms can use the polyketides they produce for pheromonal communication in the case of insects and also as protective compounds.

Polyketides are a family of natural products which are synthesized by polyketide synthase (PKS) enzymes with different biological activities and pharmacological properties. They are divided into three types: type I polyketides (macrolides produced by multimodular megasynthases), type II polyketides (aromatic molecules produced by the iterative action of dissociated enzymes), and type III polyketides (small aromatic molecules produced by fungal species) [100]. Polyketides are also found in bacteria, fungi, plants, mollusks, protists, sponges, and insects. They have notable variety in their structure and function. Some examples of polyketides antibiotics are Erythromycin, Avermectin, Nystatin, and Rifamycin [100].

4.3 Nonribosomal peptides

Nonribosomal peptides (NRPs) are peptide secondary metabolites that are synthesized by nonribosomal peptide synthetases (NRPSs) (multidomain mega-enzymes), without messenger RNAs and cell ribosomal machinery [101]. Nonribosomal peptides are naturally synthesized by bacteria, fungi, and higher eukaryotes [101]. Nonribosomal peptides are also synthesized by indigoidine (pigment). Some examples of nonribosomal peptide antibiotics are; Vancomycin, bacterium, Ramoplanin, Teicoplanins, Gramicidin, Bacitracin, Polymyxin [102].

4.4 Ribosomal peptides

Streptomyces azureus is produced from several strains of streptomycetes (Thiostrepton), *Escherichia coli* produced from Microcins and Bacteriocins [82].

4.5 Glucosides

Streptomyces species produced from Nojirimycin [82].

4.6 Alkaloids

Pseudoalteromonas produced by Tetrodotoxin, a neurotoxin [82].

5. Conclusion

Natural products originate as secondary metabolites Plants possess different indigenous defensive mechanisms to cope with certain environmental stresses. Secondary metabolites are natural tools used by plants to combat biotic and abiotic stresses. Microorganisms can produce several antibiotics and other pharmaceutically important drugs to treat bacterial and fungal infections. The secondary metabolites from natural products help us to understand their classes, sources, pharmacological importance and examples associated with the secondary metabolites derived from plants, fungi, and bacteria.

Author details

Stella Omokhefe Bruce
Faculty of Pharmaceutical Sciences, Department of Pharmacognosy and Traditional Medicine, Nnamdi Azikiwe University, Awka, Nigeria

*Address all correspondence to: stellaobruce@yahoo.com

IntechOpen

© 2022 The Author(s). Licensee IntechOpen. This chapter is distributed under the terms of the Creative Commons Attribution License (http://creativecommons.org/licenses/by/3.0), which permits unrestricted use, distribution, and reproduction in any medium, provided the original work is properly cited.

References

[1] Návarová H, Bernsdorff F, Döring AC, Zeier J. Pipecolic acid, any endogenous mediator of defense amplification and priming, is a critical regulator of inducible plant immunity. Plant Cell. 2012;**24**(12):5123-5141

[2] Bourgaud F, Gravot A, Milesi S, Gontier E. Production of plant secondary metabolites: A historical perspective. Plant Science. 2001;**161**(5):839-851

[3] Bruce SO, Onyegbule FA. Biosynthesis of natural products. In: Zepka LQ, do TC, Jacob-Lopes E, editors. Bioactive Compounds - Biosynthesis, Characterization and Applications. London: IntechOpen; 2021. DOI: 10.5772/intechopen.97660. Available from: https://www.intechopen.com/chapters/76931

[4] Seigler DS. Plant Secondary Metabolism. New York: Springer US; 1998

[5] Korkina L, Kostyuk V, Potapovich A, Mayer W, Talib N, De Luca C. Secondary plant metabolites for sun protective cosmetics: From pre-selection to product formulation. Cosmetics. 2018;**5**(2):32

[6] Kumar P, Mina U. Life Sciences: Fundamentals and Practice. 3rd ed. New Delhi: Pathfinder Academy; 2013

[7] O'Connor SE. In: Mander L, Lui H-W, Alkaloids. In book: Comprehensive Natural Products II Chemistry and Biology. Vol. 1. Oxford: Elsevier; 2010. pp. 977-1007. DOI: 10.1016/B978-008045382-8.00013-7

[8] Croteau R, Kutchan TM, Lewis NG. Natural products (secondary metabolites). In: Civjan N, editor. Natural Products in Chemical Biology. Hoboken, New Jersey: Wiley; 2012. pp. 1250-1319

[9] Kukula-Koch WA, Widelski J. Chapter 9 - Alkaloids. In: Delgoda R, editor. Pharmacognosy. Fundamentals, Applications and Strategies. Lublin, Poland: Academic Press; 2017. pp. 163-198. DOI: 10.1016/B978-0-12-802104-0.00009-3

[10] Twyman RM, Stöger E, Christou P. Molecular farming. In: Encyclopedia of Applied Plant Sciences. 2nd ed. Vol. 2. Amsterdam: Elsevier Science B.V; 2003. pp. 77-82

[11] Das PR, Eun J-B. Tea antioxidants in terms of phenolic and nonphenolic metabolites. In: Preedy VR, editor. Pathology: Oxidative Stress and Dietary Antioxidants. 1st ed. London: Academic Press; 2020. pp. 357-367

[12] Bruce SO, Onyegbule FA, Ihekwereme CP. Evaluation of hepato-protective and anti-microbial activities of ethanol extracts and fractions of *Picralima nitida* seed and pod. Journal of Phytomedicine and Therapeutic. 2016;**1**(2):1-21

[13] Ayad R, Akkal S. Phytochemistry and biological activities of Algerian *Centaurea* and related genera. In: Atta-ur-Rahman, editor. Bioactive Natural Products, Studies in Natural Products Chemistry. Vol. 63. Amsterdam, The Netherlands: Elsevier; 2019. pp. 357-414

[14] Saranraj P, Behera SS, Ray RC. Chapter 7 - Traditional foods from tropical root and tuber crops: Innovations and challenges. In: Galanakis CM, editor. Innovations in Traditional Foods. Chania, Greece: Woodhead Publishing; 2019. pp. 159-191. DOI: 10.1016/B978-0-12-814887-7.00007-1

[15] Gan RY, Chan CL, Yang QQ, Li HB, Zhang D, Ge YY, et al. Bioactive

compounds and beneficial functions of sprouted grains. In: Feng H, Nemzer B, DeVries JW, editors. Sprouted Grains. United States: Woodhead Publishing and AACC International Press; 2019. pp. 191-246

[16] Tiwari R, Rana CS. Plant secondary metabolites : A review. International Journal of Engineering Research and General Science. 2015;**3**(5):661-667. ISSN 2091-2730

[17] Bruce SO, Onyegbule FA, Ezugwu CO. Pharmacognostic, physicochemical and phytochemical evaluation of the leaves of *Fadogia cienkowskii* Schweinf (Rubiaceae). Journal of Pharmacognosy and Phytotherapy. 2019;**11**(3):52-60

[18] Maimone T. Classic Terpene Syntheses I. In: An introduction to Terpenes. Baran Lab; 2002. pp. 1-18

[19] Mazid M, Khan TA, Mohammad F. Role of secondary metabolites in defense mechanisms of plants. Biology and Medicine. 2011;**3**(2):232-249

[20] Kennedy DO, Wightman EL. Herbal extracts and phytochemicals: Plant secondary metabolites and the enhancement of human brain function. Advances in Nutrition. 2011;**1**:32-50

[21] Wang G, Tang W, Bidigare RR. Terpenoids as therapeutic drugs and pharmaceutical agents. In: Zhang L, Demain AL, editors. Natural Products. Vol. 12. Totowa, NJ: Humana Press; 2013. pp. 153-175

[22] Onyegbule FA, Bruce SO, Onyekwe ON, Onyealisi OL, Okoye PC. Evaluation of the in vivo antiplasmodial activity of ethanol leaf extract and fractions of Jatropha gossypifolia in Plasmodium berghei infected mice. Journal of Medicinal Plant Research. 2019;**13**(11):269-279

[23] Cho KS, Lim Y, Lee K, Lee J, Lee JH, Lee I. Terpenes from forests and human health. Toxicological Research. 2017;**33**(2):97-106

[24] Bouvier F, Rahier A, Camara B. Biogenesis, molecular regulation and function of plant isoprenoids. Progress in Lipid Research. 2005;**44**:357-429

[25] Weng JK, Philippe RN, Noel JP. The rise of chemodiversity in plants. Science. 2012;**336**:1667-1670

[26] Roba K. The role of terpene (secondary metabolite). Holeta, Ethiopia: Holeta honeybee research center. Natural Products Chemistry & Research; 2020;**9**(8):411

[27] Kabera J, Semana E, Mussa AR, He X. Plant secondary metabolites: Biosynthesis, classification, function and pharmacological classification, function and pharmacological properties. Journal of Pharmacy and Pharmacology. 2014;**2**(7):377-392

[28] Castells AA. The Role of Terpenes in the Defensive Responses of Conifers against Herbivores and Pathogens. Spain: Universitat Autònoma de Barcelona; 2015. pp. 1-185

[29] Ishida M, Hara M, Fukino N, Kakizaki T, Morimitsu Y. Glucosinolate metabolism, functionality and breeding for the improvement of Brassicaceae vegetables. Breeding Science. 2014;**64**(1): 48-59

[30] Bone K, Mills S. Principles of herbal pharmacology. In: Principles and Practice of Phytotherapy. Modern Herbal Medicine. 2nd ed. London, United Kingdom: Churchill Livingstone; 2013. pp. 962-967

[31] Bruce SO, Onyemailu VO, Orji CE. Evaluation of The antiulcer activity and GC-MS spectroscopic analysis of the crude ethanolic

extract of *Peuraria Phaseoloide* Leaf (Roxb) Benth. (FABACEAE). World Journal of Pharmaceutical Research. 2021;**10**(7):39-59

[32] Gerber M. Oxidative stress, antioxidants and cancer. In: Sen C, Packer L, Hänninen O, editors. Handbook of Oxidants and Antioxidants in Exercise. Vol. 1220. Amsterdam: Elsevier; 2000

[33] Paluszczak J, Baer-dubowska W. DNA methylation as a target of cancer chemoprevention by dietary polyphenols. In: Polyphenols in Human Health and Disease. Elsevier; 2014. pp. 1385-1392. DOI: 10.1016/B978-0-12-398456-2.00105-5

[34] Halkier BA, Gershenzon J. Biology and biochemistry of glucosinolates. Annual Reviews of Plant Biology. 2006;**57**:303-333

[35] Baenas N, Villaño D, Cristina G-V, Moreno DA. Optimizing elicitation and seed priming to enrich broccoli and radish sprouts in glucosinolates. Food Chemistry. 2014;**204**:314-319

[36] Angelino D, Dosz EB, Sun J, Hoeflinger JL, Van Tassell ML, Chen P, et al. Myrosinase-dependent and – independent formation and control of isothiocyanate products of glucosinolate hydrolysis. Frontiers in Plant Science. 2015;**6**:831

[37] Agerbirk N, Olsen CE. Glucosinolate structures in evolution. Phytochemistry. 2012;**77**:16-45

[38] Gunatilaka AA. Fungal Secondary Metabolites. Tucson, Arizona: **Leslie** Office of Arid Lands Studies, Southwest Center for Natural Products Research and Commercialization. The University of Arizona; 2010. DOI: 10.1036/1097-8542.YB100063

[39] Onyemailu VO, Bruce SO, Iloh ES. UV-Visible and FTIR Spectroscopic Analysis of The Crude Ethanolic Extract of *Peuraria phaseoloide* Leaf (Roxb) Benth. (FABACEAE). International Journal of Modern. Pharmaceutical Research. 2021;**5**(3):148-153

[40] Quin MB, Flynn CM, Schmidt-Dannert C. Traversing the fungal terpenome. Natural Product Reports. 2014;31(10):1449-1473

[41] Boruta T. Uncovering the repertoire of fungal secondary metabolites: From Fleming's laboratory to the International Space Station. Bioengineered. 2018;**9**(1):12-16

[42] king R, Marahiel MA. Biosynthesis of nonribosomal peptides 1. Annual Review of Microbiology. 2004;**58**:453-488

[43] Liu J, Zhang J, Shi Y, Grimsgaard S, Alraek T, Fønnebø V. Chinese red yeast rice (Monascus purpureus) for primary hyperlipidemia: a meta-analysis of randomized controlled trials. Chinese Medicine. 2006;**1**(1):4

[44] Zhao ZJ, Pan YZ, Liu QJ, Li XH. Exposure assessment of lovastatin in Pu-erh tea. International Journal of Food Microbiology. 2013;**164**(1):26-31

[45] Uys H, Berk M. A controlled double blind study of zuclopenthixol acetate compared with clothiapine in acute psychosis including mania and exacerbation of chronic psychosis. European Neuropsychopharmacology. 1996;**6**:60

[46] Bruce SO, Usifoh SF, Nduka SO, Anetoh MU, Isidienu CP. A retrospective study of antimalarial drug utilization in a secondary healthcare institution in Nigeria. World Journal of Pharmaceutical Research. 2019;**8**(13):271-281

[47] Demain AL, Fang A. The natural functions of secondary metabolites. Advances in Biochemical Engineering/ Biotechnology. 2000;**69**:1-39

[48] Hicks J, Yu JH, Keller N, Adams TH. Aspergillus sporulation and mycotoxin production both require inactivation of the FadA G-alpha protein-dependent signaling pathway. The EMBO Journal. 1997;**16**:4916-4923

[49] Bruce SO, Ugwu RN, Onu JN, Iloh ES, Onwunyili AR. Pharmacognostic, antimicrobial and hepatoprotective activities of the sub-fractions of *Picralima nitida* (Durand and Hook) (APOCYNACEAE) seeds. World Journal of Pharmaceutical Sciences. 2021;**9**(8):77-91

[50] Stone MJ, Williams DH. On the evolution of functional secondary metabolites (natural products). Molecular Microbiology. 1992;**6**:29-34

[51] Ihekwereme CP, Bruce SO, Orji CE, Ibe CI, Iloh ES. Aqueous extracts of *Ocimum gratissimum* and *Anacardium occidentale* synergises in anti-diarrhoeal property. International Journal of Modern Pharmaceutical Research (IJMR). 2020;**4**(4):06-11

[52] Sekiguchi J, Gaucher GM. Conidiogenesis and secondary metabolism in *Penicillium urticae*. Applied and Environmental Microbiology. 1977;**33**:147-158

[53] Calvo AM, Gardner HW, Keller NP. Genetic connection between fatty acid metabolism and sporulation in *Aspergillus nidulans*. The Journal of Biological Chemistry. 2001;**276**:20766-20774

[54] Champe SP, El-Zayat AAE. Isolation of a sexual sporulation hormone from *Aspergillus nidulans*. Journal of Bacteriology. 1989;**171**:3982-3988

[55] Champe SP, Rao P, Chang A. An endogenous inducer of sexual development in *Aspergillus nidulans*. Journal of General Microbiology. 1987;**133**:1383-1388

[56] Mazur P, Nakanishi K, El-Zayat AAE, Champe SP. Structure and synthesis of sporogenic psi factors from *Aspergillus nidulans*. Journal of the Chemical Society, Chemical Communications. 1991;**20**:1486-1487

[57] Alspaugh JA, Perfect JR, Heitman J. *Cryptococcus neoformans* mating and virulence are regulated by the G-protein alpha subunit GPA1 and cAMP. Genes & Development. 1997;**11**:3206-3217

[58] Kawamura C, Tsujimoto T, Tsuge T. Targeted disruption of a melanin biosynthesis gene affects conidial development and UV tolerance in the Japanese pear pathotype of *Alternaria alternata*. Molecular Plant-Microbe Interactions. 1999;**12**:59-63

[59] Trail F, Mahanti N, Linz J. Molecular biology of aflatoxin biosynthesis. Microbiology. 1995;**141**:755-765

[60] Bird BA, Remaley AT, Campbell IM. Brevianamides A and B are formed only after conidiation has begun in solid cultures of *Penicillium brevicompactum*. Applied and Environmental Microbiology. 1981;**42**:521-525

[61] Ramaswamy A. Ecological analysis of secondary metabolite production in *Aspergillus* spp. Master's thesis. College Station: Office of Graduate Studies of Texas A & M University; 2002

[62] Sim SC. Characterization of Genes in the Sterigmatocystin Gene Cluster and Their Role in Fitness of *Aspergillus nidulans.* Master's thesis. College Station: Office of Graduate Studies of Texas A & M University; 2001

[63] Hadley G, Harrold CE. The sporulation of *Penicillium notatum* westling in submerged liquid cultures. Journal of Experimental Botany. 1958;**9**:418-428

[64] Park D, Robinson PM. Sporulation in *Geotrichum candidum*. Br. Mycol. Soc. 1969;**52**:213-222

[65] Onyegbule FA, Okoli OG, Bruce SO. In vivo evaluation of the antimalarial activity of the aqueous ethanol extract of Monodora myristica seed in Albino Mice. International Journal of Science and Research (IJSR). 2019;**8**(6):1530-1538

[66] Wolf JC, Mirocha CJ. Regulation of sexual reproduction in *Gibberella zeae* (*Fusarium roseum* 'Graminearum') by F-2 (zearalenone). Canadian Journal of Microbiology. 1973;**19**:725-734

[67] Shimizu K, Keller NP. Genetic involvement of cAMP-dependent protein kinase in a G protein signaling pathway regulating morphological and chemical transitions in *Aspergillus nidulans*. Genetics. 2001;**157**:591-600

[68] Schimmel TG, Coffman AD, Parsons SJ. Effect of butyrolactone I on the producing fungus, *Aspergillus terreus*. Applied and Environmental Microbiology. 1998;**64**:3703-3712

[69] Ana M. Calvo, Richard A. Wilson, Jin Woo Bok, Nancy P. Keller. Relationship between secondary metabolism and fungal development microbiology and molecular biology reviews. ASM Journals. 2020;66(3):447-459

[70] Moore BS, Hopke JN. Discovery of a new bacterial polyketide biosynthetic pathway. Chembiochem. 2001;**2**:35-38

[71] Rokem JS, Lantz AE, Nielsen J. Systems biology of antibiotic production by microorganisms. Natural Product Reports. 2007;**24**:1262

[72] Katsuyama Y, Funa N, Miyahisa I, Horinouchi S. Synthesis of unnatural flavonoids and stilbenes by exploiting the plant biosynthetic pathway in *Escherichia coli*. Chemistry & Biology. 2007;**14**:613-621

[73] Chopra I, Roberts M. Tetracycline antibiotics: Mode of action, applications, molecular biology, and epidemiology of bacterial resistance tetracycline antibiotics: Mode of action, applications, molecular biology, and epidemiology of bacterial resistance. Microbiology and Molecular Biology Reviews. 2001;**65**:232-260

[74] Onyegbule FA, Ezenwa CJ, Bruce SO, Umeokoli BO. Standardization, chemical composition and antipyretic evaluation of methanol leaf extract and fractions of chrysophyllum albidum (Sapotaceae). Tropical Journal of Natural Product Research. 2020;**4**(6):216-222

[75] Tacar O, Sriamornsak P, Dass CR. Doxorubicin: An update on anticancer molecular action, toxicity and novel drug delivery systems. The Journal of Pharmacy and Pharmacology. 2013;**65**:157-170

[76] Shushni MAM, Singh R, Mentel R, Lindequist U. Balticolid: A new 12-membered macrolide with antiviral activity from an *Ascomycetous* fungus of marine origin. Marine Drugs. 2011;**9**:844-851

[77] Li J, Kim SG, Blenis J. Rapamycin: One drug, many effects. Cell Metabolism. 2014;**19**:373-379

[78] Van de Donk NWCJ, Kamphuis MMJ, Lokhorst HM, Bloem AC. The cholesterol lowering drug lovastatin induces cell death in myeloma plasma cells. Leukemia. 2002;**16**:1362-1371

[79] Hleba L, Charousova I, Cisarova M, Kovacik A, Kormanec J, Medo J, et al. Rapid identification of *Streptomyces* tetracycline producers by MALDI-TOF mass spectrometry. Journal of Environmental Science and Health, Part A. 2018;**169**(12):940-947

[80] Parthasarathy R, Sathiyabama M. Lovastatin-producing endophytic fungus isolated from a medicinal plant *Solanum xanthocarpum*. Natural Product Research. 2015;**29**:2282-2286

[81] Huang Q, Lu G, Shen HM, Chung MCM, Ong CN. Anti-cancer properties of anthraquinones from rhubarb. Medicinal Research Reviews. 2006;**27**:609-630

[82] Gokulan K, Khare S, Cerniglia C. Metabolic pathways: Production of secondary metabolites of bacteria. In: Encyclopedia of Food Microbiology. London: Academic Press; 2020. pp. 561-569

[83] Karuppiah V, Sun W, Li Z. Natural products of actinobacteria derived from marine organisms. In: Studies in Natural Products Chemistry. Vol. 48. Amsterdam, Netherlands: Elsevier; 2016. pp. 417-446

[84] Hadla M, Halabi MA. Fundamentals of quorum sensing, analytical methods and applications in membrane bioreactors 1st Edition. In: Chormey D, Bakirdere S, Turan N, Engin G, editors. Comprehensive Analytical Chemistry. Vol. 81. Elsevier; 2018. pp. 1-308

[85] Bruce SO, Nwafor OI, Omoirri MA, Adione NM, Onyeka IP, Ezeoru VC. GC-MS, FTIR and Antiulcer screening of aqueous seed extract and oil of *Nigella sativa* in Wistar rats. Journal of Drug Delivery and Therapeutics. 2021;**11**(6):48-60

[86] Watve MG, Tickoo R, Jog MM, Bhole BD. How many antibiotics are produced by the genus *Streptomyces*? Archives of Microbiology. 2001;**176**:386-390

[87] Okoye VO, Bruce SO, Onyegbule FA. Phytochemical screening and pharmacognostic properties of *Peuraria phaseoloides* leaves (roxb) benth (fabaceae). International Journal of Public Health, Pharmacy and Pharmacology. 2020;**5**(2):11-24

[88] Kinashi H. Giant linear plasmids in *Streptomyces*: A treasure trove of antibiotic biosynthetic clusters. Journal of Antibiotics (Tokyo). 2011;**64**:19-25

[89] Baltz RH. Renaissance in antibacterial discovery from actinomycetes. Current Opinion in Pharmacology. 2008;**8**:557-563

[90] Zotchev SB. Marine actinomycetes as an emerging resource for the drug development pipelines. Journal of Biotechnology. 2012;**158**:168-175

[91] Clardy J, Fischbach MA, Walsh CT. New antibiotics from bacterial natural products. Nature Biotechnology. 2006;**24**:1541-1550

[92] Sansinenea E, Ortiz A. Secondary metabolites of soil *Bacillus* spp. Biotechnology Letters. 2011;**33**:1523-1538

[93] Bruce SO, Onyegbule FA, Ezugwu CO, Nweke ID, Ezenwelu CR, Nwafor FI. Chemical composition, hepatoprotective and antioxidant activity of the crude extract and fractions of the leaves of *Fadogia Cienkowskii* Schweinf (Rubiaceae). Tropical Journal of Natural Product Research. 2021;**5**(4):720-731

[94] Wenzel SC, Muller R. Myxobacteria—'Microbial factories' for the production of bioactive secondary metabolites. Molecular BioSystems. 2009;**5**:567-574

[95] Gross H, Loper JE. Genomics of secondary metabolite production by *Pseudomonas* spp. Natural Product Reports. 2009;**26**(11):1408-1446

[96] Witting K, Sussmuth RD. Discovery of antibacterials and other bioactive compounds from microorganisms-evaluating methodologies for discovery and generation of non-ribosomal peptide antibiotics. Current Drug Targets. 2011;**12**(11):1547-1559

[97] Kohli GS, John U, Figueroa RI, Rhodes LL, Harwood DT, Groth M, et al. Polyketide synthesis genes associated with toxin production in two species of *Gambierdiscus* (Dinophyceae). BMC Genomics. 2015;**16**(1):410

[98] Florian P, Monika H. Polyketides in insects: Ecological role of these widespread chemicals and evolutionary aspects of their biogenesis. Biological Reviews. 2008;**83**:209-226

[99] Adele C, Guido C, Guido V, Angelo F. Shaping the polypropionate biosynthesis in the solar-powered mollusc *Elysia viridis*. Chembiochem. 2008;**10**:315-322

[100] Monfil VO, Casas-Flores S. Molecular mechanisms of biocontrol in *Trichoderma* spp. and their applications in agriculture. In: Gupta VK, Schmoll M, Herrera-Estrella A, Upadhyay RS, Druzhinina I, Tuohy MG, editors. Biotechnology and Biology of *Trichoderma*. Dordrecht: Elsevier; 2014. pp. 429-453. DOI: 10.1016/B978-0-444-59576-8.00032-1

[101] Ding K, Dai LX. Organic Chemistry - Breakthroughs and Perspectives. Weinheim: Wiley-VCH; 2012. pp. 1-802

[102] Soltani J. Chapter 22 - Secondary metabolite diversity of the genus Aspergillus: Recent advances. In: Gupta VK, editor. New and Future Developments in Microbial Biotechnology and Bioengineering. Elsevier; 2016. pp. 275-295. DOI: 10.1016/B978-0-444-63505-1.00035-X

Chapter 5

Metabolomics and Genetic Engineering for Secondary Metabolites Discovery

Ahmed M. Shuikan, Wael N. Hozzein, Rakan M. Alshuwaykan and Ibrahim A. Arif

Abstract

Since 1940s, microbial secondary metabolites (SMs) have attracted the attention of the scientific community. As a result, intensive researches have been conducted in order to discover and identify novel microbial secondary metabolites. Since, the discovery of novel secondary metabolites has been decreasing significantly due to many factors such as 1) unculturable microbes 2) traditional detection techniques 3) not all SMs expressed in the lab. As a result, searching for new techniques which can overcome the previous challenges was one of the most priority objectives. Therefore, the development of omics-based techniques such as genomics and metabolomic have revealed the potential of discovering novel SMs which were coded in the microorganisms' DNA but not expressed in the lab or might be produced in undetectable amount by detecting the biosynthesis gene clusters (BGCs) that are associated with the biosynthesis of secondary metabolites. Nowadays, the integration of metabolomics and gene editing techniques such as CRISPR-Cas9 provide a successful platform for the detection and identification of known and unknown secondary metabolites also to increase secondary metabolites production.

Keywords: metabolomics, genetic engineering, secondary metabolites identification, genomic, CRISPR-Cas9, production of secondary metabolites, microorganisms, gene editing

1. Introduction

Since the discovery of penicillin in the 1940s, microbial secondary metabolites (SMs) have attracted the attention of scientists all over the world. In fact, penicillin discovery has been shown to be a promising solution for many kinds of infections. As a result, the scientific world starts to search for other products that are produced by microbes that can be utilized for treating a different disease or can be useful for any aspect of our life. Therefore, the period between the 1940s – 1960s The "golden period of SM discovery" [1, 2] is referred to as "the golden era of SM discovery." During the golden era, several SMs were discovered, characterized, and reported, and they are still used today. Unfortunately, after the golden era, the development of authorized novel chemical scaffolds of secondary metabolites has declined dramatically [1] the decrease in the microbial secondary metabolites detection and identification could be due to 1) almost 99% of the microbial community unculturable [2], due to the difficulty to identify their optimal medium compositions, which means that the majority of SMs are definitely unidentified, 2) the scientists have been focused on specific groups of microorganisms such as *Actinobacteria* which resulting to the identification of known compounds and do not develop a new methodology for screening in other microorganisms.

All biochemical reactions carried out by organisms is called metabolism and all products resulting from metabolism is called metabolites. In fact, there are two kinds of metabolites resulting from the biochemical reactions that are called primary and secondary metabolites. The difference between primary and secondary is that primary metabolites are found in all living cells able to divide while secondary metabolites are present only incidentally and are not affect the organism's life immediately. Microbial SMs is low molecular mass products with an unusual chemical structure that are produced by microorganisms usually during the late growth phase and are not essential for the growth and development of the microbe but are associated with some other functions such as competition, interactions, defense, and others [3, 4]. In fact, SMs have shown a variety of biological activities that can be utilized in different aspects such as antitumor agents, immunosuppressive agents, antimicrobial agents, antiparasitic agents, anthelmintic, and food industry etc. An example for the importance of SMs in our life is the discovery of immunosuppression such as cyclosporine A, which plays a significant role in establishing the organ transplant field.

Nowadays, Over 2 million SMs have been found based on their vast diversity in structure, function, and biosynthesis (**Table 1**). Plants (about 80%) and microbes (approximately 20%) are the primary sources of secondary metabolites discovered [3]. Actinobacteria and fungi have been found to create the bulk of SMs discovered to date [5]. Nowadays, omics-based techniques such as genomics, metabolomics, proteomics, and transcriptomics have overcome the problem of

Source	All known compounds	Bioactive
Plant kingdom	600,000–700,000	150,000–200,000
Microbes	Over 50,000	22,000–23,000
Higher plants	500,000–600,000	~100,000
Animal kingdom	300,000–400,000	50,000–100,000
Protozoa	Several hundreds	100–200
Vertebrates	200,000–250,000	50,000–70,000
Marine animals	20,000–25,000	7000–8000
Invertebrates	~100,000	NA
Algae, lichens	3000–5000	1500–2000
Insects, worms	8000–10,000	800–1000

NA–Data not available.
Source: Bérdy [5].

Table 1.
Approximate number of identified natural metabolites.

identification of unculturable microbes and have revealed that microorganisms have the potential to produce more secondary metabolites than were originally expected [6, 7]. By conducting omics techniques scientists were able to detect SMs that are coded by clustered genes present on chromosomal DNA directly without doing microbial culturing.

Due to the development of the genomic and bioinformatic field, scientists are now able to access extensive genetic information and enable genome mining of relevant Biosynthesis gene cluster (BGCs) with the potential for valuable SM production [8]. therefore, genetic engineering has now become widely used and moving beyond traditional tools which open a new era in the detection of novel secondary metabolites [9]. In fact, by using bioinformatic analysis that analyzes the putative secondary metabolites genes cluster in the sequenced genome, scientists were able to predict new SMs that were not identified by using traditional techniques because all new revealed SMs are not produced naturally under the lab conditions or even though produced but in very low amount that the traditional techniques were unable to identify them [10, 11]. Metabolomics aims to characterize and identify SMs in natural and engineered biosystems.

Metabolomics based techniques such as mass spectrometry (MS) and nuclear magnetic resonance (NMR) is accurate that can measure as low molecular weight compounds as possible. In fact, mass spectrometry (MS) and nuclear magnetic resonance (NMR) have been reported as significant analytical techniques to detect secondary metabolites under specific conditions [12]. This chapter provides an overview of metabolomics and genetic engineering techniques especially the CRISPR-Cas9 technique for the discovery and production enhancement of microbial secondary metabolites.

2. Genetic engineering for SMs detection

The genes associated with the biosynthesis of secondary metabolites is named biosynthesis gene cluster (BGCs). In fact, BGCs include all genetic information required for secondary metabolites regulation, assembly, modification, and biosynthesis [13]. As mentioned previously, not all microorganisms can be cultured in the laboratory resulting in not all SMs can be expressed by using traditional techniques (culturing and detection) also a lot of microbes contains silent or cryptic genes in their genome that are responsible for the production of secondary metabolites. In fact, these silent BGCs have potentially significant in the discovery of novel secondary metabolites [13–16].

Nowadays, instead of traditional detection techniques, genetic engineering tools are utilized for the identification of novel biosynthesis gene cluster BGCs [9]. However, genetic engineering can be used in both heterologous and homologous hosts. While gene manipulation in a homologous host allow the retention of factors necessary for the production of SMs, also gene manipulation in a heterologous host enable activation of BGCs obtained from unculturable microorganism [17].

In fact, a variety of genetic engineering techniques have been developed in order to induce the expression of all genes of interest. Therefore, in metabolomic production field, several genome techniques have been utilized in order to detect and enhance secondary metabolites production such as clustered regulatory interspaced short palindromic repeat (CRISPR-Cas9), zinc finger nucleases (ZFNs), and transcriptional

	CRISPR/Cas9	Zinc finger nucleases (ZFNs)	Transcription factors like effector nucleases (TALENs)
Protein engineering steps	It does not necessitate any protein engineering steps and is very easy to test several times. Grna	It requires complex to test gRNA	TALENs need protein engineering steps to test gRNA
Mode of action	It operates by inserting double-strand breaks or single-strand DNA nicks into the target DNA (Case9 nickase)	It can induce double-strand breaks in target DNA	Induces DSBs in target DNA
Cloning	Not Required	Required	Required
Structural proteins	CRISP R is made up of a single monomeric protein as well as chimeric RNA	ZFNs are dimeric proteins that only require one protein component to function	TALENs are also dimeric and require a protein component to function
Mutation rate	It has been discovered that there is a low rate of mutation	High mutation rate observed in plants	When compared to CRISPR, the mutation rate is high
Components	crRNA, Cas9 proteins	Zn-finger domains Non- specific FOKI nuclease domain	Zn-finger domains Non-specific folk nuclease domain
Length of target sequence (bp)	20–22	18–24	24–59
Target recognition efficiency	High	High	High
Level of experiment	Easy and very fast procedure	Complicated procedure that necessitates protein engineering expertise	Relatively easy procedure
Methylated DNA cleavage	In human cells, it can cleave methylated DNA. This is an area of particular concern for plants, as it has received little attention	Unable to do so	There are many unanswered questions about TALENs' ability to cleave methylated DNA
Multiplexing	CRISPR's main advantage is that multiple genes can be edited at the same time. Only Cas9 was required	This is extremely difficult to achieve using ZFNs	Using TALENs to obtain multiplexed genes is extremely difficult. Because it necessitates distinct dimeric proteins for each target

Source: Shuikan [11].

Table 2.
Comparing different genomic engineering techniques used in metabolomics.

activator-like effector nucleases (TALENs) [18, 19]. While each technique has its advantages and disadvantages (**Table 2**), CRISPR-Cas9 has been reported to be the most promising and significant technique that can be used in the discovery and enhancement of SMs production [9, 17, 20, 21].

2.1 Gene insertion/deleting

Gene insertion or deletion is useful not only in biosynthesis gene clusters activation but also for novel SMs discovery [22]. In fact, several silent biosynthesis gene clusters have been refactored by replacing the biosynthesis gene clusters promoter to yield natural products such as secondary metabolites [23–26].

Nowadays, the promising technique has been developed in the genetic engineering field that is multiplexed CRISPR-Cas9 and transformation-associated recombination (TAR)-mediated promoter engineering method (mCRISTAR) [21, 27–30]. mCRISTAR actually combined the advantages of TAR technique and CRISPR-Cas9 technique. Basically, mCRISTAR mode of action is that CRISPR-Cas9 breaks the double-stranded in the promoter region of the biosynthesis gene cluster (BGCs), then the fragments produced are reassembled by TAR with synthetic gene-cluster specific promoter cassettes [21].

2.2 Gene cloning

Basically, gene cloning consists of some steps include 1) determining the suitable heterologous host 2) cloning the target gene, 3) transferring the gene into the suitable host, 4) expression of the gene in the suitable host system, 5) optimization of production [31].

However, many new and useful cloning techniques have been introduced such as transformation assisted recombination (TRA), Cas9-assisted targeting of chromosome segments (CATCH), and TAR-CRISPR [20, 32, 33]. CATCH is a cloning tool that uses the CRISPR-Cas9 system for direct BGCs cloning into the host. However, compared to PCR and restriction enzyme cloning techniques, CATCH is appeared to be more useful for direct cloning of large genes clusters. Whether, TAR technique has been utilized for about a decade in the cloning of large BGCs, but the TAR technique is associated with low cloning efficiency [20, 33]. To address this challenge TAR and CRISPR-Cas9 have been coupled resulting in a new approach called TAR-CRISPR [33]. Therefore, TAR-CRISPR is different than mCRISTAR as discussed earlier. It is yeast-based method, while mCRISTAR uses CRISPR-Cas9 to breaks the double-stranded in the promoter region of the BGC, and the fragments produced are reassembled by TAR with synthetic gene-cluster specific promoter cassettes. As a result, by coupling CRISPR with TAR significant increase of clone efficiency has been reported [33]. In fact, TAR-CRISTAR cloning will allow for the development of BGC cloning and SM production in the future.

While gene-editing techniques play a significant role in the detection and production of microbial secondary metabolites, metabolomics is also important in the identification and characterization of secondary metabolites produced by native or genetically modified microorganisms.

3. Identification and characterization of secondary metabolites

The identification and characterization of secondary metabolites are important. Metabolomic often requires abroad array of instrumentation such as ELSD for detecting lipids, coulometric array detectors for detecting redox compounds, and fluorescent spectrometer for detecting aromatic compounds, whereas other omics techniques such as genomics, transcriptomics, or proteomics are often conducted by a single instrument.

In microbial secondary metabolites investigation, the experiments are mainly conducted in two different approaches, targeted or untargeted metabolites identification [34]. As its name, targeted metabolites experiment aims to identify a specific group of SMs that are already known. Whereas, the untargeted secondary metabolites experiment aims to identify the large scale of SMs produced by microorganisms including novel and known metabolites [35].

Nowadays, two general technologies have been utilized as primary tools in metabolomic, mass spectrometry (MS), and nuclear magnetic resonance (NMR) [4, 36, 37].

These high-throughput tools provide broad coverage of many classes of secondary metabolites, including amino acids, lipids, sugars, organic acids, and others.

In fact, nuclear magnetic resonance (NMR) and mass spectrometry (MS) has been used to identify both targeted and untargeted secondary metabolites [38]. They are often complementary to each other. Mass spectrometry (MS) provides information of molecules whereas, nuclear magnetic resonance (NMR) is utilized to differentiate between structural isomers [39]. In fact, MS is more sensitive than NMR and able to detect the large scale of metabolites, while NMR is highly quantitative and reproducible and require larger sample amount for analysis than MS [40, 41].

4. Data analysis

In fact, the major challenges in metabolomic experiments are the huge amount of information obtained from either NMR spectroscopy or MS [7, 37]. The extraction of the significant information generated by NMR and MS is crucial by using computer software in order to organize the vast amount of data [40, 42].

Because studying individual metabolites is impractical for visualizing changes between groups of metabolites, univariate statistical approaches can be utilized to understand the results. Principal component analysis (PCA) is one of the most extensively used statistical approaches [39, 43, 44]. The data can be simplified using principle component analysis. CA without losing its core feature. In fact, principal component analysis PCA provides information on multivariate differences among secondary metabolites while, different univariant statistical tests such as non-parametric Wilcoxon signed-rank test, Kruskal–Wallis test, and the parametric.

Student's t-test and ANOVA can be utilized to analyze isolated metabolites [45].

Nowadays, most metabolites can be identified, due to the development of many bioinformatics software. There are two types of metabolites identification that are applied including 1) definitive identification and 2) putative identification [7]. Many different metabolomics databases are available online some of them are used for NMR such as METLIN (http://metlin.scripps.edu), Biological Magnetic Resonance Databank (http://www.bmrb.wisc.edu/metabolomics/), and METLIN (http://metlin.scripps.edu) while the others are used for MS such as Mass Bank (http://www.massbank.jp), http://csbdb.mpimp-golm.mpg.de/csbdb/gmd/gmd.html"http://csbdb.mpimp-golm.mpg.de/csbdb/gmd/gmd.html), the Glom Metabolite Database (GMD, NIST (http://www.nist.gov/srd/nist1a.htm), METLI and MMCD (http://mmcd.nmrfam.wisc.edu) [46].

5. Conclusion

Microorganisms are one of the most significant sources of SMs that play important roles in many aspects of our life including pharmaceutical, biomedical and food applications. The integration between genetic engineering and metabolomic provides a powerful platform for the production, detection, and characterization of known and unknown secondary metabolites. However, the combination between CRISPR-Cas9 and metabolomics may improve the efficiency of microbial SMs discovery. Thus, the need of the hour is a comprehensive and sensitive technique that has the ability to provide comprehensive information of any secondary metabolites under all conditions.

Author details

Ahmed M. Shuikan[1]*, Wael N. Hozzein[2], Rakan M. Alshuwaykan[3] and Ibrahim A. Arif[1]

1 Botany and Microbiology Department, College of Science, King Saud University, Riyadh, Saudi Arabia

2 Bioproducts Research Chair, Zoology Department, College of Science, King Saud University, Riyadh, Saudi Arabia

3 Surgery Department, King Khalid Civilian Hospital, Tabuk, Saudi Arabia

*Address all correspondence to: ahmedshuikan@gmail.com

IntechOpen

© 2022 The Author(s). Licensee IntechOpen. This chapter is distributed under the terms of the Creative Commons Attribution License (http://creativecommons.org/licenses/by/3.0), which permits unrestricted use, distribution, and reproduction in any medium, provided the original work is properly cited. (cc) BY

References

[1] Li JWH, Vederas JC. Drug discovery and natural products: End of an era or an endless frontier? Biomeditsinskaya Khimiya. 2011;**57**(2):148-160

[2] Pelaez F. The historical delivery of antibiotics from microbial natural products - Can history repeat? Biochemical Pharmacology. 2006;**71**(7):981-990

[3] McMurry JE. Organic chemistry with biological applications. In: Secondary Metabolites: An Introduction to Natural Products Chemistry. Stamford, USA: Cengage Learning Ltd; 2015. pp. 1016-1046

[4] Berg M, Vanaerschot M, Jankevics A, Cuypers B, Breitling R, Dujardin J-C. LC-MS metabolomics from study design to data-analysis – Using a versatile pathogen as a test case. Computational and Structural Biotechnology Journal. 2013;**4**:e201301002

[5] Bérdy J. Bioactive microbial metabolites. The Journal of Antibiotics. 2005;**58**(1):1-26

[6] Putri SP, Nakayama Y, Matsuda F, Uchikata T, Kobayashi S, Matsubara A, et al. Current metabolomics: Practical applications. Journal of Bioscience and Bioengineering. 2013;**115**:579-589

[7] Go EP. Database resources in metabolomics: An overview. Journal of Neuroimmune Pharmacology. 2010;**5**(1):18-30

[8] Blin K, Kim HU, Medema MH, Weber T. Recent development of antiSMASH and other computational approaches to mine secondary metabolite biosynthetic gene clusters. Briefings in Bioinformatics. 2019;**20**:1103-1113

[9] Tong Y, Weber T, Lee SY. CRISPR/Cas-based genome engineering in natural product discovery. Natural Product Reports. 2019;**36**:1262-1280

[10] Lim FY, Sanchez JF, Wang CCC, Keller NP. Toward awakening cryptic secondary metabolite gene clusters in filamentous fungi. Methods in Enzymology. 2012;**517**:303-324

[11] Shuikan AM, Hozzein WN, Alzharani MM, Sandouka MN, Al Yousef SA, Alharbi SA, et al. Enhancement and identification of microbial secondary metabolites. In: Extremophilic Microbes and Metabolites - Diversity, Bioprespecting and Biotechnological Applications. London: IntechOpen; 2020

[12] Lenders J, Frédérich M, De Tullio P. Nuclear magnetic resonance: A key metabolomics platform in the drug discovery process. Drug Discovery Today: Technologies. 2015;**13**:39-46

[13] Bino RJ, Hall RD, Fiehn O, Kopka J, Saito K, Draper J, et al. Potential of metabolomics as a functional genomics tool. Trends in Plant Science. 2004;**9**(9):418-425

[14] Tran PN, Yen MR, Chiang CY, Lin HC, Chen PY. Detecting and prioritizing biosynthetic gene clusters for bioactive compounds in bacteria and fungi. Applied Microbiology and Biotechnology. 2019;**103**:3277-3287

[15] Valayil JM. Activation of microbial silent gene clusters: Genomics driven drug discovery approaches. Biochem Anal Biochem. 2016;**5**:276

[16] Rutledge PJ, Challis GL. Discovery of microbial natural products by

activation of silent biosynthetic gene clusters. Nature Reviews. Microbiology. 2015;**13**:509-523

[17] Zhang MM, Wang Y, Ang EL, Zhao H. Engineering microbial hosts for production of bacterial natural products. Natural Product Reports. 2016;**33**:963-987

[18] Miller JC, Holmes MC, Wang J, Guschin DY, Lee YL, Rupniewski I, et al. An improved zinc-finger nuclease architecture for highly specific genome editing. Nature Biotechnology. 2007;**25**:778-785

[19] Jankele R, Svoboda P. TAL effectors: Tools for DNA targeting. Briefings in Functional Genomics. 2014;**13**:409-419

[20] Yamanaka K, Reynolds KA, Kersten RD, Ryan KS, Gonzalez DJ, Nizet V, et al. Direct cloning and refactoring of a silent lipopeptide biosynthetic gene cluster yields the antibiotic taromycin A. Proceedings of the National Academy of Sciences. 2014;**111**:1957-1962

[21] Kang HS, Charlop-Powers Z, Brady SF. Multiplexed CRISPR/Cas9-and TAR-mediated promoter engineering of natural product biosynthetic gene clusters in yeast. ACS Synthetic Biology. 2016;**5**:1002-1010

[22] Voytas DF. Plant genome engineering with sequence-specific nucleases. Annual Review of Plant Biology. 2013;**64**:327-350

[23] Lee NC, Larionov V, Kouprina N. Highly efficient CRISPR/Cas9-mediated TAR cloning of genes and chromosomal loci from complex genomes in yeast. Nucleic Acids Research. 2015;**43**:e55-e55

[24] Horbal L, Marques F, Nadmid S, Mendes MV, Luzhetskyy A. Secondary metabolites overproduction through transcriptional gene cluster refactoring. Metabolic Engineering. 2018;**49**:299-315

[25] Shao Z, Rao G, Li C, Abil Z, Luo Y, Zhao H. Refactoring the silent spectinabilin gene cluster using a plug-and-play scaffold. ACS Synthetic Biology. 2013;**2**:662-669

[26] Bauman KD, Li J, Murata K, Mantovani SM, Dahesh S, Nizet V, et al. Refactoring the cryptic streptophenazine biosynthetic gene cluster unites phenazine, polyketide, and nonribosomal peptide biochemistry. Cell Chemical Biology. 2019;**26**:724-736

[27] Pohl C, Kiel JAKW, Driessen AJM, Bovenberg RAL, Nygard Y. CRISPR/Cas9 based genome editing of Penicillium chrysogenum. ACS Synthetic Biology. 2016;**5**:754-764

[28] Sander JD, Joung JK. CRISPR-Cas systems for editing, regulating and targeting genomes. Nature Biotechnology. 2014;**32**:347-355

[29] Zhang MM, Wong FT, Wang Y, Luo S, Lim YH, Heng E, et al. CRISPR–Cas9 strategy for activation of silent Streptomyces biosynthetic gene clusters. Nature Chemical Biology. 2017;**13**:607

[30] Li L, Zheng G, Chen J, Ge M, Jiang W, Lu Y. Multiplexed sitespecific genome engineering for overproducing bioactive secondary metabolites in actinomycetes. Metabolic Engineering. 2017;**40**:80-92

[31] Greunke C, Duell ER, D'Agostino PM, Glöckle A, Lamm K, Gulder TAM. Direct pathway cloning (DiPaC) to unlock natural product biosynthetic potential. Metabolic Engineering. 2018;**47**:334-345

[32] Jiang W, Zhao X, Gabrieli T, Lou C, Ebenstein Y, Zhu TF. Cas9-assisted

targeting of chromosome segments CATCH enables one step targeted cloning of large gene clusters. Nature Communications. 2015;**6**:810

[33] Bonet B, Teufel R, Crüsemann M, Ziemert N, Moore BS. Direct capture and heterologous expression of Salinispora natural product genes for the biosynthesis of enterocin. Journal of Natural Products. 2015;**78**:539-542

[34] Breitling R, Ceniceros A, Jankevics A, Takano E. Metabolomics for secondary metabolite research. Metabolites. 2013;**3**:1076-1083

[35] Wu C, Kim HK, van Wezel GP, Choi YH. Metabolomics in the natural products field—A gateway to novel antibiotics. Drug Discovery Today: Technologies. 2015;**13**:11-17

[36] Dunn WB, Broadhurst DI, Atherton HJ, Goodacre R, Griffin JL. Systems level studies of mammalian metabolomes: The roles of mass spectrometry and nuclear magnetic resonance spectroscopy. Chemical Society Reviews. 2011;**40**(1):387-426

[37] Midelfart A, Dybdahl A, Gribbestad IS. Metabolic analysis of the rabbit cornea by proton nuclear magnetic resonance spectroscopy. Ophthalmic Research. 1996;**28**(5):319-329

[38] Dettmer K, Aronov PA, Hammock BD. Mass spectrometry based metabolomics. Mass Spectrometry Reviews. 2007;**26**(1):51-78

[39] Alia A, Ganapathy S, de Groot HJ. Magic angle spinning (MAS) NMR: A new tool to study the spatial and electronic structure of photosynthetic complexes. Photosynthesis Research. 2009;**102**(2-3):415-425

[40] Dunn WB, Broadhurst D, Begley P, Zelena E, Francis-McIntyre S, Anderson N, et al. Procedures for large-scale metabolic profiling of serum and plasma using gas chromatography and liquid chromatography coupled to mass spectrometry. Nature Protocols. 2011;**6**(7):1060-1083

[41] Johnson CH, Gonzalez FJ. Challenges and opportunities of metabolomics. Journal of Cellular Physiology. 2012;**227**(8):2975-2981

[42] Lu W, Bennett BD, Rabinowitz JD. Analytical strategies for LC-MS-based targeted metabolomics. Journal of Chromatography. B, Analytical Technologies in the Biomedical and Life Sciences. 2008;**871**(2):236-242

[43] H. H. Analysis of a complex of statistical variables into principal components. Journal of Education & Psychology. 1933;**24**:417-441

[44] Young SP, Wallace GR. Metabolomic analysis of human disease and its application to the eye. Journal of Ocular Biology, Diseases, and Informatics. 2009;**2**(4):235-242

[45] Broadhurst DI, Kell DB. Statistical strategies for avoiding false discoveries in metabolomics and related experiments. Metabolomics. 2006;**2**(4):171-196

[46] Brown M, Dunn WB, Dobson P, Patel Y, Winder CL, Francis-McIntyre S, et al. Mass spectrometry tools and metabolite-specific databases for molecular identification in metabolomics. Analyst. 2009;**134**(7): 1322-1332

Chapter 6

In Vitro Cultures for the Production of Secondary Metabolites

Grazia Maria Scarpa, Vanda Prota, Nicola Schianchi and Federica Manunta

Abstract

Plants' secondary metabolism is an important source of medicinal and industrial products. Even though natural ecosystems are still the most important font of this kind of substance, excessive harvesting of spontaneous flora can act as a direct cause of biodiversity loss. Different technologies are used for *in vitro* production which, in addition to being useful for safeguarding biodiversity, make available to industry substances that are difficult to produce *in vivo*. Moreover, the growing demand for secondary metabolites encourages the use of new biotechnology tools to create new, more productive *in vitro* transgenic plant cultures.

Keywords: medicinal plants, metabolites, vitro, secondary metabolism, medicinal plants, elicitor, cell multiplication

1. Introduction

Several problems might arise when producing secondary metabolites using both spontaneous and cultivated plants or parts of plants. If the material for extraction is collected by spontaneous plants, the major risk is related to the impoverishment of resources and biodiversity, consequently. Although natural ecosystems are usually rich in officinal plants that can be used by humans, an excessive collection of spontaneous flora can act as a direct cause of biodiversity loss [1, 2]. Currently, it is estimated that at least 50,000 plant species are used, which in the majority of cases grow spontaneously, however, sometimes products come from specific cultivation. Based on what was reported by the 2020 edition of the State of the World's Plants and Fungi [3], climate change is threatening two-fifths of the plants currently known; this value is doubled compared to what observed in 2016 and, among these, species are included many medicinal plants used both as a natural remedy and for drug production. According to such data collection that involved 210 scientists and 42 countries, over 140,000 plants should be classified as under extinction threat, including 730 medicinal plants. Among known species, 5500 medicinal plants can be found and approximately 13% of these are under extinction threat [4]. Concerning the most vulnerable plants, we can mention *Brugmansia sanguinea* (Ruiz & Pav.) D. Don, used in medicine to treat cardiovascular disorders, which can only be found as a cultivated plant. Fate similar to

Nepenthes khasiana Hook. f., mostly used to treat skin problems as well as *Warburgia salutaris* (G. Bertol.) Chiov., indicated when respiratory problems occur [3, 5].

1.1 Historical notes

A large number of species belonging to the plant kingdom have always coexisted on Earth, over the years they have created a great heritage of biodiversity. Plants have always been a primary source of sustenance for herbivorous and omnivorous animals including the human species, the latter, however, over time, has realized the possibility of using plant biomass to also obtain substances to be utilized in various effective ways, for example as medication or food supplements.

Western medical culture can be traced back to the Sumerian Nippur tablets of 3000 BC on which the names of medicinal herbs are reported. The first known writing on the subject is a papyrus (1552 BC), dating back to an Egyptian dynasty. It features numerous herbal formulas and, between magic and medicine, even invocations to ward off disease and a catalog of plants, minerals, magical amulets, and useful spells. It is based on more than 500 plants, nearly a third of which are still found in today's Western pharmacopoeias.

The most famous Egyptian physician was Imhotep (Memphis around 2500 BC) whose "materia medica" included practices to reduce head and thoracic trauma, wound care, prevention, treatment of infections, and principles of hygiene.

The first Chinese manual of materia medica, Shennong Ben Cao Jing (Emperor Shennong's classic Materia medica), written in the first century, describes 365 medicines, 252 derived from herbs.

Ancient literature also provided the manuscript "Recipes for fifty-two foods," the longest medical text found in the Chinese tomb of Mawangdui, (168 BC), the Wushi'er Bingfang (9950 characters). It along with others shows the early development of Chinese medicine while subsequent generations have developed Yaoxing Lun, a "Treatise on the Nature of Medicinal Herbs."

Ayurveda is the traditional medicine in India that emphasizes plant-based treatments, hygiene, and the balance of the state of the body. The Indian Materia Medica included knowledge of plants, the place of its growth, the methods of conservation, and the duration of the collected materials; includes also directions for extracting juices, powders, cold infusions, and extracts.

Later in Greece, it was Hippocrates, a philosopher known as the father of medicine, who in 460 BC founded a school focused on the necessity to discover the causes of disease to combat them. His treatises, aphorisms, and prognostics, in addition to describing 265 drugs, supported the importance of diet for the treatment of diseases.

Theophrastus (390–280 BC), a disciple of Aristotle's, historically known as the "father of botany," wrote the treatise Historia Plantarium, the first attempt to classify plants and botanical morphology in Greece with details of medicinal herbs and concoctions based on them.

Later Galen, philosopher, physician, pharmacist, and prolific writer of medical matters, collected the medical knowledge of his time in an extensive report and wrote on the structure of organs, the impulse and its association with respiration, arteries, and blood circulation, and the uses of the "Theriac" "In treatises such as on Theriac to Piso, on Theriac to Pamphilius, and on Antidotes, Galen identified in the Teriaca a compound of 64 ingredients, which can be defined as a polypharmaceutical, suitable for treating every known disease." His work rediscovered in the fifteenth century became the authority on medicine and healing for the next two centuries.

The Greek physician Dioscorides treated medical questions in five volumes, entitled Περὶ ὕλης ἰατρικῆς in Greek and De Materia Medica in Latin; they include about 500 plants and direct observations of the plants and the effects that the various drugs have had on patients. De Materia Medica was the first extensive drug system comprising a 1000 natural drugs (products for most basic plants). The classification used by Dioscorides is of an elementary type even if he uses a botanical taxonomy. The books written by Dioscorides on medicinal herbs of history are considered the precursors of the modern pharmacopeia remaining in use until the 1600s.

2. Secondary metabolism

Active principles synthesized through secondary metabolites act as a defense strategy, playing an active role in plant ecophysiology against herbivores, attacks by pathogens but also as a response to abiotic stress, and competition with other plants; at the same time, they play a crucial role in attracting beneficial organisms, such as symbionts and pollinators. Recently, several studies on "secondary metabolism" highlighted additional features related to these molecules, which make them essential for the organism that produces them as they provide useful information on quality and on specific features of a range of raw materials, both of animal and vegetal origin as well as on food produced with them [6, 7]. As a matter of fact, the secondary metabolites pool is often influenced by specific environmental conditions, for instance, in the case of essential oil profile; for this reason, secondary metabolite products in essential oils may provide important support in acquiring valuable information on their origin.

Unlike primary metabolites that are stable in concentration and chemical structure, ensuring cell structural and functional integrity, secondary metabolites show a "high degree of freedom" as far as these aspects are concerned [8, 9].

Due to an enormous diversity in structure and intraspecific variability, biosynthesis in secondary metabolites is limited to definite groups of plants and thus they are not ubiquitous. Synthesis in secondary metabolites was selected when during evolution such compounds managed to respond to specific needs by vegetal organisms [10]. This is the case, for instance, of the variation of scents and colors in flowers to attract pollinators and promote and increase efficiency in pollination [11].

Secondary metabolism-derived molecules are released in the environment through different mechanisms, among others we can mention volatilization that leads to a dispersion of substances such as ethylene and sesquiterpenes that can be absorbed by surrounding plants directly through the soil or atmosphere; lisciviation, instead, promotes the release of substances, such as sugar, amino acids, alkaloids, fatty acids, terpenoids, and phenolic acids, from the aerial part of the plant through hydrosolubility caused by rain or fog. Other mechanisms promoting dispersion are 40 exudation and decomposition.

The activity of substances released also depends on the physiological and nutritional status of plants and environmental abiotic factors, such as light, rain, and temperature [12].

During the nineteenth century, chemists showed interest in the study of secondary metabolism and metabolites, concerning especially drugs, poison, aromatizers, and industrial products, all representing as a whole the final products of metabolic pathways or networks of these; actually, more than 200,000 are known to date.

Recently, potential roles of secondary products at the cell level that have been identified are—plant growth regulation, gene expression modulation, and compounds involved in signal transduction [13, 14]. Hence, while for centuries secondary metabolites have been used in traditional medicine, nowadays, they act as valuable pharmaceutical, cosmetic, chemical compounds, and nutraceuticals in the recent past [15].

2.1 Secondary metabolites in natural environment

Active principles can be divided into three big molecule families based on the biosynthesis pathways from which they are originated—terpenoids and steroids, alkaloids and phenolics [15].

2.1.1 Terpenoids

These are the most recurrent compounds; lipid molecules synthesized starting from acetyl CoA or from glycolysis intermediates reaching a total of 35,000, abundant in essential oils, resins, rubber, volatile molecules, scented, colorless, soluble in oil or highly lipophilic solutions, and inflammable. They function as protectors for wood tissues, exert antibacterial effects, are responsible for insect attraction and repulsion, as well as represent the base material for vegetal hormones or pigments (chlorophyll and carotenoids) synthesis; they also take part in the mitochondrial electron transport and plastoquinone.

2.1.2 Alkaloids

These molecules, which accumulate nitrogen becoming an important source of it, are produced by approximately 20% of plants; more than 20,000 different alkaloids are known and are synthesized principally from amino acids.

They play an important role as an advanced chemical defense system of plants under predators' pressure (larvae, insects, herbivores, mammals). They work as antibiotics and pesticides with a deterrent action to prevent plants from being ingested.

Alkaloids used as drugs, poison, with stimulating and narcotizing effects were used even by Greek and Romans, such as atropine (*Atropa belladonna* L), cocaine (*Erytroxylon coca* Lam. leaves), morphine and opium (*Papaver somniferum* L. fruits), nicotine (*Nicotiana tabacum* L. leaves), and strychnine (*Strychnos nux-vomica* L. seeds).

2.1.3 Phenolic compounds

Secondary plant metabolites belonging to the big family of polyphenols [16], having mostly hydrosoluble characteristics. They represent one of the main classes of secondary metabolites that includes a wide range of highly heterogeneous substances having all in common an aromatic ring. They are formed through the biosynthesis pathway of shikimic or mevalonic acids; a total of 15,000 are known and represent a group of substances easily occurring in superior plants; the most common cinnamic acid derivatives are caffeic, p-Coumaric, ferulic, gallic, and synaptic acids.

Compounds of different colors accumulate especially in aerial plant organs (stems, leaves, flowers, and fruits) rather than in roots; such a preferred location is related to a light-induced effect on phenolic metabolism; besides, phenolic compounds play a protective role against UV that are successfully absorbed and accumulated into leaves epidermis to avoid damage caused to cell DNA [16]. They influence the color, generally yellow, of flowers and fruits where they can be found as glycosides diluted in cell juice except for anthocyanidines and their glucosides (anthocyanins) that are red, purple, or blue depending on the pH of cell juice [17]. The flavonoids content in plants depends not only on the genotype but is also closely related to environmental conditions especially by light radiation such as UV; the latter, in fact, induces a significant increase of flavonoids in leaves [18, 19].

Flavonoids and phenolic acids are the most important antioxidants in the diet and can be found also in tea, wine, and beer [8].

They are considered pharmacologically active compounds having anti-inflammatory activity, active against liver injury due to hepatotoxicity, and acting as antitumoral, antimicrobials, antivirals, enzyme inhibitors, antioxidants, protect against capillary fragility, as well as playing a role as insect repellents and signaling in plant-organism interactions.

In the recent past, the most common use involving the antioxidant properties has been represented by the "scavenger" activity exerted by a series of enzymes, such as dismutase, superoxide, catalase, glutathione peroxidase; they play a role in halting the radical reaction cascade causing acceleration of cell senescence processes.

Among multiple biological activities exerted by these secondary metabolism molecules, we highlight the role of antioxidants against aging, such as in the case of cocoa (*Theobroma cacao* L.), coffee (*Coffea arabica* L.), tea (*Camellia sinensis* L.). The content of phenolic compounds in vegetal tissues varies based on the species, variety, specific organ considered, physiological status, and pedoclimatic conditions.

The high variety of phenolic structures shows the same amount of function diversification—they can play a role as low molecular weight flower pigments, antibiotics, and anti-UV screens.

Likewise, elicitation on a secondary metabolic pathway by a pathogen can lead to *ex novo* production and accumulation of phytoalexins in a plant. This event is exploited through some biotechnological applications in which elicitors are used to stimulate the production of secondary metabolites.

2.2 Applications in food

Antioxidants can be defined as any substance that is able to delay or significantly inhibit oxidation in a specific substrate even if it shows a really low concentration compared to the oxidable substrate [20]. Nutrition plays a crucial role in ensuring the efficacy of antioxidant enzyme defenses—many essential oligoelements, such as selenium, copper, manganese, and zinc, are involved in the molecular structure or in the catalytic activity of these enzymes. The main antioxidant compounds in food are—ascorbic acid (vitamin C), tocopherols (vitamin E), carotenoids, flavonoids.

Over the years, pharmaceutical companies have been focusing on antioxidant compounds from food to promote healthy properties of food as available data show that an increase in oxidant intake from natural sources, specifically from fruit and vegetables, may have a beneficial effect on disease prevention. Their production can be effectively achieved through *in vitro* cultures.

3. Production through *in vitro* culture

Secondary metabolites can be produced *in vivo* from spontaneous or cultivated plants or *in vitro*. Unlike primary metabolites, an accumulation can be detected in vacuoles, they are not ubiquitous and synthesis depend on the development stage of the plant. Production conventional methods from vegetal tissues include different extracting methodologies through solvents, steam, or supercritical CO_2 [21]. *In vivo* culture refers to plants grown in the natural environment or cultivated in non-sterile conditions. Instead, the definition "*in vitro* culture" refers to the culture of explants, tissues, or isolated cells on the artificial substrate, under controlled conditions, in a sterile environment. In *in vitro* culture, the same metabolites that plants naturally produce can be accumulated through physiological stimulation, stress, or hormones. The development of methodologies such as vegetal tissue culture, enzyme production, and fermentation technologies gave a significant contribution to the production of this kind of molecule [21].

In vitro secondary metabolite production is based on a procedure in two separate phases—mass production and secondary metabolites synthesis. These phases are performed separately and are independent each other; at the same time, they have different requirements and can be optimized separately [22, 23].

Cultures of vegetal tissues (**Figure 1**) or isolated cells (**Figure 2**) are inoculated in sterile conditions starting from explants, such as leaves, stems, meristems, roots, buds, callus (**Figure 3**) both for multiplication and secondary metabolite production. Production can take place in more than one tissue.

3.1 Biomass production

Depending on the species, biomass production can be initiated from an undifferentiated callus or cell suspension. In other cases, sprouts, roots, and somatic embryos can be cultured. Using differentiated tissues or organs is crucial when the requested metabolite is produced in specific plant tissue or organ or also in specialized glands

Figure 1.
Culture of shoots on liquid substrate.

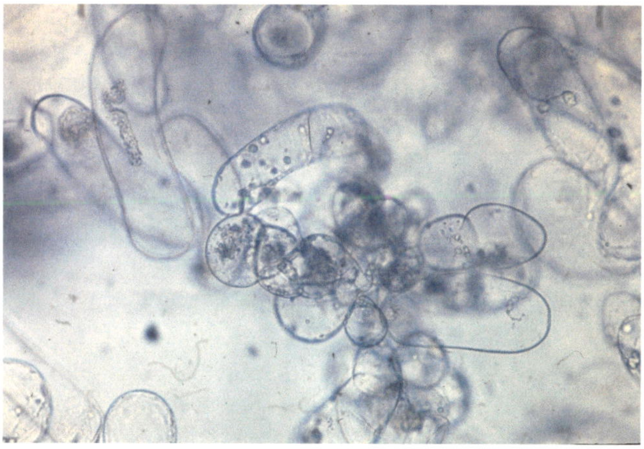

Figure 2.
Cell suspension culture.

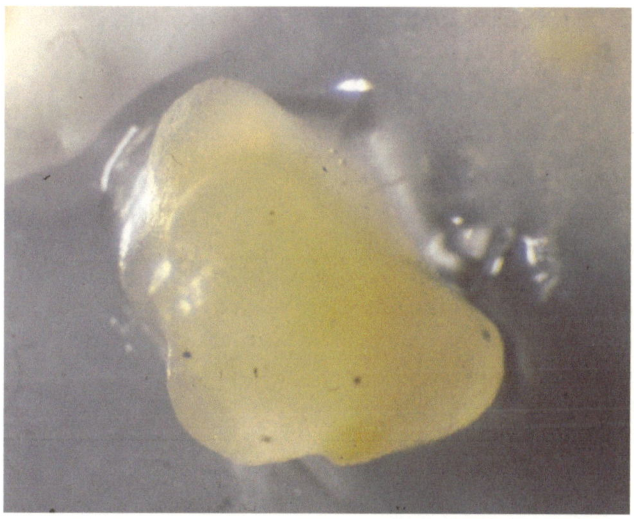

Figure 3.
Isolated callus on solidified medium.

such as in the case of essential oils [24, 25]. Although different studies showed efficacy in secondary metabolite production through cultures of differentiated tissues and callus, the technique mostly used is cell suspension [22, 23, 26, 27]. The latter is a culture of cells isolated in a liquid medium that exploits cell totipotency for large-scale production. Each cell, in fact, keeps the biosynthesis ability of the plant and under the right conditions can produce metabolites identical to the ones produced by the mother plant. Furthermore, it can be noticed that cell cultures have greater and faster potential application to the market compared to other production methods [25, 28]. This technique ensures the continuous production of metabolites of interest while offering an elevated quality standard and product uniformity. In addition, it is possible through biotechnology applications to produce new metabolites not synthesized by the mother plant [29, 30]. Currently, different metabolites with an interesting market value are produced using cell suspension culture, such as taxol [31, 32], resveratrol [33], artemisinin [34], ginsenosides [35], raubasine [36].

Among differentiated tissues, hairy roots should be highlighted as they enable the production of secondary metabolites from a considerable number of plants.

Hairy roots are formed in nature on plants following an infection caused by *Agrobacterium rhizogenes*. This bacterium carries genes that encode phenotypic mutation inducing the formation of hairy roots. After infection, a DNA (T-DNA) segment is transferred in the plant genome through the root-inducing (Ri) plasmid [37].

Agrobacterium can transfer genetic information to plants inducing transformations. Once the infection takes place, a plasmid fragment called T-DNA can be integrated into the plant nuclear DNA where genes are integrated. The composition and organization of T-DNA sequences vary considerably. As some cT-DNA genes show strong growth effects when expressed in other species, they can also influence the growth of natural transformants. However, there is still a need to fully identify the mechanisms through which these genes alter growth models and their regulation by promoters and plant transcription factors [38]. Among the advantages of such a technique, we can mention the high level of cell differentiation, rapid growth, relatively easy production, genetic, and biochemical stability. It should also be taken into consideration the potential accumulation of secondary metabolites in the aerial part of the plant. However, technical problems might arise in cultivation systems for the market [37].

3.2 Immobilization

The process can involve both cells and elicitors. They are bound inside a matrix through trapping, absorption, or covalent bonding. The system must be integrated with an adequate substrate as in the case of cultures of suspended cells, as well as regulating chemical and physical parameters, such as pH and temperature.

In a system of this kind, secondary metabolites must be released by cells in the culture media naturally or through induced secretion. One of the advantages of this methodology is the potential stabilization of a continuous production process through the adoption of a specific system of bioreactors.

4. Substrates

In vitro production includes cultural techniques on explants, tissues, or vegetal cells under controlled conditions supported by a substrate that plays a vital role for plants; in fact, it acts as the "place" where all elements needed for plant survival are located [39]. The explant sometimes represented by the cell alone has to be able to regenerate; thus, the substrate provides it with all substances needed, such as macro and microelements, vitamins, carbon sources, growth regulators, and in the case of solid substrates jellifying agents, usual agar in quantities that can vary between 0.7–0.8 g/L, agarose, and starch.

Generally, the substrate contains mineral elements formed by macro and microelements and an organic component formed by vitamins, amino acids, and other nitrogen components as well as carbohydrates. There are different substrates that can play a specific role in achieving different objectives, as a consequence of the concentration of specific substances contained in them, such as those indicated, for example, in **Table 1**.

Macro and microelements	MS (mg/L)	WPM (mg/L)	B5 (mg/L)	NN (mg/L)
Ammonium nitrate	1650.000	400.000		720.000
Boric acid	6.200	6.200	3.000	10.000
Anhydrous calcium chloride	332.200	72.500	113.24	
Cobalt chloride hexahydrate $6H_2O$	0.025		0.025	
Tripotassium phosphate	170.000	170.000	130.500	68.000
Potassium iodide	0.830		0.750	
Sodium molybdate $2H_2O$	0.250	0.250	0.250	0.250
Calcium nitrate		386.000		
Potassium nitrate	1900.000		2500.000	950.000
Ammonium sulfate			134.000	
Iron sulfate·$7H_2O$	27.800	27.800	27.850	27.850
Anhydrous magnesium sulfate	180.700	180.700	122.09	90.340
Manganese sulfate·H_2O	16.900	22.300	10.000	18.940
Potassium sulfate		990.000		
Copper sulfate $5H_2O$	0.025	0.250	0.025	0.025
Zinc sulfate $7H_2O$	8.600	8.600	2.000	10.000
VITAMINS				
Folate				0.500
Nicotinic acid	0.500	0.500	1.000	5.000
Biotin				0.050
Myo-inositol	100.000	100.000	100.000	100.000
Pyridoxine·HCl	0.500	0.500	1.000	0.500
Thiamine·HCl	0.100	1.00	10.000	0.500
OTHER ADDITIVES				
Disodium EDTA (·$2H_2O$)	37.260	37.300	37.250	37.250
Glycine	2.000	2.000		2.000

MS = Murashige & Skoog [40]; WPM = woody plant medium [41]; B5 [42]; NN = Nitsch & Nitsch [42].

Table 1.
Composition of main substrates for in vitro *culture.*

The choice of an appropriate substrate should be based on the following [43]:

- the type of ions contained
- macroelements balance
- total ionic concentration of medium

Microelements are used in small quantities; lack of such elements causes specific symptoms as they intervene in plant metabolism; they are integrated into enzymes.

Some microelements can influence the production of secondary metabolites, acting as elicitors [44].

Hormones carry out an essential role as growth regulators in plants [45, 46]. The need to add growth regulators to substrates is based on the fact that normal tissues or small organs placed *in vitro* are not able to synthesize a sufficient quantity of them.

Among known hormones mostly utilized we can find:

1. Auxin: Natural auxin is Indole-3-acetic acid (IAA, 3-IAA); in substrates for *in vitro* culture mostly synthetic compounds are used with an auxin-like function such as:

    - IBA (Indole-3-butyric acid), the most commonly used;
    - NAA (1-Naphthaleneacetic acid);
    - 2,4 D (2,4-Dichlorophenoxyacetic acid).

    Natural auxins are added in variable quantities (0.01–10 mg/l) and the synthetic ones are added in quantities between 0.001 and 10 mg/l, determining—elongation and tissue distension, cell division, adventitious roots formation [47, 48].

2. Cytokines: Natural cytokinins are as follows:

    - Kinetin (N6-Furfuryladenine, 6-Furfurylaminopurine)
    - Zeatin [6-(4-Hydroxy-3-methylbut-2-enylamino)purine]
    - 2Ip [N6-(2-Isopentenyl)adenine]

    Cytokines are used in concentrations between 1 and 10 mg/l to stimulate cell division, stimulate adventitious buds production from tissues or from callus, and growth of somatic embryos, to induce the development of axillary buds. In addition, cytokines inhibit root development [49].

3. Gibberellins: Among gibberellins, the most used is GA3 (gibberellic acid) which promotes internode elongation, meristem, and bud development while inhibiting the formation of roots; thus, it is employed in subsequent phases after planting [50, 51].

    pH: Another factor essential for a cultural substrate is pH as its value influences —salt solubility, elements absorption, and substrate solidification; for these reasons, the pH range is quite limited ranging from 5.2 to 5.8. As far as secondary metabolites are concerned, optimal ranges are established both for pH and temperature according to the cultured species.

5. Elicitors

To achieve secondary metabolite production, elicitation is one of the most important strategies and is used to increase productivity; it takes place through the addition

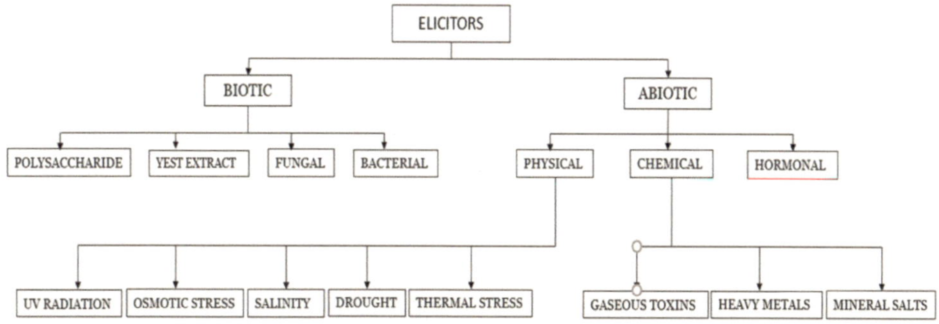

Figure 4.
Classification of elicitor based on different features.

of compounds called elicitors—they can be defined as stress-inducing compounds that induce or improve biosynthesis of specific compounds when a specific amount is applied to a living system [52, 53].

Elicitors can be biotic, such as jasmonic acid, hydrolyzed casein, cellulase, macerozyme, yeast extract, fungal extract, chitin; in addition, chemical compounds usually synthesized from pathogens; abiotic elicitors that include nonorganic substances and can be divided into physical, chemical, and hormonal factors (**Figure 4**) [53, 54].

6. Growth curve

In cell suspension, depending on environmental parameters and on bioreactor features, the development of cells cultured on the liquid substrate is based on specific phases illustrated in **Figure 5**. The graph shows time (horizontal axis) and cell number (vertical axis). At the beginning a slow-growth phase is shown, known as lag phase followed by a phase in which cell concentration grows based on a logarithmic scale,

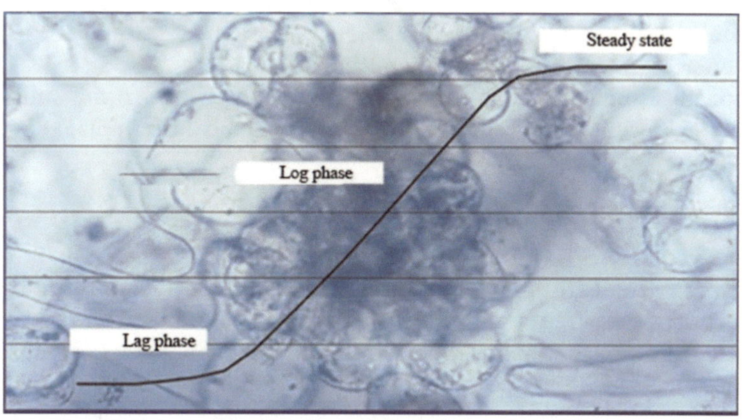

Figure 5.
Growth curve of a cell suspension culture.

log phase; then a second slow-growth phase occurs followed by a phase in which the culture is numerically constant indicated as a plateau or steady state.

During the latency phase, reduced growth and an accumulation of substances useful for cell development occur, while during the exponential growth phase a considerable biomass increase can be observed. In a discontinue culture, in the case that cells accumulate metabolites in vacuoles, the biomass is removed at the end of the exponential phase; during the stationary phase a balance occurs between new cells and dead cells, then secondary metabolites are excreted in culture media. In this case, the collection is carried out by replacing from time to time or continuously the culture media.

7. Production increase

In the production of high-value secondary metabolites, a good strategy is offered by the use of technologies that ensure elevated yield and stable over time. It should be underlined that the production of secondary metabolites from plants is genotype-dependent and this fact influences both metabolite type and quantity. Mother plants can be selected to a first selection to identify plants that ensure also *in vivo* a higher production of compound needed. Once the *in vitro* culture is stabilized, both from cells and organs, hyperproductive lines can be selected [23]. Selection is carried out through *in vitro* growth analysis on cell lines or organs, evaluating the multiplication degree but also assessing the quantity and quality of metabolite produced through chromatography and spectroscopic analysis [23].

The output can also be increased through conventional systems or metabolic engineering methodologies [22, 55].

7.1 Biosynthesis pathways

By using metabolic engineering, the biosynthesis pathways can be studied more efficiently [56, 57] through studying gene overexpression that alterates pathways. The study design includes analysis of enzyme reactions and biosynthesis processes at genetic, transcriptomic, and proteomic levels; in addition, it is also studied the manipulation of genes that encode critical enzymes and those that regulate the speed in the biosynthesis pathways [58, 59]. However, to date this system is limited to experimental settings and no method has been identified yet for industrial transfer of such methodology. Currently, the study of the biosynthesis pathway in phenylpropanol seems to be one of the most promising given that this substance is involved in the biosynthesis of different secondary metabolites in plants [60, 61].

8. Conventional technologies

Culture parameters are among the factors that mostly influence secondary metabolite production—substrate composition both in terms of mineral and organic compounds; pH; characteristics of cell inoculation; physical parameters, such as temperature, light intensity, duration, shaking, and aeration [22, 23, 27]. The substrate should be selected based on the requirements of plant species. Each substrate parameter can be modified to better adjust to the species and to metabolites to be obtained by it—salt type and concentration, carbon source, growth regulators. In nature, secondary metabolites production is in response to environmental stimuli, or for defensive purposes. This mechanism can be simulated in the laboratory through the modification of the culture parameters, for example, light, temperature, or through the use of substances called elicitors. To elicitors belong both organic and inorganic molecules, such as methyl jasmonate, salicylic acid, microbial cell wall extracts (e.g., yeast extract, chitosan), inorganic salts, heavy metals, physical agents (e.g., UV radiation) among others (**Tables 2** and **3**).

Aid to the production of new secondary metabolites, or increased production of those already known and used, can come from new technologies, such as transgenic

Elicitor	Plant species	Culture	Compound	References
Ozone (O_3)	*Melissa officinalis*	Shoot	Rosmarinic acid	[62]
	Hypericum perforatum	Cell suspension	Hypericin	[63]
	Pueraria thomsnii	Cell suspension	Puerarin	[64]
pH	*Bacopa monnieri*	Shoot	Bacoside A	[65, 66]
	Withania somnifera	Hairy root	Withanolide A	[67]
	Withania somnifera	Cell suspension	Withanolide A	[68]
Sucrose	*Hypericum adenotrichum*	Seedling	Hypericin and pseudohypericin	[69]
	Corylus avellana	Cell suspension	Paclitaxel	[70]
	Bacopa monnieri	Shoot	Bacoside A	[65, 66]
	Withania somnifera	Cell suspension	Withanolide A	[68]
Ultraviolet C	*Vitis Vinifera*	Cell suspension	Stilbene	[71]
Proline	*Stevia rebaudiana*	Callus and suspension	Steviol glycoside	[72]
Polyethylene glycol	*Stevia rebaudiana*	Callus and suspension	Steviol glycoside	[72]
	Hypericum adenotrichum	Seedling	Hypericin and pseudohypericin	[69]
Jasmonic acid	*Bacopa monnieri*	Shoot	Bacoside A	[73]
	Plumbago indica	Hairy root	Plumbagin	[74]
	Plumbago rosea	Cell suspension	Plumbagin	[75]
Methyl jasmonate	*Salvia miltiorrhiza*	Hairy root	Tanshinone	[76]
	Perovskia abrotanoides	Adventitious roots	Cryptotanshinone and tanshinone IIA	[77]
	Vitis vinifera	Cell suspension	Stilbene	[71]
	Bacopa monnieri	Shoot	Bacoside	[73]
	Salvia officinalis	Shoot	Diterpenoid	[78]
	Silybum marianum	Cell suspension	Silymarin	[79]
	Salvia castanea	Hairy root	Tanshinone	[80]
	Gymnema sylvestre	Cell suspension	Gymnemic acid	[81]
	Withania somnifera	Hairy roots	Withanolide A, withanone, and withaferin A	[82]

Elicitor	Plant species	Culture	Compound	References
	Andrographis paniculata	Cell suspension	Andrographolide	[83]
	Vitis vinifera	Cell suspension	trans-Resveratrol	[84]
	Taverniera cuneifolia	Root	Glycyrrhizic acid	[85]
Gibberellic acid	*Salvia miltiorrhiza*	Hairy root	Tanshinones	[86]
	Echinacea pupurea	Hairy root	Caffeic acid derivatives	[87]
Salicylic acid	*Salvia miltiorrhiza*	Hairy root	Tanshinone	[76]
	Vitis vinifera	Cell suspension	Stilbene	[71]
	Digitalis purpurea	Shoot	Digitoxin	[88]
	Hypericum hirsutum	Shoot	Hypericin and pseudohypericin	[89]
	Gymnema sylvestre	Cell suspension	Gymnemic acid	[81]
	Withania somnifera	Hairy root	Withanolide A, withanone, and withaferin A	[82]
	Datura metel	Root	Hyoscyamine and scopolamine	[90]
	Glycyrrhiza uralensis	Adventitious root	Glycyrrhizic acid	[91]
Sodium salicylate	*Salvia officinalis*	Shoot	Carnosol	[92]
Sodium chloride	*Catharanthus roseus*	Embryogenic tissues	Vinblastine and vincristine	[93]
Sorbitol	*Perovskia abrotanoides*	Adventitious roots	Cryptotanshinone and tanshinone IIA	[77]
Silver (Ag)	*Perovskia abrotanoides*	Adventitious roots	Cryptotanshinone and tanshinone IIA	[77]
	Vitis vinifera	Cell suspension	Resveratrol	[94]
	Salvia castanea	Hairy root	Tanshinone	[80]
	Datura metel	Hairy root	Atropine	[95]
Cadmium (Cd)	*Vitis vinifera*	Cell suspension	Resveratrol	[94]
	Datura stramonium	Root	Sesquiterpenoid	[96]
Cobalt (Co)	*Vitis vinifera*	Cell suspension	Resveratrol	[94]
Copper (Cu)	*Ammi majus*	Shoot	Xanthotoxin	[97]
	Bacopa monnieri	Shoot	Bacoside	[73]
	Datura stramonium	Root	Sesquiterpenoid	[96]

Table 2.
Abiotic elicitors.

cultures. Several works have demonstrated the safety of these technologies, and their effectiveness, at low cost, for the production of secondary metabolites for medicine and industry [103].

Elicitor	Plant species	Culture	Compounds	References
Chitin	*Hypericum perforatum*	Shoot	Hypericin and pseudohypericin	[98]
	Hypericum perforatum	Cell suspension	Phenylpropanoid and naphtodianthrone	[99]
	Vitis vinifera	Cell suspension	trans-Resveratrol and viniferins	[83]
Pectin	*Hypericum perforatum*	Shoot	Hypericin and pseudohypericin	[98]
Dextran	*Hypericum perforatum*	Shoot	Hypericin and pseudohypericin	[98]
Yeast extract	*Perovskia abrotanoides*	Adventitious roots	Cryptotanshinone and tanshinone IIA	[77]
	Plumbago rosea	Cell suspension	Plumbagin	[75]
	Silybum marianum	Cell suspension	Silymarin	[79]
Trichoderma atroviride	*Salvia miltiorrhiza*	Hairy root	Tanshinone	[100]
Protomyces gravidus	*Ambrosia artemisiifolia*	Hairy root	Thiarubrine A	[101]
Claviceps purpurea	*Azadirachta indica*	Hairy root	Azadirachtin	[102]
Mucor hiemalis	*Taverniera cuneifolia*	Root	Glycyrrhizic acid	[85]
Fusarium oxysporum	*Hypericum perforatum*	Cell suspension	Phenylpropanoid and naphtodianthrone	[99]
Phoma exigua	*Hypericum perforatum*	Cell suspension	Phenylpropanoid and naphtodianthrone	[99]
Botrytis cinerea	*Hypericum perforatum*	Cell suspension	Phenylpropanoid and naphtodianthrone	[99]
Aspergillus niger	*Gymnema sylvestre*	Cell suspension	Gymnemic acid	[55]
Saccharomyces cerevisiae	*Gymnema sylvestre*	Cell suspension	Gymnemic acid	[55]
Agrobacterium rhizogenes	*Gymnema sylvestre*	Cell suspension	Gymnemic acid	[55]
Bacillus subtilis	*Gymnema sylvestre*	Cell suspension	Gymnemic acid	[55]
Escherichia coli	*Gymnema sylvestre*	Cell suspension	Gymnemic acid	[55]
	Datura metel	Hairy root	Atropine	[95]
Bacillus cereus	*Datura metel*	Hairy root	Atropine	[95]
Staphylococcus aureus	*Datura metel*	Hairy root	Atropine	[95]
Rhizobium leguminosarum	*Taverniera cuneifolia*	Root	Gymnemic acid	[85]

Table 3.
Biotic elicitors.

9. Potential applications in agriculture

Focusing on biodiversity can be useful to strengthen food security and human nutrition aiming at promoting general sustainable development. Traditional crops represent an important biodiversity source and carry out a key role in preserving the identity of specific production areas as well the consumer behavior and transfer of cultural heritage to next generations. However, these cultures and foods require to be preserved from genetic erosion that can determine tragic effects on biodiversity, environmental sustainability, and rural economies.

As a matter of fact, this methodology based exclusively on a phenotypic evaluation does not allow to easily distinguish between genotype and effects on the environment. Recent methodologies based on gene markers enable us to identify species, cultivars, and autochthone varieties easily and rapidly.

Elevated costs and technical problems that might arise when the relationship between phenotype features and gene expression is studied, make the application of these methodologies often difficult. Recently, secondary metabolite analysis has been proposed as a crucial tool to identify a specific species; the metabolic profile, in fact, can lead to the identification of a huge quantity of local autochthone varieties, acting against globalization of agriculture production and being at the same time a tool to identify metabolites useful in traditional project characterization.

10. Importance of modern biotechnologies secondary metabolites production

The parts of plants to be used for therapy, nutrition, and other activities can be obtained from spontaneous or cultivated plants; the choice of production method is mostly determined by economic factors is affordable to collect spontaneous plants when abundant and costs are relatively low, however, in case of high collection costs and lack of spontaneous plants, cultivation can be less expensive [17]. Furthermore, a lot of spontaneous plants are collected without any control and are currently under extinction threat; just a small percentage is cultivated [104]—all these factors are of concern due to the decrease and loss of gene diversity and environmental degradation. Advantages of open field cultivations are related not only to the fact that they give a solution to a lack of vegetal material available in nature, but also to the fact that the wild plant often offers a highly heterogeneous which might be at the same time inadequate in terms of continuous supply and quality standards. Production of secondary metabolites from cell cultures is a valuable option for molecules that have elevated extraction costs and low output from plant material coming from cultivation [105, 106].

For these reasons and due to the current increased demand for natural food products and drugs of natural origin, the employment of biotechnological artificial culture systems might be a good alternative to conventional cultivations for *in vitro* production of secondary metabolites as well as a viable option to replace industrial biosynthesis products. These issues, together with the need to increase the production of plant materials with uniform quality standards, are encouraging pharmaceutical companies to innovate research aiming also at gene and cell technologies indicated as biotechnologies [107].

On one hand, *in vitro* cultivation systems give us the chance to exploit cells, tissues, organs, or organisms as a whole also through gene manipulation to obtain desired compounds [25]; on the other hand, they play a potential role in terms of large-scale productions, production from secondary metabolic pathways.

Plant tissue culture is based on the principle that the same substances found in nature inside an organ, a fruit, or other plant tissues can be induced to accumulate in undifferentiated cells while keeping gene information and the ability to produce that range of active principles detected in the mother plant [108].

Multiple factors influencing *in vitro* secondary metabolite synthesis can be found: type of raw material, environmental and climate conditions, culture media, the quantity of carbohydrates contained that influences biomass, type and quantity of hormones, light (optimal light quantity and intensity is a prerequisite for maximum expression of metabolites), temperature. Substrate composition strongly influences secondary metabolite production, especially for what concerns salt and growth regulators besides subsequent glucose addition that might increase accumulation if compared to cultures in which a fix concentration is used [109].

Plant cell cultures are defined also as "chemical factories for secondary metabolites" [25] and represent to date a viable alternative to the cultivation of pharmaceutical plants both from *in vitro* and non *in vitro* origin.

The most important reason for pharmaceutical companies to obtain valuable secondary metabolites in this way is due to the fact that conventional cultivations in fields of pharmaceutical plants of some species are time-consuming, expensive, and generate a reduced output.

Some large-scale protocols of productions for the market have been set up for extractions of berberine, shikonin, and *Ginseng* saponins [25, 109] by using bioreactors. Berberine is produced *in vitro* from two members of *Ranuncolaceae* (*Thalictrum minus* and *Coptis japonica*); shikonin is produced *in vitro* from *Lithospermum erythrorhizon* in quantities 800 times higher than quantities obtained from plant roots; saponins are produced *in vitro* from *Panax ginseng*.

Further research was performed on other secondary metabolites such as flavoring agents (i.e., vanillin produced in bioreactors from calluses explanted from *Vanilla planifolia* by the company ESCAgenetic Corporation—San Carlos—CA, USA), food coloring (e.g., anthocyanins from *Euphorbia milii*), drugs (e.g., taxol), different essential oils and natural insect repellents [25].

11. Conclusions

Although for the production of food from plants there is an increasing tendency toward natural agriculture, in the production of substances intended for industry, in particular the medicinal industry, a cultivated or spontaneous plant cannot always guarantee a constant and high-quality product. Pollution problems, climate change, and the political unsafe of some harvesting and cultivation areas also make production uncertain. In this situation, the production of secondary metabolites *in vitro* ensures a safe and constant making of the substances of interest.

New technologies, always evolving, can give an even greater push toward *in vitro* culture, since they guarantee safe products, at lower costs, often difficult to obtain in nature.

Acknowledgements

The present work is co-financed by the RAS (Autonomous Region of Sardinia), under the Advanced Technologies for LANds management and Tools for Innovative Development of an EcoSustainable agriculture.

Conflicts of interest

The authors declare no conflict of interest.

Author details

Grazia Maria Scarpa*, Vanda Prota, Nicola Schianchi and Federica Manunta
Department of Agriculture, University of Sassari, Sassari, Italy

*Address all correspondence to: grazia@uniss.it

IntechOpen

© 2022 The Author(s). Licensee IntechOpen. This chapter is distributed under the terms of the Creative Commons Attribution License (http://creativecommons.org/licenses/by/3.0), which permits unrestricted use, distribution, and reproduction in any medium, provided the original work is properly cited.

References

[1] Vines G. Herbal Harvests with a Future: Towards Sustainable Sources for Medicinal Plants. Salisbury, UK: Plantlife International; 2004. p. 12

[2] Carlen C, Vouillamoz J, Baroffio C, Simonnet X, Quennoz M. Genetic resources, conservation and breeding. Neunkirchen am Brand, Germany: Society for Medicinal Plant and Natural Product Research, Permanent Committee on Breeding and Cultivation of Medicinal Plants; 2009. p. 52

[3] Antonelli A, Fry C, Smith R, Simmonds M, Kersey P, Pritchard H, et al. State of the World's Plants and Fungi 2020. Royal Botanic Gardens: Kew; 2020. DOI: 10.34885/172

[4] Schippmann U, Leaman DJ, Cunningham AB. Impact of cultivation and gathering of medicinal plants on biodiversity: Global trends and issues. In: Biodiversity and the Ecosystem Approach in Agriculture, Forestry and Fisheries. Satellite event on the occasion of the Ninth Regular Session of the Commission of Genetic Resources for Food and Agriculture; Rome; 12–13 October 2002. Rome: Inter-Departmental Working Group on Biological Diversity for Food and Agriculture, FAO; 2002. pp. 1-21

[5] Allen D, Bilz M, Leaman DJ, Miller RM, Timoshyna A, Window J. European Red List of Medicinal Plants. Luxembourg: European Commission; 2016. DOI: 10.2779/907382

[6] Füllbeck M, Michalsky E, Dunkel M, Preissner R. Natural products: Sources and databases. Natural Product Reports. 2006;**23**(3):347-356

[7] Hartmann T. From waste products to ecochemicals: Fifty years research of plant secondary metabolism. Phytochemistry. 2007;**68**(22-24): 2831-2846

[8] Maffei M. Metabolismo e prodotti secondari delle piante. Alimentari e Forestali, Utete; 1999. D.D. n. 23043

[9] Collin HA. Secondary product formation in plant tissue cultures. Plant Growth Regulation. 2001;**34**(1): 119-134

[10] Pichersky E, Gang DR. Genetics and biochemistry of secondary metabolites in plants: An evolutionary perspective. Trends in Plant Science. 2000;**5**:439-445

[11] Mol J, Grotewold E, Koes R. How genes paint flowers and seeds. Trends in Plant Science. 1998;**3**:212-217

[12] Rovira AD. Plant root exudates. The Botanical Review. 1969;**35**(1):35-57

[13] Berlin J. Formation of secondary metabolites in cultured plant cells and its impact on pharmacy. In: Medicinal and Aromatic Plants I. Berlin, Heidelberg: Springer; 1988. pp. 37-59

[14] Briskin DP. Medicinal plants and phytomedicines. Linking plant biochemistry and physiology to human health. Plant Physiology. 2000;**124**(2): 507-514

[15] Borman JF, Teramura AH. Effects of UV-B radiation on terrestrial plants. In: Young AR, Bjorn LO, Moan J, Nultsch W, editors. Environmental UV Photobiology. New York: Plenum; 1993. pp. 427-471. ISBN 0-306-44443-7

[16] Pietta P. Frutta e verdure: Non solo vitamine, minerali e fibra. Verona:

UNAPROA; 2005. Available from: www.unaproa.com

[17] Samuelsson G. Drugs of Natural Origin: A Textbook of Pharmacognosy. Stockholm: 5th Swedish Pharmaceutical Press; 2004

[18] Bornman JF, Reuber S, Cen YP, Weissenbock G. Ultraviolet radiation as a stress factor and the role of protective pigments. In: Plants and UV-B: Responses to Environmental Change. Cambridge: Cambridge University Press; 1997

[19] Lercari B, Bertram L. Photomorphogenic response to UV radiation IV: A comparative study of UVB effcts on growths and pigment accumulation in etiolated and de-etiolated hypocotyls of wild-type and aurea mutant of tomato (*Lycopersicon esculenteum* Mill.). Plant Biosystems. 1997;**131**:83-92

[20] Halliwell B. Antioxidant characterization. Methodology and mechanism. Biochemical Pharmacology. 1997;**49**:1341,1348

[21] Thirumurugan D, Cholarajan A, Raja SS, Vijayakumar R. An introductory chapter: Secondary metabolites. In: Secondary Metabolites – Sources and Applications. 2018. pp. 1-21. DOI: 10.5772/intechopen.79766. Available from: https://www.intechopen.com/chapters/62731

[22] Isah T, Umar S, Mujib A, Sharma MP, Rajasekharan PE, Zafar N, et al. Secondary metabolism of pharmaceuticals in the plant in vitro cultures: Strategies, approaches, and limitations to achieving higher yield. Plant Cell Tissue and Organ Culture. 2018;**132**:239-265. DOI: 10.1007/s11240-017-1332-2

[23] Murthy HN, Lee EJ, Paek KY. Production of secondary metabolites from cell and organ cultures: Strategies and approaches for biomass improvement and metabolite accumulation. Plant Cell Tissue and Organ Culture. 2014;**118**:1-16. DOI: 10.1007/s11240-014-0467-7

[24] Karuppusamy S. A review on trends in production of secondary metabolites from higher plants by in vitro tissue, organ and cell cultures. Journal of Medicinal Plants Research. 2009;**3**: 1222-1239

[25] Rao SR, Ravishankar GA. Plant cell cultures: Chemical factories of secondary metabolites. Biotechnology Advances. 2002;**20**:101-153. DOI: 10.1016/S0734-9750(02)00007-1

[26] Yue W, Ming Q-L, Lin B, Rahman K, Zheng C-J, Han T, et al. Medicinal plant cell suspension cultures: Pharmaceutical applications and high-yielding strategies for the desired secondary metabolites. Critical Reviews in Biotechnology. 2016; **36**:215-232. DOI: 10.3109/07388551.2014.923986

[27] Ochoa-Villarreal M, Howat S, Hong S, Jang MO, Jin Y-W, Lee E-K, et al. Plant cell culture strategies for the production of natural products. BMB Reports. 2016;**49**:149-158. DOI: 10.5483/BMBRep.2016.49.3.264

[28] Xu J, Ge X, Dolan MC. Towards high-yield production of pharmaceutical proteins with plant cell suspension cultures. Biotechnology Advances. 2011; **29**:278-299. DOI: 10.1016/j.biotechadv.2011.01.002

[29] Zhang X, Ye M, Dong Y, Hu HB, Tao SJ, Yin J, et al. Biotransformation of bufadienolides by cell suspension cultures of *Saussurea involucrata*. Phytochemistry. 2011;**72**:1779-1785. DOI: 10.1016/j.phytochem.2011.05.004

[30] de Pádua RM, Meitinger N, Filho JDS, Waibel R, Gmeiner P, Braga FC, et al. Biotransformation of 21-O-acetyl-deoxycorticosterone by cell suspension cultures of *Digitalis lanata* (strain W. 1.4). Steroids. 2012;**77**: 1373-1380. DOI: 10.1016/j.steroids.2012.07.016

[31] Patil RA, Lenka SK, Normanly J, Walker EL, Roberts SC. Methyl jasmonate represses growth and affects cell cycle progression in cultured Taxus cells. Plant Cell Reports. 2014;**33**:1479-1492. DOI: 10.1007/s00299-014-1632-5

[32] Sharma K, Zafar R. Optimization of methyl jasmonate and β-cyclodextrin for enhanced production of taraxerol and taraxasterol in *Taraxacum officinale* Weber cultures. Plant Physiology and Biochemistry. 2016;**103**:24-30. DOI: 10.1016/j.plaphy.2016.02.029

[33] Cai Z, Knorr D, Smetanska I. Enhanced anthocyanins & resveratrol production in *Vitis vinifera* cell suspension culture by indanoyl-isoleucine, N-linolenoyl-l-glutamine, and insect saliva. Enzyme and Microbial Technology. 2012;**50**:29-34. DOI: 10.1016/j.enzmictec.2011.09.001

[34] Baldi A, Dixit VK. Yield enhancement strategies for artemisinin production by suspension cultures of *Artemisia annua*. Bioresource and Technology. 2008;**99**:4609-4614. DOI: 10.1016/j.biortech.2007.06.061

[35] Jeong C, Murthy HN, Hahn E, Paek K. Improved production of ginsenosides in suspension cultures of ginseng by medium replenishment strategy. Journal of Bioscience and Bioengineering. 2008;**105**:288-291. DOI: 10.1263/jbb.105.288

[36] Ten Hoopen HJG, Vinke JL, Moreno PRH, Verpoorte R, Heijnen JJ. Influence of temperature on growth and ajmalicine production by *Catharantus roseus* suspension cultures. Enzyme and Microbial Technology. 2002;**30**:56-65. DOI: 10.1016/S0141-0229(01)00456-2

[37] Chandra S, Chandra R. Engineering secondary metabolite production in hairy roots. Phytochemical Reviews. 2011;**10**:371-395. DOI: 10.1007/s11101-011-9210-8

[38] Otten L. Natural agrobacterium-mediated transformation in the genus Nicotiana. In: Ivanov N, Sierro N, Peitsch M, editors. The Tobacco Plant Genome. Compendium of Plant Genomes. Cham: Springer; 2020. DOI: 10.1007/978-3-030-29493-9_12

[39] George EF. Plant Propagation by Tissue Culture. Part 2—In Practice. 2nd ed. Edington: Exegetics; 1996

[40] Murashige T, Skoog F. A revised medium for rapid growth and bioassays with tobacco tissue cultures. Physiologia Plantarum. 1962;**15**(3):473-497

[41] Lloyd G, Mccown B. Commercially-feasible micropropagation of mountain laurel, Kalmia latifolia, by use of shoot-tip culture. Combined Proceedings, International Plant Propagators' Society. 1980;**30**:421-427

[42] Gamborg OL, Miller R, Ojima K. Nutrient requirements of suspension cultures of soybean root cells. Experimental Cell Research. 1968;**50**(1): 151-158

[43] Coelho VAT, Rodas CL, Coelho LC, Carvalho JG, Almeida EFA, Figueiredo MA. Caracterização de sintomas visuais de deficiências de macronutrientes e boro em plantas de gengibre. Revista Brasileira de Horticultura Ornamental. 2012;**18**:47-55

[44] Welch RM, Shuman R. Micronutrient nutrition of plants. Plant Science. 1995;**14**(1):49-82. DOI: 10.1080/07352689509701922

[45] Gross D, Parthier B. Novel natural substances acting in plant growth regulation. Journal of Plant Growth Regulation. 1994;**13**:93-114

[46] Liu JH, Reid DM. Auxin and ethylene-stimulated adventitious rooting in relation to tissue sensitivity to auxin and ethylene production in sunflower hypocotyls. Journal of Experimental Botany. 1992;**43**:1191-1198

[47] Aloni R. The induction of vascular tissues by auxin and cytokinin. In: Davies PJ, editor. Plant Hormones. Dordrecht: Kluwer Academic Publishers; 1995. pp. 531-546

[48] Tamas IA. Hormonal regulation of apical dominance. In: Davies PJ, editor. Plant Hormones. Dordrecht: Kluwer Academic Publishers; 1995. pp. 572-597

[49] Krikorian AD. Hormones in tissue culture and micropropagation. In: Davies PJ, editor. Plant Hormones. Dordrecht: Kluwer Academic Publishers; 1995. pp. 774-796

[50] Lance B, Reid DM, Thorpe TA. Endogenous gibberellins and growth of tobacco callus cultures. Physiologia Plantarum. 1976;**36**:287-292

[51] Ziv M, Ariel T. Bud proliferation and plant regeneration in liquid-cultured Philodendron treated with ancymidol and paclobutrazol. Journal of Plant Growth Regulation. 1991;**10**:53-57

[52] Radman R, Saez T, Bucke C, Keshavarz T. Elicitation of plant and microbial systems. Biotechnology and Applied Biochemistry. 2003;**37**: 91-102

[53] Zhao J, Davis LC, Verpoort R. Elicitor signal transduction leading to production of plant secondary metabolites. Biotechnology Advances. 2005;**23**:283-333

[54] Poornananda MN, Al-Khayri JM. Abiotic and Biotic Elicitors—Role in Secondary metabolites Production through In Vitro Culture of Medicinal Plants. 2016. Chapter 10. pp. 259-289

[55] Khani S, Barar J, Movafeghi A, Omidi Y. Production of anticancer secondary metabolites: Impacts of bioprocess engineering. Biotechnological Production of Secondary Metabolites. Bentham Science Publishers. 2012;**13**: 215-240

[56] Verpoorte R, Alfermann AW. Metabolic Engineering of the Plant Secondary Metabolism. New York: Springer; 2000. pp. 1-29

[57] O'Connor SE. Engineering of secondary metabolism. Annual Review of Genetics. 2015;**49**:71-94

[58] Cusido RM, Onrubia M, Sabater-Jara AB, Moyano E, Bonfill M, Goossens A, et al. A rational approach to improving the biotechnological production of taxanes in plant cell cultures of Taxus spp. Biotechnology Advances. 2014; **32**(6):1157-1167

[59] Lu X, Tang K, Li P. Plant metabolic engineering strategies for the production of pharmaceutical terpenoids. Frontiers in Plant Science. 2016;7:1647. DOI: 10.3389/fpls.2016.01647

[60] Dudareva N, Andersson S, Orlova I, Gatto N, Reichelt M, Rhodes D, et al. The nonmevalonate pathway supports both monoterpene and sesquiterpene formation in snapdragon flowers. Proceedings of the National Academy of

Sciences USA. 2005;**102**:933-938. DOI: 10.1073/pnas.0407360102

[61] Nanda S, Mohanty JN, Mishra R, Joshi RK. Metabolic engineering of phenyl propanoids in plants. In: Jha S, editor. Transgenesis and Secondary Metabolism: Part of the Series Reference Series in Phytochemistry. New York: Springer; 2016. pp. 1-26

[62] Tonelli M, Pellegrini E, D'Angiolillo F, Nali C, Pistelli L, Lorenzini G. Ozone-elicited secondary metabolites in shoot cultures of *Melissa officinalis* L. Plant Cell Tissue and Organ Culture. 2015;**120**:617-629

[63] Xu M, Yang B, Dong J, Lu D, Jin H, Sun L, et al. Enhancing hypericin production of *Hypericum perforatum* cell suspension culture by ozone exposure. Biotechnology Progress. 2011;**27**: 1101-1106

[64] Sun L, Su H, Zhu Y, Xu M. Involvement of abscisic acid in ozone–induced puerarin production of *Pueraria thomsnii* Benth. suspension cell cultures. Plant Cell Reports. 2012;**31**:179-185

[65] Naik PM, Manohar SH, Praveen N, Murthy HN. Effects of sucrose and pH levels on in vitro shoot regeneration from leaf explants of *Bacopa monnieri* and accumulation of bacoside A in regenerated shoots. Plant Cell Tissue and Organ Culture. 2010;**100**: 235-239

[66] Praveen N, Murthy HN. Synthesis of withanolide A depends on carbon source and medium pH in hairy root cultures of *Withania somnifera*. Industrial Crops and Products. 2012;**35**:241-243

[67] Praveen N, Murthy HN. Establishment of cell suspension cultures of Withania somnifera for the production of withanolide A. Bioresource Technology. 2010;**101**: 6735-6739

[68] Omer Y, Bengi E. Effects of sucrose and polyethylene glycol on hypericins content in Hypericum adenotrichum. Eurasian Journal of Biosciences. 2013;7: 101-110

[69] Sara-Alsadat R, Ahmad M, Mokhtar JJ. Paclitaxel production is enhanced in suspension–cultured hazel (*Corylus avellana* L.) cells by using a combination of sugar, precursor, and elicitor. Engineering in Life Sciences. 2015;**15**:234-242

[70] Xu A, Zhan JC, Huang WD. Effects of ultraviolet C, methyl jasmonate and salicylic acid, alone or in combination, on stilbene biosynthesis in cell suspension cultures of *Vitis vinifera* L. cv. Cabernet Sauvignon. Plant Cell Tissue and Organ Culture. 2015;**122**:197-211

[71] Gupta P, Sharma S, Saxena S. Biomass Yield and Steviol Glycoside Production in Callus and Suspension Culture of Stevia rebaudiana Treated with Proline and Polyethylene Glycol. Applied Biochemistry and Biotechnology 2015;**176**:863–874. DOI: 10.1007/s12010-015-1616-0

[72] Munish S, Ashok A, Rajinder G, Sharada M. Enhanced bacoside production in shoot cultures of *Bacopa monnieri* under the influence of abiotic elicitors. Natural Product Research. 2015;**29**(8):745-749

[73] Gangopadhyay M, Dewanjee S, Bhattacharya S. Enhanced plumbagin production in elicited *Plumbago indica* hairy root cultures. Journal of Bioscience and Bioengineering. 2011;**111**:706-710

[74] Silja PK, Gisha GP, Satheeshkumar K. Enhanced plumbagin accumulation in embryogenic cell

suspension cultures of *Plumbago rosea* L. following elicitation. Plant Cell Tissue and Organ Culture. 2014;**119**:469-477

[75] Xiaolong H, Min S, Lijie C, Chao X, Yanjie Z, Guoyin K. Effects of methyl jasmonate and salicylic acid on tanshinone production and biosynthetic gene expression in transgenic *Salvia miltiorrhiza* hairy roots. Biotechnology and Applied Biochemistry. 2015;**62**(1): 24-31. DOI: 10.1002/bab.1236

[76] Arehzoo Z, Christina S, Florian G, Parvaneh A, Javad A, Seyed HM, et al. Effects of some elicitors on tanshinone production in adventitious root cultures of *Perovskia abrotanoides* Karel. Industrial Crops and Products. 2015;**67**: 97-102

[77] Izabela G, Halina W. The effect of methyl jasmonate on production of antioxidant compounds in shoot cultures of *Salvia officinalis* L. Herba Polonica. 2009;**55**:238-243

[78] Firouzi A, Mohammadi SA, Khosrowchahli M, Movafeghi A, Hasanloo T. Enhancement of silymarin production in cell culture of *Silybum marianum* (L) Gaertn by elicitation and precursor feeding. Journal of Herbs, Spices and Medicinal Plants. 2013;**19**(3): 262-274

[79] Li B, Wang B, Li H, Peng L, Ru M, Liang Z, et al. Establishment of *Salvia castanea* Diels f. tomentosa Stib. hairy root cultures and the promotion of tanshinone accumulation and gene expression with Ag^+, methyl jasmonate, and yeast extract elicitation. Protoplasma. 2016 Jan;**253**(1):87-100. DOI: 10.1007/s00709–015–0790–9

[80] Chodisetti B, Rao K, Gandi S, Giri A. Gymnemic acid enhancement in the suspension cultures of *Gymnema sylvestre* by using the signaling molecules—methyl jasmonate and salicylic acid. In Vitro Cellular and Developmental Biology—Plant. 2015;**51**:88-92

[81] Sivanandhan G, Dev GK, Jeyaraj M, Rajesh M, Arjunan A, Muthuselvam M, et al. Increased production of withanolide A, withanone, and withaferin A in hairy root cultures of *Withania somnifera* (L.) Dunal elicited with methyl jasmonate and salicylic acid. Plant Cell Tissue and Organ Culture. 2013;**114**:121-129

[82] Sharma SN, Jhaa Z, Sinhab RK, Gedac AK. Jasmonate–induced biosynthesis of andrographolide in Andrographis paniculata. Physiologia Plantarum. 2015;**153**:221-229

[83] Taurino M, Ingrosso I, D'amico L, Domenico SD, Nicoletti I, Corradini D, et al. Jasmonates elicit different sets of stilbenes in *Vitis vinifera* cv. Negramaro cell cultures. SpringerPlus. 2015;**4**:49

[84] Awad V, Kuvalekar A, Harsulkar A. Microbial elicitation in root cultures of *Taverniera cuneifolia* (Roth)Arn. for elevated glycyrrhizic acid production. Industrial Crops and Products. 2014;**54**: 13-16

[85] Yuan Y, Huang L, Cui G, Mao Y, He X. Effect of gibberellins and its synthetic inhibitor on metabolism of tanshinones. Chinese Journal of Experimental Traditional Medical Formulae. 2008;**6**:002

[86] Abbasi BH, Stiles AR, Saxena PK, Liu CZ. Gibberellic acid increases secondary metabolite production in *Echinacea purpurea* hairy roots. Applied Biochemistry and Biotechnology. 2012; **168**(7):2057-2066

[87] Patil JG, Ahire ML, Nitnaware KM, Panda S, Bhatt VP, Kishor PB, et al. In vitro propagation and production of

cardiotonic glycosides in shoot cultures of *Digitalis purpurea* L. by elicitation and precursor feeding. Applied Microbiology and Biotechnology. 2013;**97**:2379-2393

[88] Coste A, Vlase L, Halmagyi A, Deliu C, Coldea G. Effects of plant growth regulators and elicitors on production of secondary metabolites in shoot cultures of *Hypericum hirsutum* and *Hypericum maculatum*. Plant Cell Tissue and Organ Culture. 2011;**106**: 279-282

[89] Ajungla L, Patil PP, Barmukh RB, Nikam TD. Influence of biotic and abiotic on accumulation of hyoscyamine and scopolamine in root cultures of *Datura metel* L. Indian Journal of Biotechnology. 2009;**8**(3):317-322

[90] Li J, Wang J, Li J, Li J, Liu S, Gao W. Salicylic acid induces the change in the adventitious root of *Glycyrrhiza uralensis* Fisch.: Bioactive compounds and antioxidant enzymes. Research on Chemical Intermediates. 2015;**42**(2): 1503-1519. DOI: 10.1007/s11164–015–2099–x

[91] Kračun-Kolarević M, Dmitrović S, Filipović B, Perić M, Mišić D, Simonović A, et al. Influence of sodium salicylate on rosmarinic acid, carnosol and carnosic acid accumulation by *Salvia officinalis* L. shoots grown in vitro. Biotechnology Letters. 2015;**37**(8): 1693-1701

[92] Fatima S, Mujib A, Dipti T. NaCl amendment improves vinblastine and vincristine synthesis in *Catharanthus roseus*: A case of stress signalling as evidenced by antioxidant enzymes activities. Plant Cell Tissue and Organ Culture. 2015;**121**:445-458

[93] Cai Z, Kastell A, Speiser C, Smetanska I. Enhanced resveratrol production in *Vitis vinifera* cell suspension cultures by heavy metals without loss of cell viability. Applied Biochemistry and Biotechnology. 2013; **171**:330-340

[94] Zahra S, Mehrnaz K, Gholamreza A, Mustafa G. Improvement of atropine production by different biotic and abiotic elicitors in hairy root cultures of *Datura metel*. Turkish Journal of Biology. 2015;**39**:111-118

[95] Furze JM, Rhodes MJC, Parr AJ, Robins RJ, Withehead IM, Threlfall DR. Abiotic factors elicit sesquiterpenoid phytoalexin production but not alkaloid production in transformed root cultures of *Datura stramonium*. Plant Cell Reports. 1991;**10**(3):111-114

[96] Purohit M, Pande D, Datta A, Srivastava PS. Enhanced xanthotoxin content in regenerating cultures of *Ammi majus* and micropropagtion. Planta Medica. 1995;**61**:481-482

[97] Sonja GS, Oliver T, Stéphane M, Alain D, Claude J, Daniel H. Effects of polysaccharide elicitors on secondary metabolite production and antioxidant response in *Hypericum perforatum* L. shoot cultures. The Scientific World Journal. 2014;**11**:609-649

[98] Sonja GS, Oliver T, Stéphane M, Christophe H, Alain D, Brigitte C, et al. Fungal elicitor-mediated enhancement in phenylpropanoid and naphtodianthrone contents of *Hypericum perforatum* L. cell cultures. Plant Cell Tissue and Organ Culture. 2015;**122**: 213-226

[99] Qianliang M, Chunyan S, Chengjian Z, Min J, Qiaoyan Z, Hong Z, et al. Elicitors from the endophytic fungus *Trichoderma atroviride* promote *Salvia miltiorrhiza* hairy root growth and tanshinone biosynthesis. Journal of

Experimental Botany. 2013;**64**(18): 5687-5694

[100] Bhagwath SG, Hjortsø MA. Statistical analysis of elicitation strategies for thiarubrine A production in hairy root cultures of *Ambrosia artemisiifolia*. Journal of Biotechnology. 2000;**80**:159-167

[101] Satdive RK, Fulzele DP, Eapen S. Enhanced production of azadirachtin by hairy root cultures of *Azadirachta indica* A. Juss by elicitation and media optimization. Journal of Biotechnology. 2007;**128**:281-289

[102] Chodisetti B, Rao K, Gandi S, Giri A. Improved gymnemic acid production in the suspension cultures of *Gymnema sylvestre* through biotic elicitation. Plant Biotechnology Reports. 2013;7:519-525

[103] Kowalczyt T, Wieczfinska J, Skała E, Śliwiński T, Sitarek P. Transgenesis as a tool for the efficient production of selected secondary metabolites from plant in vitro cultures. Plants. 2020;**2020**(9):132. DOI: 10.3390/plants9020132

[104] Canter P, Thomas H, Ernst E, 2005. Bringing medicinal plants into cultivation: Opportunities and challenges for biotechnology. Trends in Biotechnology. 2005;**23**(4): 180-185

[105] Dornenburg H, Knorr D. Challenges and opportunities for metabolite production from plant cell and tissue cultures. Food Technology. 1997;**51**(11):47-54

[106] Ravishankar GA, Venkataraman LV. Role of plant cell culture in food biotechnology: Current trends, limitations and future prospect's. In: Prakash J, Pierik RLM, editors. Plant Biotechnology: Commercial Prospects and Problems. New Delhi: Oxford IBH Press; 1993. pp. 255-274

[107] Falcone AM. La coltura in vitro applicata alle specie aromatiche. Rivista di Agricoltura Subtropicale e Tropicale. 1990;**3**:377-386

[108] Ruffoni B, Pistelli L, Bertoli A, Pistelli L. Plant cell cultures: Bioreactors for industrial production. Advances in Experimental Medicine and Biology. 2010;**698**:203-218

[109] Smith MAL. Large scale production of secondary metabolites. In: Terzi M, Cella R, Falavigna A, editors. Current Issues in Plant Molecular and Cellular Biology. Vol. 22. Springer, Dordrecht: Current Plant Science and Biotechnology in Agriculture. 1995. DOI: 10.1007/978-94-011-0307-7_93

Chapter 7

Biological Activity of Defence-Related Plant Secondary Metabolites

Ananth Anbu and Umadevi Ananth

Abstract

The message that everyone needs to know is that secondary metabolites in plants and natural products are involved in various activities. The phenolics, quinones, terpenes, flavonoids, and other thousands of low molecular weight metabolites activity is unknown. Well-understood secondary metabolites have been implicated in the defense against pathogens; the operating system of some of these has been established. In particular, to date, a relatively small number of processes have been shown to be targets of plant metabolism, including electron transport chains, mitochondrial function, and membrane integration. However, it is now emerging that other specific enzymes and processes may also be targets of specific metabolites. There is a general belief that modern genetic approaches will identify new targets and mechanisms of plant metabolism. Molecules that trigger apoptosis or autoimmunity in tumor cells, especially triterpenoids, are of particular interest in this regard. Before proceeding to specific studies in plant or human cells, we discuss whether there is a case for conducting preliminary studies on the mechanism of action in the genetic pathway system, such as yeast *Saccharomyces cerevisiae*, considering the approaches taken so far in botany and strategies that have led to success in the biomedical field.

Keywords: natural products, *Saccharomyces cerevisiae*, Defensins, pathogen

1. Introduction

Secondary metabolites in plants are commonly used to describe metabolic pathways that produce molecules or metabolites that can provide for normal growth or are only needed under certain conditions. In contrast, primary metabolites traditionally describe key household maintenance functions, such as energy production or the production of essential metabolites and macromolecules. These differences may be somewhat misleading; however, as is now known, secondary metabolites compounds plays a very important role in the biology of various organisms. In fact, it is clear that evolution would not selectively maintain the complex pathways that make up secondary metabolites if there were no competing advantages for the developing organism.

This logic, coupled with the fact that the biological function of the majority of plant and microbial secondary metabolites is poorly understood, has led to an alternative description of plant metabolites as "natural products" [1], though that description also carries some limitations. Nature produces a tremendous array of secondary metabolites or natural products, with the most diversity seen in micro-organisms and plants [2]. It is a great resource for mankind and many examples of microbial or plant metabolisms are exploited by man, for example, antibiotics and pharmaceuticals. However, we have only scratched the surface, especially since there are various natural metabolites that have applications in the field of biomedicine. It is the basis of many natural product discovery projects, for example, attempts to use metagenomics to study marine microbial diversity [3]. In contrast to these attempts to explore metabolic diversity in new key locations, plant metabolic diversity has been exploited by humans throughout history, initially using plant extracts and more recently through scientific activity to identify metabolites with specific functions and then use these products directly or as traces for therapeutic compounds [4].

2. Defense against pathogens with secondary metabolism

Knowledge of how these molecules affect the exploitation of natural materials is often followed by an understanding of the role of metabolism in the producing organism. In plants, well-understood secondary metabolism is involved in pathogen protection or perception and signaling. In terms of pathogen protection, fungal diseases pose a major threat to plant health, with estimates below of 13,000 phytopathogenic fungal species in the United States alone. Therefore, it is not surprising that plants have developed comprehensive protection mechanisms against fungal pathogens, with chemical protection being one of the key weapons in the plant arsenal [5]. Although thousands of different molecular companies are believed to play a role in plant protection against bacterial and fungal pathogens, the mechanism of action of relatively few has been the subject of extensive research.

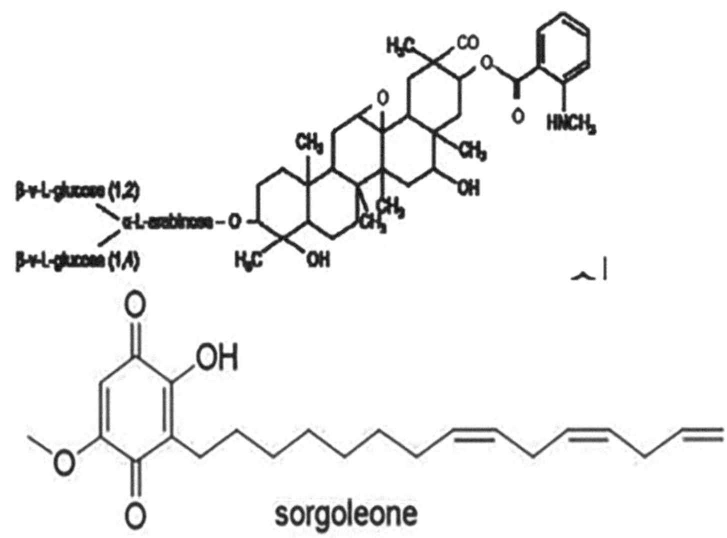

Figure 1.
Natural product can be localized into plant tissue or secreted externally.

Plant defense molecules may be pre-formed in plant tissues (**Figure 1**) or synthesized in response to the pathogenic attack, resulting in variations leading to the terms phyto antiseptics and phytoalexins, respectively [6]. This difference does not provide any specific information about the chemical structure or mechanism of metabolism and in some cases, misleading defense molecules are pre-manufactured but concentrated in high concentrations at the site of infection are reasonably considered to be phyto antiseptics or phytoalexins. In practice, when studying the range of possible biological functions involved in metabolism, the chemical structure of natural products is more relevant than the exact time produced at the plant.

3. The role of secondary metabolites in plant-microbial signaling

In terms of signal, the most comprehensible metabolites are flavonoids involved in symbiotic lentil-rhizobia interactions that lead to the formation of nitrogen-fixing nodules in root tissues [7]. Collectively, plants produce more than 5000 different flavonoids, with only a small subgroup involved in specific interactions with Rhizobia. This interaction begins with the secretion of signal flavonoids at the root exudates, followed by the bacterial understanding of the signal and direct contact with the bacterial nodule transcriptional activator. This triggers a series of events that create convenient rhizobial infection of the plant root and nitrogen-fixing nodules.

The other major beneficial plant-microbe interaction that occurs in nature is the formation of mycorrhizal roots. Once again, there is a facilitated infection of plant roots, this time by arbuscular mycorrhiza fungi, which develop specialized structures called arbuscles within the root for nutrient exchange between the plant and fungus. Although a role for signaling has long been postulated, it is only in recent years that the first experimental evidence demonstrating a role for a plant chemical has been obtained, showing that a particular class of sesquiterpene, the strigolactones, can induce hyphal branching, an important step in the symbiosis [8].

As an added twist, several studies have shown that certain strigolactones actually play a role in regulating plant hormones and spruce branches in the plant, thus regulating processes above and below the ground [9].

4. Allelopathic reactions

Allelopathy is defined as the inhibition of the growth of one species by chemicals produced by another species, and although this is a matter of controversy in the scientific literature, this concept has been generally accepted in recent years [10]. This definition is significantly shorter than the original use of the word, which may involve both positive and negative interactions, but it is also a reflection of the importance of allelopathy between domestic and introduced plant species, especially when introduced species can invade and displace native plants. Engineering mills, especially those with grains, have a considerable interest in controlling weeds in their own surroundings using allelopathy in agriculture.

The basic premise of allelopathy is that plants secrete phytotoxic metabolites in their surroundings (primarily rhizosphere) and inhibit the growth of plants that are susceptible to these metabolites. This process can be reasonably classified as protective or signal and, in fact, molecules such as strigolactones may have dual roles.

Allelopathy is believed to have an evolutionary dimension, so long-term coexisting plants have developed co-adaptation and tolerance mechanisms, whereas ecologically separated plants may not have these tolerance or resistance mechanisms. The various molecules present in the root glands are known to have phytotoxic properties at biologically related concentrations (**Figure 1**).

The majority are phenolics, including simple phenolics, flavonoids, and quinones; terbenes, monoterpenoids, sesquiterpene lactones, diterbenes and Benzoxazinoids or glucosinolates. An important feature when considering plant protection against microbial or insect pathogens, signaling and allopathy is that overall classes of molecules are also included in these cases. Our ability to determine whether, specific metabolites may first be lost in evolutionary history as signal molecules, as protection against pathogens, or as phytotoxic agents to enhance competitiveness. With regard to the exploitation of these natural products (lead) as herbicides, plant protection products, or drugs, it is now an important quest to understand their mechanism of action in targeted and non-targeted organisms.

5. Process of plant natural products

5.1 Identifying the targets of plant metabolites

Despite the vast number of biological reactions in biological structures and cells, a relatively small number are exploited by man. For example, 270 herbicides in commercial use target only 17 different processes and medicinal and agricultural fungicides target only six different processes [11]. As the synthesis of natural substances in plants runs into many thousands of different molecules, many new inhibitors of cellular functions can be identified. This belief drives most of the research on plant natural products and their mode of action. Although many plant metabolites have been described chemically and have played many roles in signaling, defense, and allelopathy, the exact action of some has been determined in no detail. In cases where attempts have been made to determine how chemicals cause their effects, the interpretation of the results is often complicated by several goals, including difficulty in separating primary and secondary effects and difficulties in determining whether data obtained from *in-vivo* studies are related to *in-vivo*. To a large extent, these difficulties are indicative of limitations with the test methods used, and there is certainly a case for conducting studies in genetically controllable systems such as yeast. Nevertheless, it was possible to identify key processes that are normally targeted by plant metabolites and specific enzymes that can be inhibited by specific metabolites.

5.2 Inhibition of specific enzymes

Plant secondary metabolites can inhibit specific enzymes in plants or other organisms, such as fungi or animals. In some cases, it appears to be the only function of the metabolism, while in others, it forms part of a set of enzyme inhibitory effects. It should be noted, however, that the uniqueness of some of the findings and the biological relevance of *in-vivo* are questionable. An example of this is the inhibition of various enzyme reactions, including the plant hormone biosynthetic enzymes, catalase, maltase, and phosphatase by phenolics and phenolic acids [12].

Sesquiterpenes are one of the largest families of plant natural products and have many common effects associated with this type of molecule. It is believed that some sesquiterpenes inhibit the activity of enzymes containing sulfhydryl-containing enzymes (e.g., phosphor-fructokinase) and that this may be due to the general apoptotic effects of plant sesquiterpenes on animal cells, but more detailed investigations are needed in this area. In contrast to those common effects, quinone sorgoleone (**Figure 1**) inhibits the enzyme 4-hydroxyphenylpyruvate dioxygenase (HPPD) [11]. Plastoquinone and ultimately chloroplast synthesis require HPPD activity and are targeted to sulcotrione and other herbicides [13]. Other quinones, such as juglone made from the walnut tree, can also inhibit HPPD activity.

Another example is the steroidal alkaloid tomatidine, which in particular inhibits the C24 sterol methyltransferase reaction, which is essential for the synthesis of the essential fungal membrane sterol, ergosterol. This anti-fungal metabolism is

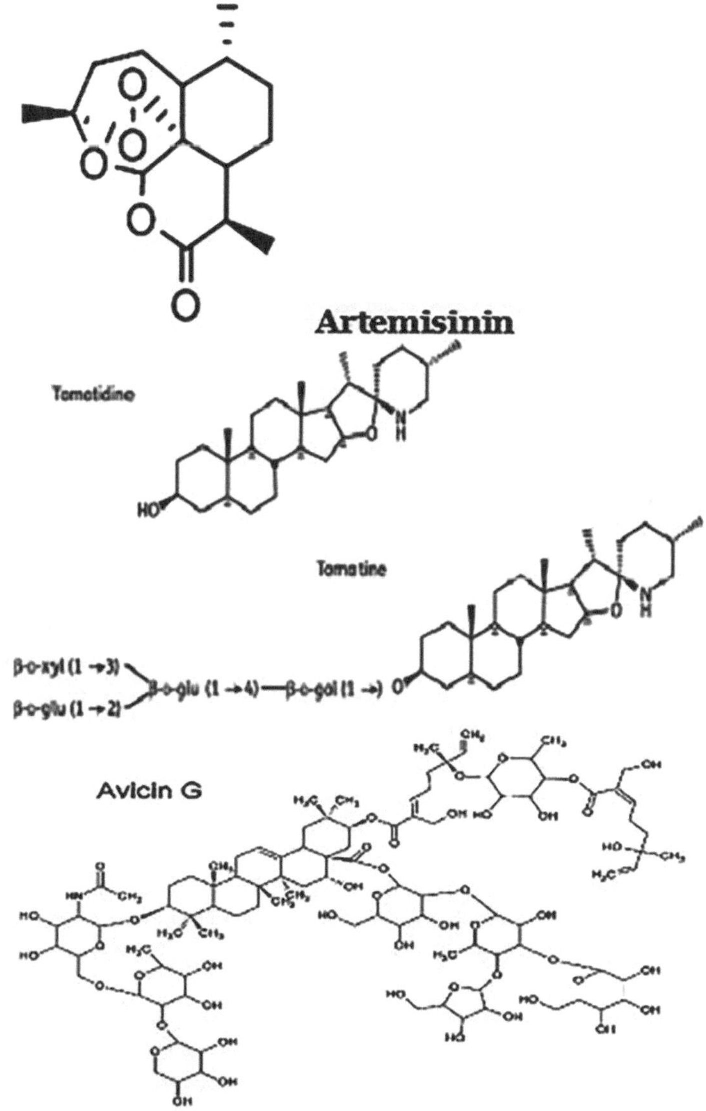

Figure 2.
Structures of some plant secondary metabolites.

synthesized in tomatoes in a glycosylated form called α-tomatine and is closed to the steroidal alkaloid tomatidine by fungal enzymes during plant infection (**Figure 2**). Studies with yeast *Saccharomyces cerevisiae* have unique modes of action with tomatidine, which is 50 times more potent than α-tomatine and tomaditin action α-tomatine.

Interestingly, the importance of C24 sterol methyltransferase for ergosterol biosynthesis has already been recognized and commercial fungicides such as fenpropimorph target the same enzyme. The fact that the enzymes in question have already been identified, and used as pharmacological targets in both sorgoleone/HPPT and tomatidine/C24 sterol methyltransferase confirms the technique of identifying new enzyme targets of plant natural products as intervention drugs or chemicals. Some new natural ingredients or enzymes are under investigation in this regard. For example, 1,4-cineole (monoterpene) inhibits the synthesis of asparagine and quassinoids (diterpenes) are believed to inhibit membrane NADH oxidase [14].

6. Inhibition of electron transport systems

6.1 Target of photosynthesis and respiration

Photosynthesis is centrally important for plant health; It is, therefore, a clear target for natural and synthetic inhibitory molecules. At least 59 different herbicides target Photo System II (PSII), primarily by interfering with electron transport [13]. PS-II quinone was found to be the main target of sorgoleone, the same metabolite that inhibits the enzyme HPPD (above). Sorgoleone is believed to compete with plastoquinone for binding to D1 proteins in PS-II [15] and is secreted in droplets from the root hairs, which accumulate in the soil around the plant roots at 10–100 μM.

The imbalance between the number of herbicides and natural metabolites that inhibit photosynthesis is surprising and suggests that there may be many more natural inhibitors of photosynthesis yet to be identified. Respiration is another important function of the cell based on electron transport chains and also is the target of inhibitory molecules. The clearest example is probably the cyanogenic glycosides that are produced by more than 200 different types of plants. These are synthesized by converting amino acid precursors to oximes, which are then glycosylated. The hydrolysis of cyanogenic glycosides in response to tissue damage produces hydrogen cyanide (HCN), a potent respiratory toxin [6]. Glucosinolates are molecules associated with the evolution that is synthesized only by a subgroup of organisms, mainly within the order capparalase, including the agriculturally important Brassicaceae family [16].

The hydrolysis of glucosinolates yields isothiocyanates, thiocyanates, and nitriles and although the fungal pattern of these metabolites has not been demonstrated, cyanide moiety is said to be the target of some of these metabolites. Other low molecular weight natural products are also believed to target respiration, but in many cases, it is difficult to establish definitively and studies with isolated mitochondria have sometimes produced conflicting results. Therefore, although some phenolic acids inhibit the absorption of iodine by the mitochondria, the concentrations of phenolics appear to be unreliable, while there are suggestions that phenolics may inhibit electron transport in the b/c1 cytochrome complex and those phenolics actually induce respiration in some cases [17].

7. Biological activity of the antimalarial drug artemisinin

The use of sesquiterpene lactone artemicin has been reported to have a variety of physiological effects on target cells, including disruption of mitochondrial function [18]. Artemisinin is a natural product synthesized by the Chinese plant *Artemesia annua* (sweet wormwood), which means that this molecule and its derivatives are now part of the front-line anti-malarial treatment. The effect of artemisinin on plant cells is unknown, but several studies have attempted to determine why this metabolism is toxic to the malaria parasite *Plasmodium falciparum* and other protozoa. Osteomycin is abnormal in possessing an endoperoxide moiety essential for cell function (**Figure 2**).

Artemisinin *Plasmodium falciparum* inhibits the absorption of oxygen, indicating that it may be the target respiratory chain [19]. In a new strategy, Li and colleagues explored the mechanism of action of artemisinin using a yeast sample and using yeast genetics to disrupt the mitochondrial membrane capacity of artemisinin [20]. Their work pointed out that the electron transport chain actually activates the mitochondrial depolarizing function of artemisinin. In contrast, Nagamun and colleagues found that artemisinin could not affect the mitochondrial membrane ability of another protozoan parasite *Toxoplasma gondii*, suggesting that mitochondria were not a primary target in the *T. gondii* [21]. In fact, there is now strong evidence that calcium affects homeostasis in the target species of artemisinin.

Eukaryotic cells use Ca^{2+} as a second messenger and generally maintain very low cytoplasmic concentrations of Ca^{2+} by dividing Ca^{2+} into segments, such as the endoplasmic reticulum. One of the key enzymes in this process is the sarcoplasmic/endoplasmic reticulum Ca^{2+} -ATPase (SERCA). Heterologues host, *Xenopus lewis* using early work, demonstrated that artemisinin inhibits *P. falciparum* function [22] and recent experiments revealing the *T. gondii* in *S.cerevisiae* demonstrated that the *T. gondii* enzyme was inhibited by artemisinin. Physiological tests in many protozoa are consistent with the effects of artemisinin on calcium homeostasis, suggesting that it may account for a significant portion of the biological activity of this metabolite [23]. There are conflicting opinions on the biological functions of artemisinin with further studies to determine whether the malaria parasite *P. falciparum* and in fact plant organisms affect the mitochondria as the primary or secondary target of artemisinin. Artemisinin appears to cause specific nonspecific effects such as the production of free radicals and immune stimulation, and it is absolutely plausible that, like sorgoleone, artemisinin has more than one target or effect. The interactions between mitochondrial function, calcium signaling, and apoptosis should go unnoticed, and the effects that appear pleiotropic may actually become part of the same process.

8. Disruption of plasma membrane integrity

8.1 Importance of the fungal membrane as a target

As previously highlighted, secondary metabolites plays an important role protection of plants against fungal pathogens. This is an important area of interest in modern agriculture and medicine with the fungal cell membrane for clinical medicine and agro-fungal drugs. The fungal membrane has unique features, especially sterol ergosterol other than cholesterol or stigmasterol, which is present in animal and plant membranes, respectively. Other differences include the presence of specific lipids on the outer leaf of the membrane. Common antifungal compounds include amphotericin B, which binds to ergosterol, which leads to pore formation, azoles, and morpholine, which inhibit ergosterol biosynthesis. Evolution has failed to observe this effect on fungi and plants but develops different types of antifungal defense metabolites that target the membranes of phytopathogenic fungi. The well-understood of these are defensins and saponins.

8.2 Plant defenses have specific binding sites in fungal membranes

Defensins are the most basic, cysteine-rich peptides, typically 40–45 amino acids in length, produced by plants, insects and other animals as antimicrobial defensins molecules [24]. Molecular phylogenetic analysis while the evolutionary roots of these molecules were probably in plants, there was a significant functional difference in the family of cationic antimicrobial peptides (cAMPs) defensins by evolution [25]. A variety of defensins has been reported to have antiviral, antibacterial, and antifungal activity. The prevailing opinion is that the positive charge of peptides mediates specific non-binding with phospholipids, which leads to pore formation and loss of membrane integrity. Although this is a common feature of cAMPs, in recent years it has emerged that specific interactions play a role in the functioning of some cAMPs. For example, human α-defensins have been shown to inactivate adenovirus by binding directly to the viral protein, and endogenous targets for cAMPS have been identified in some bacteria. It is already known that plant defensins and some insect defensins have a specific binding target and mode of action. This discovery initially came from work using *S. cerevisiae* as a model to study the antifungal activity of plant defensins. Yeast strain DmAMP1, a mutation in a gene required for the synthesis of sphingolipids, altered sensitivity to plant defensins [26]. Sphingolipids are commonly found on the outer leaf of eukaryotic membranes and resemble phospholipids, except that the vertebrae are not made of diacylglyceride.

Many variants of Sphingolipids have some unique structures in different fungi in eukaryotic membranes. In a series of studies, some plant and insect defensins bind to different fungal sphingolipids or different nuclei in the same sphingolipids. Following binding, membrane infiltration occurs, but it is not yet known whether this is the result of the signal layer or the biophysical effect. However, it is clear that plant defensins do not particularly penetrate fungal membranes, creating pores and destroying membrane integrity. Interestingly, in *Candida albicans*, the anti-fungal activity of a plant called RsAFP2, which binds to glucosylceramide, was found to

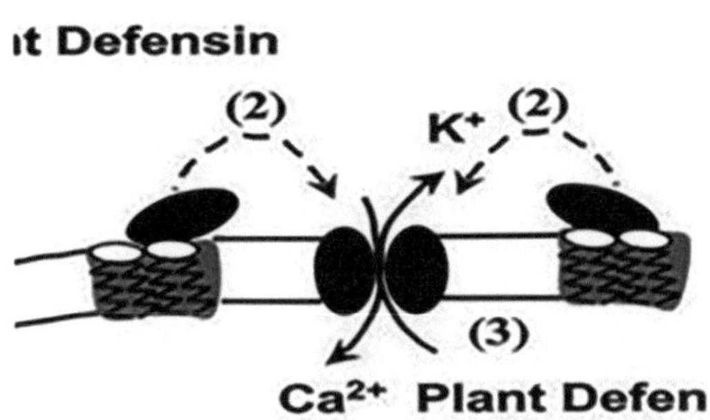

Figure 3.
The antifungal activity of plant defensins is specific and involves receptors and signal transduction pathways.

involve the production of reactive oxygen species (ROS) suggesting that binding to the membrane ligand initiated a signal transduction cascade that culminated in the production of ROS and membrane infiltration (**Figure 3**).

Despite advances in the study of plant defensins, some serious questions and challenges remain to be resolved. First, most extensive work has been done with a limited number of specific defensins, and it remains to be determined whether this is the only procedure. Second, it is not known what signal transmission paths are activated in response to defensins. Third, it is not yet clear whether defensins are internalized after binding to or with sphingolipids. Working with human cAMPs is said to be at least absorbed by some bacteria and reported to be absorbed by the fungal cells of pea defensins [27].

8.3 Lysis of fungal membranes by saponins

Saponins are a structurally different class of secondary metabolites found in different plants. For example, a survey lists more than 200 plants that isolated saponins between 1998 and 2003. The basic structure of all saponins consists of the polar core and the polar glycosyl group or groups, which give the molecules ambiguous properties. Typically, saponins are classified as triterpenoid or steroidal, with a subset of steroidal alkaloids (steroidal glycoalkaloids) depending on the structure of the hydrophobic center. However, some authors consider steroidal glycolic colloids to be a unique natural product, and recently, a new saponin classification has been proposed into 11 different families depending on the structure of the spine. Saponins are present in significant concentrations in many traditional herbal medicines and a variety of beneficial functions, including common ingredients such as ginseng, are often attributed to the components of saponin.

Within plants, saponins are believed to provide protection against phytopathogenic fungi because they have powerful antifungal activity, are usually accommodated in the epidermal layers of plant tissues and have been shown to play a protective role in many pathogenic interactions. The amphibian nature of saponins represents a mechanism of action and it has been demonstrated that saponins penetrate fungal membranes. The proposed mechanism is that the hydrophobic core enters the outer membrane, forming a compound with ergosterol. Subsequent interaction between polar glycocytic sidechains leads to aggregation, pore formation, and loss of membrane integrity [28]. The ability to penetrate membranes has been demonstrated *in vitro* and *in vivo* in sample membranes of *S.cerevisiae* to explore the anti-fungal activity of the steroidal glycoalkaloid saponin. However, the study also showed that the alkaloid α-tomatine did not penetrate the membranes of the α-tomatine, which is more potent than the α-tomatine, in fact inhibiting ergosterol biosynthesis.

However, the study also showed that the aglycone of α-tomatine did not penetrate the membranes of the α-tomatine, was more potent than the α-tomatine, and actually inhibited ergosterol biosynthesis. Furthermore, several studies have proposed additional functions for α-tomatine and its derivatives. β2-tomatine (created by removing sugar from sugar α-tomatine) has been found to be capable of suppressing plant defense response, and α-tomatine has been reported to induce projected cell death in fungi called *Fusarium oxysporum*, lack of membrane penetration. Finally, studies with potato steroidal glycoalkaloids (saponins), α-chaconine and α-solanine, have identified various toxic effects on animal systems that differ from membrane penetrating activity. In conclusion, although membrane penetration activity contributes to the antifungal activity of saponins, saponins may have other biological properties, including beneficial roles in human health [29].

9. Anti-tumor activity of plant natural products

Among the various properties associated with saponins, the ability of some saponin products to inhibit the growth of tumor cells *in vitro* attracts much attention. In fact, many saponins have been reported to have such activity, increasing the likelihood of developing novel saponin-based anti-cancer drugs [30]. Some commentators have questioned the relevance of these *in vitro* data and want to prove that their validity covers specific endogenous targets and is not related to membrane penetration. Significant progress in this direction is due to the synthesis of two different groups of legume triterpenoid saponins, avicins, *Acacia victoria* (**Figure 2**), and soyasaponin from soybean plants. This work was triggered by reports that Avicenna had apoptotic activity against human tumor cells. Several studies have established that it is mediated by mitochondrial dysfunction, indicating that the effects are twofold—disruption of the outer membrane energy and closure of the voltage-dependent ion channel (VDAC) in the mitochondrial membrane. The link between saponin-induced mitochondrial disfunction and apoptosis was further supported by a recent report showing that treatment of HeLa cells with soyasaponin products led to apoptosis via the mitochondrial pathway [31].

Other endogenous targets for specific avicin have also been reported; however, pro-apoptotic effects may involve multiple targets or indicate that different avicins have specific target processes. In the East model, evidence was obtained for the modulation or inhibition of RO-based signaling and CAMP/PKA signal transmission pathways. A more direct link to apoptosis or automation was obtained from studies with avicin D, where avicin D activates AMP-activated kinase (AMPK), thereby inhibiting mTORC1 and downstream targets. Although many studies of plant natural products have reported a "pro-apoptotic" function, it is worth noting here that there is not much difference between apoptosis and autoimmunity in the plant literature in general. In fact, although the end result is the same, the paths and processes involved are completely different and this is a topic that will require closer attention in the future. Autophagy is given more importance by discovering that the production of β-group soysasaponins reduces mTORC activity, this time apparently by activating another kinase, Akt. The general significance of apoptotic pathways as a target for plant natural metabolism is that other saponin non-metabolites, such as sesquiterpenoid helenalin, which inhibits telomerase, have proliferative effects on mammalian cells [32]. Some of the more than 5000 different flavonoids that occur naturally in plants have effects associated with apoptosis in the future. Nutmeg flavonoid, (−) catechins, for example, inhibit seed germination and cause cell death in sensitive species. This effect appears to involve the generation of reactive oxygen species and may also be linked to calcium signaling or homeostasis. Again, the relationship between ROS, calcium homeostasis, mitochondrial function, autoimmunity, and apoptosis [33] should be kept in mind.

10. Conclusion

Plant natural products, especially those involved in the protection against pathogens, can lead to biotechnological applications. However, beyond the phytochemical list and general studies, there is a need to go for experiments to identify specific screens and functional patterns. It should take two forms, identifying the biological role of metabolism in the plant and determining the effects of metabolism on other organisms. The latter is a compelling argument for the use of unbiased genetic or proteomic methods and cell-based assessments to avoid confusion with specific nontargets. Finally, once the candidate goals have been identified, it is necessary to carry out detailed structural and functional studies of the interaction in the actual hosts. However, for preliminary screens and analyzes, plant natural product scientists must follow their biomedical counterparts in eukaryotes and *S. cerevisiae* using chemical genetics and molecular techniques.

Conflicts of interest

The author declares no conflict of interest.

Author details

Ananth Anbu[1*] and Umadevi Ananth[2]

1 Department of Microbiology, Srinivasan College of Arts and Science, Perambalur, Tamilnadu, India

2 Department of Microbiology, Sri Saradha College of Women, Perambalur, Tamilnadu, India

*Address all correspondence to: uananth4@gmail.com

IntechOpen

© 2022 The Author(s). Licensee IntechOpen. This chapter is distributed under the terms of the Creative Commons Attribution License (http://creativecommons.org/licenses/by/3.0), which permits unrestricted use, distribution, and reproduction in any medium, provided the original work is properly cited.

References

[1] Field B, Jordan F, Osbourn A. First encounters-deployment of defence-related natural products by plants. The New Phytologist. 2006;**172**:193-207

[2] Wink M. Evolution of secondary metabolites from an ecological and molecular phylogenetic perspective. Phytochemistry. 2003;**64**:3-19

[3] Kennedy J, Marchesi JR, Dobson AD. Metagenomic approaches to exploit the biotechnological potential of the microbial consortia of marine sponges. Applied Microbiology and Biotechnology. 2007;**75**:11-20

[4] Balunas MJ, Kinghorn AD. Drug discovery from medicinal plants. Life Sciences. 2005;**78**:431-441

[5] Maor R, Shirasu K. The arms race continues: Battle strategies between plants and fungal pathogens. Current Opinion in Microbiology. 2005;**8**:399-404

[6] Morrissey JP, Osbourn AE. Fungal resistance to plant antibiotics as a mechanism of pathogenesis. Microbiology and Molecular Biology Reviews. 1999;**63**: 708-724

[7] Perret X, Staehelin C, Broughton WJ. Molecular basis of symbiotic promiscuity. Microbiology and Molecular Biology Reviews. 2000;**64**:180-201

[8] Akiyama K, Hayashi H. Strigolactones: Chemical signals for fungal symbionts and parasitic weeds in plant roots. Annals of Botany (London). 2006;**97**:925-931

[9] Klee H. Plant biology: Hormones branch out. Nature. 2008;**455**:176-177

[10] Macias FA, Molinillo JM, Varela RM, Galindo JC. Allelopathy - a natural alternative for weed control. Pest Management Science. 2007;**63**:327-348

[11] Thevissen K, Kristensen HH, Thomma BP, Cammue BP, Francois IE. Therapeutic potential of antifungal plant and insect defensins. Drug Discovery Today. 2007;**12**:966-971

[12] Meazza G, Scheffler BE, Tellez MR, Rimando AM, Romagni JG, Duke SO. The inhibitory activity of natural products on plant p-hydroxyphenylpyruvate dioxygenase. Phytochemistry. 2002;**60**: 281-288

[13] Cole D, Pallet K, Rodgers M. Discovering new modes of action for herbicides and the impact of genomics. Pesticide Outlook. 2000;**11**:223-229

[14] Simons V, Morrissey JP, Latijnhouwers M, Csukai M, Cleaver A, Yarrow C. Osbourn A. Dual effects of plant steroidal alkaloids on *Saccharomyces cerevisiae*. Antimicrobial Agents and Chemotherapy. 2006;**50**:2732-2740

[15] Rimando AM, Dayan FE, Czarnota MA, Weston LA, Duke SO. A new photosystem II electron transfer inhibitor from sorghum bicolor. Journal of Natural Products. 1998;**61**:1456

[16] Halkier BA, Gershenzon J. Biology and biochemistry of glucosinolates. Annual Review of Plant Biology. 2006;**57**:303-333

[17] Macias FA, Molinillo JM, Varela RM, Galindo JC. Allelopathy - a natural alternative for weed control. Pest Management Science. 2007;**63**:327-348

[18] Golenser J, Waknine JH, Krugliak M, Hunt NH, Grau GE. Current perspectives on the mechanism of action of artemisinins. International Journal for Parasitology. 2006;**36**:1427-1441

[19] Krungkrai J. The multiple roles of the mitochondrion of the malarial parasite. Parasitology 2004;**129**:511-524

[20] Li W, Mo W, Shen D, Sun L, Wang J, Lu S. Yeast model uncovers dual roles of mitochondria in action of artemisinin. PLoS Genetics. 2005a;**1**:e36

[21] Nagamune K, Moreno SN, Sibley LD. Artemisinin-resistant mutants of *Toxoplasma gondii* have altered calcium homeostasis. Antimicrobial Agents and Chemotherapy. 2007b;**51**:3816-3823

[22] Eckstein-Ludwig U, Webb RJ, Van Goethem ID, East JM, Lee AG, Kimura M. Artemisinins target the SERCA of *Plasmodium falciparum*. Nature. 2003;**424**:957-961

[23] Krishna S, Woodrow CJ, Staines HM, Haynes RK, Mercereau-Puijalon O. Re-evaluation of how artemisinins work in light of emerging evidence of in vitro resistance. Trends in Molecular Medicine. 2006;**12**:200-205

[24] Wong JH, Xia L, Ng TB. A review of defensins of diverse origins. Current Protein & Peptide Science. 2007;**8**:446-459

[25] Thomma BP, Cammue BP, Thevissen K. Plant defensins. Planta. 2002;**216**:193-202

[26] Thevissen K, Idkowiak-Baldys J, Im YJ, Takemoto J, Francois IE, Ferket KK. SKN1, a novel plant defensin-sensitivity gene in *Saccharomyces cerevisiae*, is implicated in sphingolipid biosynthesis. FEBS Letters. 2005;**579**:1973-1977

[27] Cabral KM, Almeida MS, Valente AP, Almeida FC, Kurtenbach E. Production of the active antifungal *Pisum sativum* defensin 1 (Psd1) in *Pichia pastoris*: overcoming the inefficiency of the STE13 protease. Protein Expression and Purification. 2003;**31**:115-122

[28] Armah CN, Mackie AR, Roy C, Price K, Osbourn AE, Bowyer P, et al. The membrane-permeabilizing effect of avenacin A-1 involves the reorganization of bilayer cholesterol. Biophysical Journal. 1999;**76**:281-290

[29] Ito S, Ihara T, Tamura H, Tanaka S, Ikeda T, Kajihara H. Alpha-Tomatine, the major saponin in tomato, induces programmed cell death mediated by reactive oxygen species in the fungal pathogen *Fusarium oxysporum*. FEBS Letters. 2007;**581**:3217-3222

[30] Rao AV, Gurfinkel DM. The bioactivity of saponins: Triterpenoid and steroidal glycosides. Drug Metabolism and Drug Interactions. 2000;**17**:211-235

[31] Xiao JX, Huang GQ, Zhang SH. Soyasaponins inhibit the proliferation of Hela cells by inducing apoptosis. Experimental and Toxicologic Pathology. 2007;**59**:35-42

[32] Huang PR, Yeh YM, Wang TC. Potent inhibition of human telomerase by helenalin. Cancer Letters. 2005;**227**: 169-174

[33] Iwashina T. Flavonoid function and activity to plants and other organisms. Biological Sciences in Space. 2003;**17**:24-44

Section 3

Plant Secondary Metabolites

Chapter 8

Use of Plant Secondary Metabolites to Reduce Crop Biotic and Abiotic Stresses: A Review

Ziming Yue, Varsha Singh, Josiane Argenta, Worlanyo Segbefia, Alyssa Miller and Te Ming Tseng

Abstract

Plant secondary metabolites (PSM) are small molecules of organic compounds produced in plant metabolism that have various ecological functions, such as defense against pathogens, herbivores, and neighboring plants. They can also help to reduce abiotic stresses, such as drought, salinity, temperature, and UV. This chapter reviewed the ecological functions of the PSM and how people utilize these metabolites to reduce crop biotic and abiotic stresses in agriculture. Specific topics covered in this review are (1) extraction of PSM from plant parts and its application on crops; (2) screening of crop/cover crop germplasms for high PSM content and with resistance to pathogens, herbivores, and/or neighboring plants; (3) regulation of PSM biosynthesis (including plant hormones and defense activators) to increase plant readiness for defense; (4) transcriptome and genome technology improvements in the last decade leading to valuable tools to characterize differential gene expression and gene composition in a genome, and lineage-specific gene family expansion and contraction. In addition, there is a critical need to understand how the biosynthesis and release of allelochemicals occur. Filling this knowledge gap will help us to improve and encourage sustainable weed control practices in agriculture.

Keywords: allelopathy, pathogen defense, herbivore defense, plant defense, cover crops, sustainable pest management, organic farming

1. Introduction

Plant secondary metabolites (PSM) are small organic molecules produced during plant metabolism that can function as a plant defense against herbivores, pathogens, neighboring plants, or environmental stresses [1–3]. Although proven to be incorrect, PSM [4, 5] used to be defined as (1) the part of metabolites not present in nonplant organisms or as (2) the part of plant metabolites not required for simple growth and development. These outdated PSM definitions still reflected some properties of PSM—they are widespread in the plant kingdom and are beyond the highly conserved primary metabolites, which are required in plant growth and development, such as proteins, carbohydrates, lipids, and nucleic acids. Hence, they represent plant diversity. The description of PSM often starts from the sessile property of terrestrial plants [1, 2, 6], where they cannot flee from the threat or stress from the environment and hence have to develop strategies to defend or reduce the threat or stress. PSM are their strategies.

Environmental factors, such as temperature, salinity, and water, are also called abiotic stresses [7]. The herbivores, pathogens, and neighboring plants are also called biotic stresses. Plant metabolites can be classified into primary metabolites, secondary metabolites, and plant hormones [3]. The defense function of secondary metabolites is often realized by integration with physical structures, such as cell wall, cutin, suberin, wax, and bark. According to Hartman [1], plant secondary metabolites are often lineage-specific and aid plants in interacting with the biotic and abiotic environment. For example, pine trees and mint plants often contain terpenes, peppers often contain capsaicin, and sicklepod contains anthraquinone derivatives for defense. The production of secondary metabolites can be constitutive or induced. Some plant secondary metabolites, such as anthraquinone derivatives, in sicklepod are routinely produced, and they are called constitutive secondary metabolites. The production of secondary metabolites demands a high metabolic cost on the host plant; thus, many of these compounds are not produced in large quantities until after insects have begun to feed. These secondary metabolites are called induced secondary metabolites [7].

The number of secondary metabolites reported is vast, and they have widespread applications. The most prominent application of the plant secondary metabolites is in the pharmaceutical industry, where about 25% of the drugs in use by humans are derived from medicinal plants [8]. The type and concentration(s) of the secondary molecule(s) produced by a plant are determined by the species, genotype, physiology, developmental stage, and environmental factors during its growth [2].

The application of plant secondary metabolites in agriculture is the focus of this chapter. In standard agricultural practices, the species, physiology, and development stages usually follow biological laws, and we cannot do much to change them. The genotype and environmental factors are currently where most work has been focused on in agriculture. According to Hartman [1], the functions of plant secondary metabolites could fall into three categories—(1) defense and competition involving herbivores (arthropods, vertebrates, and invertebrates), pathogens (viruses, bacteria, and fungi), and plants (allelopathy); (2) attraction and stimulation (pollination, seed dispersal, food-plant recognition, oviposition, sequestration, and symbiosis); and, (3) abiotic stresses defense. Compared to other reviews on secondary metabolites, this review chapter focuses on the agricultural applications of plant secondary metabolites, specifically categories (1) and (3).

2. Secondary metabolites as resources to reduce crop biotic stresses

2.1 Main groups of plant secondary metabolites

PSM are widely spread in the whole plant kingdom. As they are lineage-specific, the total number of PSM is much more than the number of primary metabolites [5]. PSM derive from primary metabolites using a limited number of key pathways. Their functional diversity is gained by adding diverse combination of reactive functional groups [9]. Terpenoids are the largest group of PSM and occur in all plants, including over 22,000 compounds. The simplest terpenoid is isoprene (C_5H_8), a volatile gas produced during photosynthesis in leaves. Terpenoids are classified into monoterpenoids consisting of two isoprene units, sesquiterpenoids (three units), diterpenoids (four units), and triterpenoids (six units), depending on how many isoprene units are in their structures [7]. Mint plants (Mentha spp.) produce large quantities of the monoterpenoids menthol and menthone stored in glandular trichomes on the epidermis [7]. Pyrethrins are monoterpenoid esters produced by chrysanthemum plants that act as insect neurotoxins (Saxona 1988). Gossypol (*Gossypium hirsutum*) from cotton is a diterpenoid [7]. The fresh scent of lemon and orange peel results from a class of triterpenoids called limonoids. The active ingredient of neem oil, azadirachtin, is a powerful limonoid isolated from neem trees (*Azadirachta indica*) [10]. Phenolics are another large group of PSM, which includes a wide variety of defense-related compounds, such as flavonoids, anthocyanins, phytoalexins, tannins, lignin, and furanocoumarins [7]. Flavonoids are one of the largest classes of phenolics. Soybean contains a large amount of isoflavone [7]. Tannins are water-soluble flavonoid polymers produced by plants and stored in vacuoles. Tannins are toxic to insects because they bind to salivary proteins and digestive enzymes, including trypsin and chymotrypsin, resulting in protein inactivation. Alkaloids are a large class of bitter-tasting nitrogenous compounds found in many vascular plants and include caffeine, cocaine, morphine, and nicotine [7]. Capsaicin and related capsaicinoids produced by members of the genus *Capsicum* are the active components of chili peppers and have their characteristic burning sensation in hot and spicy foods [7]. Anthraquinones are present in different plant families, such as Leguminosae, Polygonaceae, Rubiaceae, Rhamnaceae, Scrophulariaceae, Liliaceae, Verbenaceae, and Valerianaceae [11]. Anthraquinone derivatives from sicklepod (Leguminosae) have been used to repel deer from browsing soybean [12]. Chlorogenic acid (CGA) or caffeoylquinic acid (CQA) exists in all plants [13], suggesting they are among the oldest PSMs.

2.2 PSM as resources to reduce crop biotic and abiotic stresses

Crop biotic stresses come from microbial pathogens, nematodes, insects, and mammalian herbivores. Crop abiotic stresses come from drought, salinity, temperature, ultraviolet, etc. Plant secondary metabolites can help to reduce these stresses. For example, some secondary metabolites containing benzene rings can absorb ultraviolet (UV) light and release the energy in the visible light range as fluorescence to avoid crop damage from UV light.

3. Use of secondary metabolites to reduce biotic and abiotic stresses

3.1 Extraction of secondary metabolites

Secondary metabolites have a defense function in plants [1, 2]. The simplest way to utilize secondary metabolites for crop protection is to extract the secondary metabolites and apply them to crops for protection against pathogens, insects, and mammalian herbivores.

3.1.1 Secondary metabolites used as a deer repellent

Deer is the primary pest in row crop production in the US. This was first concerned in the 1960s and gradually confirmed by the agricultural community during the following 40 years [14, 15]. The annual loss of row crops in the US was estimated to be up to $4.53 billion [14]. Deer repellent is one of the primary strategies to solve crop deer damage. Among them, deer repellent with putrescent egg solids as active ingredients occurred in the 1990s and still dominates the deer repellent market today. Deer acceptance of food is dependent on the concentration of secondary metabolites present [16]. They usually avoid plants containing high concentrations of terpenes, tannins [17], and gossypol (cotton). Sicklepod (*Senna obtusifolia* L.) is one of the southern US's top ten most troublesome weeds [18]. It belongs to the Leguminosae family and is famous for its high concentrations of anthraquinone derivatives [19], another group of secondary metabolites. Anthraquinone was reported as a mammalian animal repellent since the 1940s [20, 21]. To protect soybean damage from deer, deer repellents were developed using sicklepod fruits [12]. After several modifications of the extraction protocol, the sicklepod extract matched the deer repelling efficacy of Liquid Fence® Deer & Rabbit Repellent, a popular commercial deer repellent with putrescent egg solids as active ingredients. Besides the anthraquinone derivatives, some other plant secondary metabolites were used as deer repellents, such as capsaicin in pepper plants, and monoterpenoids menthol and menthone in peppermint (the active ingredients in Deer Out™, a commercial deer repellent).

3.1.2 Secondary metabolites as insecticides

One of the best examples of secondary metabolites used as an insecticide was the development of the popular insecticide bifenthrin. The pyrethrins from chrysanthemum (*Chrysanthemum cinerariaefolium*) flower extract were used to develop this insecticide. The safety of this product is, however, questionable. Sesbania extracts developed using a similar extraction method were applied on soybean leaves and exposed to soybean loopers in a 40 mm rearing cup for 24 hours. The looper mortality reached 60% in cups containing sesbania extract-treated soybean leaves.

3.2 Germplasm screening for secondary metabolites

3.2.1 Cotton germplasm screening for gossypol

Gossypol is a unique diterpenoid in the cotton genus Gossypium. Cotton germplasm is not as big as soybean and rice, but variations in gossypol content in cotton leaves are still significant. Low gossypol variety suffering heavy insect defoliation was observed (Dr. Saha personal communication). Unlike food crops, genetically modified cotton is not debated so critically, so Bt-based GMO method was adopted early to prevent insect defoliation. Gossypol screening is still a cultivar selection and breeding direction to defend insects and nematodes.

3.2.2 Allelopathic crop screening

Allelopathy is another term introduced to the science of plant ecology to describe the addition of chemical compounds (toxic or nontoxic) from a plant into the environment that affects the germination, growth, health, development, and population biology or behavior of another plant species [22]. Weeds are considered the most severe biotic constraint on crop production, with yield losses ranging from 45 to 95%, depending on environmental conditions and agronomic practices [23].

3.2.2.1 Rice allelopathy

Rice (*Oryza sativa*) is the most important grain crop cultivated in the world. More than half of the world's population has rice as their primary food source [24]. Weed infestation is the main reason for rice yield loss. The most common weeds found in rice fields worldwide are Echinochloa species, such as *Echinochloa cruss-galli* and *Echinochloa colona,* and weedy rice species (*Oryza sativa*) [25]. According to Oerke [26], weed species account for more than one-third of the losses in global rice production. Therefore, using integrated pest management (IPM), including the use of allelopathic varieties, can be an important tool to control weed species and manage weed resistance to synthetic herbicides.

A diversity of allelochemical compounds, such as fatty acids, phenolic acids, indoles, steroids, and others were found to be released by different parts of the plants, in root exudates, and rice soil [27]. Yet, rice inhibits weed growth primarily by secreting momilactone B, a diterpene produced from geranylgeranyl diphosphate (GGPP) [28]. It has been shown that momilactones A and B released by allelopathic rice varieties inhibit shoot and root growth of *E. crus-galli* (**Figure 1**). Additionally, weed species growing near rice deficient in momilactone biosynthesis produced more biomass when compared to the ones growing near wild-type rice [29].

The rice germplasm has a large variation when testing for allelopathy. However, it was found that among the Brazilian and Asian cultivars tested, only about 3–4% showed greater allelopathic potential [30]. Similar results were found when testing allelopathic cultivars able to suppress the growth of weeds, such as *E. crus-galli*, *Cyperus difformis,* and several aquatic weeds [31]. Thus, allelopathy is still an area to be investigated since this information can be used to improve rice production.

3.2.2.2 Cotton allelopathy

1. Weeds are a continuous hazard to agriculture in the United States, costing farmers up to $20 billion each year [32]. Herbicide resistance in weeds influences the long-term effectiveness of weed management practices globally [33]. Pesticide residues in food and the environment, as a result, are a significant public health hazard [34]. The use of weed suppressive traits in crop types, commonly known as allelopathy, is one of the potential weed control techniques in cotton production [35]. Several studies have reported using allelopathic crop varieties in weed management, including rice, wheat, sunflower, and canola [36, 37].

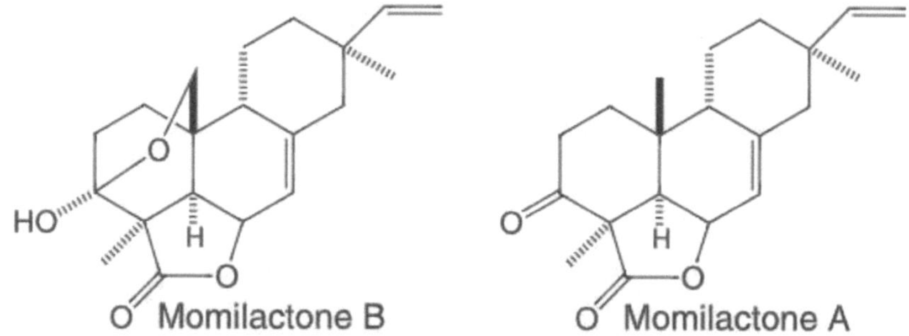

Figure 1.
Chemical structure of momilactones A and B, allelopathic molecules released by rice plants.

However, there is limited research on the direct allelopathic effect of cotton on weeds. A few research studies have established that cotton produces allelochemicals, which can impede the growth of pigweeds in other investigations [38]. According to preliminary studies on cotton allelopathy [39], cotton root showed significant quantities of four phenolic chemicals, including p-hydroxybenzoic acid, ferulic acid, gallic acid, and vanillin. A greenhouse study was conducted using eleven cotton chromosome substitution (CS) lines for allelopathy screening against Palmer amaranth (*Amaranthus palmeri*) (PA) [40]. The cotton lines were tested using a modified stair-step assay. Reductions in PA height and chlorophyll concentration were measured for each cotton line. Variations in PA height among the CS lines were more prominent 21 days after establishment. CS-B22sh and T26lo were most effective in reducing Palmer amaranth height; 77 and 68% height reduction, respectively. A multivariate cluster analysis revealed that CS-B22sh and CS-T26lo were clustered in one group, suggesting similar allelopathic potential against Palmer amaranth. Allelochemicals, produced by the allelopathic cotton CS lines, are a potential bioherbicide and a possible alternative to synthetic herbicides.

3.2.2.3 Sweetpotato allelopathy

Sweetpotato [*Ipomoea batatas* (L.) Lam.] is a nutrition-rich food with high fiber, vitamins, and antioxidants. Weed management is a major concern for sweetpotato producers [41] as weeds result in significant crop yield loss and higher production costs [42]. Being a plant of vine nature, sweetpotato grows close to the soil surface, and hand-weeding is one of the most effective mechanical options for weed management in sweetpotato fields [43]. To maintain and promote crop productivity and reduce labor requirements, chemical herbicides have been widely applied for weed control. However, long-term and large-scale herbicide applications have increased the number of herbicide-resistant weeds, environmental issues, loss of biodiversity, and threats to ecosystem safety [44]. Allelopathy can be a possible strategy for integrated, sustainable, and ecological weed management. Allelopathic properties of sweetpotato have been demonstrated to reduce the growth and development of weeds, such as alfalfa, yellow nutsedge, Palmer amaranth, and *Mikania micrantha* (**Figure 2**) [45, 46]. Alfalfa root growth was inhibited by methanol and aqueous extracts from sweetpotato leaves, stems, and roots [47]. Aqueous extracts from sweetpotato leaves or roots reduced the biomass, root and shoot length, and inhibited the germination of *Lactuca sativa* [42]. Leaf leachates from sweetpotato cultivars, Sinyulmi, Sinhwangmi, Purple, and Jami demonstrated an inhibitory effect on alfalfa [47]. Palmer amaranth growth was inhibited when they were irrigated with water-containing root exudates from different sweetpotato varieties [46].

Some sweetpotato varieties produce several allelochemicals, such as coumarin, chlorogenic acid, caffeic acid, hydroxycinnamic and trans-cinnamic acids [48] which were weed suppressive in rice. In terms of concentration, sweetpotato leaves were found to have the highest concentration of phenolic compounds, followed by stems and roots [47]. The allelopathic effect of sweetpotato on cowpea was reported when cowpea was grown as the following crop on the same field due to the presence of leaf litters and decaying residues of sweetpotato. Allelopathic varieties with the potential to suppress weed growth may be useful for breeding cultivars designed for organic production systems.

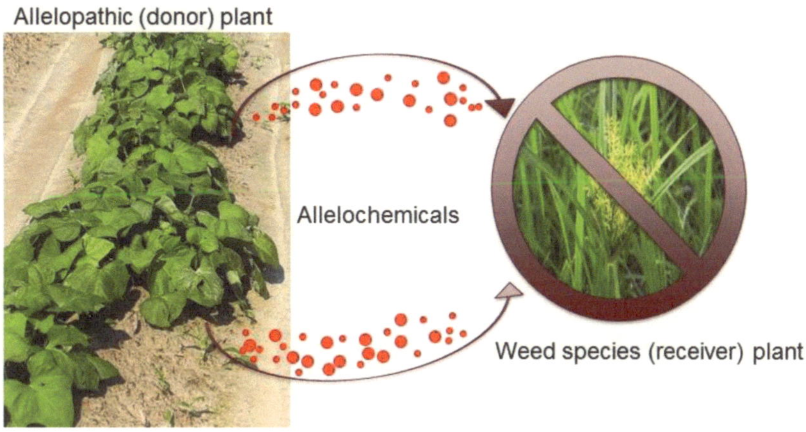

Figure 2.
Allelochemicals are released by above- and belowground parts of sweetpotato (donor) plants suppressing the surrounding receiver plants.

3.2.2.4 Sorghum allelopathy

Sorghum [*Sorghum bicolor* (L.)] is an annual grass belonging to the family Poaceae and subfamily Panicoideae. It originated in Africa and migrated to other continents [49]. Accounting for more than 22% of the world's sorghum production, the United States leads in the production globally [50]. Sorghum has wide-ranging utilization as food, fodder, technology, and construction [51]. The importance of sorghum is increasing globally due to its high functional value and ability to acclimatize to changing environmental conditions, especially drought [52]. Allelopathic or weed suppressive potential of sorghum has been documented in the past four decades. Several allelochemicals, such as sorgoleone and its analogs (**Figure 3**), phenolic acids, and their aldehyde derivatives, determine sorghum's allelopathic potential [53]. The amount of allelochemical production depends on the plant part and age of the sorghum plant, environmental conditions, and the receiver plant. Sorgoleone, a lipophilic secondary metabolite, is the primary allelochemical produced by sorghum which consists of a quinone ring and aliphatic chain [54]. Its analogs contain aliphatic side chains or additional methoxy groups in the ring [55, 56].

Phenolic acids (**Figure 4**) with phytotoxic activities, such as gallic, syringic, p-hydroxybenzoic, benzoic, vanillic, p-coumaric, and benzoic acids are also produced by sorghum [57]. However, the amount of production of these compounds depends on the type of cultivar [58] and the development stage of the sorghum plant [59].

Weed suppressing potential of sorghum on several weed species has been explored by using it as a cover crop, intercrop, crop rotation, sorghum water extract, soil incorporation of sorghum residue, and allele-herbicides derived from sorghum [60, 61]. Sorghum extracts can be combined with lower herbicide doses to effectively manage the weeds and reduce the overall herbicide introduction into the environment. Sorghum residues combined with 50% of the labeled rate of trifluralin were effective in preventing yield loss in broad beans [62]. Aqueous extracts from Brassica–sunflower–sorghum reduced weed biomass of several species, such as Purple nutsedge, bermudagrass, crowfoot grass, horse purslane, field bindweed, jungle rice, and goosegrass. The extent of suppression was comparable

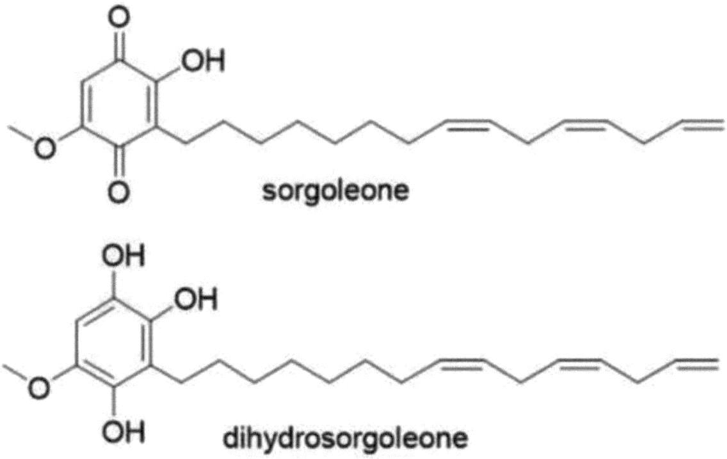

Figure 3.
Structures of sorgoleone and dihydrosorgoleone (reduced analog).

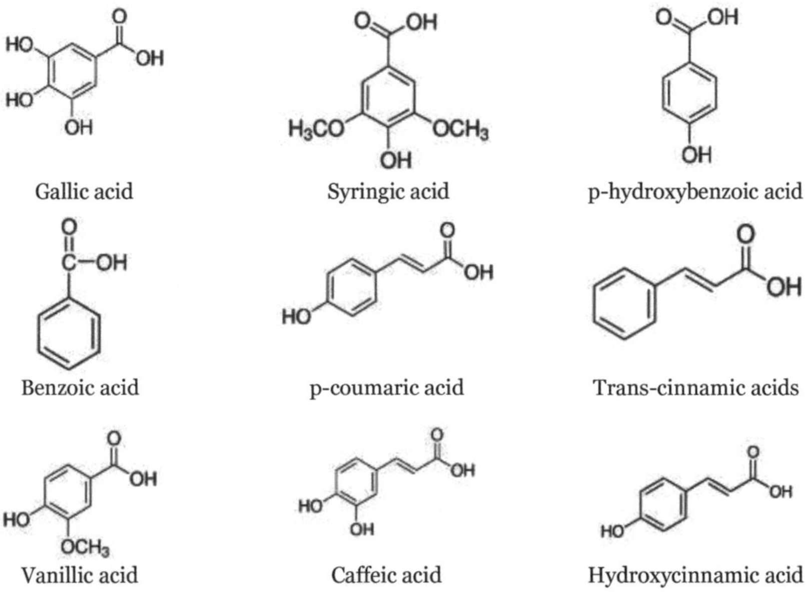

Figure 4.
Chemical structure of allelopathic phenolic compounds.

with the full rate of atrazine or S-metolachlor with half rate of atrazine [63]. Sorghum water extract combined with a reduced rate of herbicides such as isoproturon and metsulfuron-methyl demonstrated similar weed control as the full rate of these herbicides in the wheat field [64, 65]. A combination of water extracts from sunflower, rice, and sorghum can reduce the rates by 27–67% for herbicides such as ethoxysulfuron, butachlor, and pretilachlor in rice fields [66]. The utilization of allelopathy in agriculture can be a more sustainable and cost-effective strategy for weed management.

3.2.2.5 Allelopathic cover crops

The method of using cover crops in agricultural fields has been a widespread practice among a broad range of farms. Cover crops are crops that are grown prior to harvested crops to help increase the potential of the harvested crops [67]. In agricultural systems, the practice of using cover crops is shown to improve the quality of the soil by virtue of incorporating crop residues (organic matter) [68], Using a cover crop approach can also be beneficial via enhanced hydro-availability, decrease evaporation from the soil, as well as escalate the biodiversity of the soil.

An additionally impactful use for using the cover crop method in agricultural systems is its ability to suppress weeds due to either physical biomass of the terminated cover crop essentially smothering the weedy plants, physical shading of the cover crop causing inhibition of sunlight to the weeds, as well as via the production of allelochemicals from the cover crop. Allelochemicals are the product of allelopathy, which is positive or negative impact of one plant (the allelopathic plant) on another plant. Allelochemicals can increase or decrease the nutrient availability to surrounding plants by virtue of the symbiotic microbes [69]. It is appropriately thought that the use of cover crops with allelopathic properties in an agricultural field can have positively novel effects on the growth, ability to thrive, and production yields of so-called “cash crops”.

During a study in a semiarid area of Texas, USA, during a 3-year period, cotton that was cultivated following cover crop termination showed a shorter plant height and seed and lint yields. Simultaneously, the plant density did not affect the cover crops. Benzoxaziones concentrations in the soil were 2 to 3-fold higher under the cover crop treatments than in the fallow (control) plot. Though allelopathy may not be the only factor to cause these findings, it is likely to have played a significant role [70].

During a study on non-chemical weed suppression in vegetable fields, it was shown that there was a correlation between the quantity of cover crop biomass with the level of weed suppression (**Figures 5** and **6**). An 8 t ha^{-1} or greater cover crop biomass is possibly a significant enough level to have sufficient weed suppression [71]. Although this level of weed suppression may not have everything to do with allelopathy from the cover crops, it certainly played a critical role [72].

In a study focused on weed germination and the growth of IdaGold mustard, a seed germination bioassay technique was used. Phenol (allelochemical) concentrations were measured during this study. The total concentration of phenols in the soil was negatively correlated with the level of weed germination (**Figure 7**). Also, there were low concentrations of phenol in the soil that contained live microbes (<20 ng). Additionally, the germination rates were lower when compared to a nonmicrobe-containing soil with the same concentrations of phenol [73].

Numerous studies have demonstrated the weed suppressive property of allelopathic cover crops, which is a piece of good news for farmers [74]. There is a need for more research on the possible positive growth effects of allelopathic cover crops on the cash crops’ ability to thrive.

3.3 Secondary metabolites biosynthesis regulation

While PSMs have a constitutive part, i.e., routinely produced, they are also induced. This is mainly reflected in pathogen-induced resistance (including PSM production)

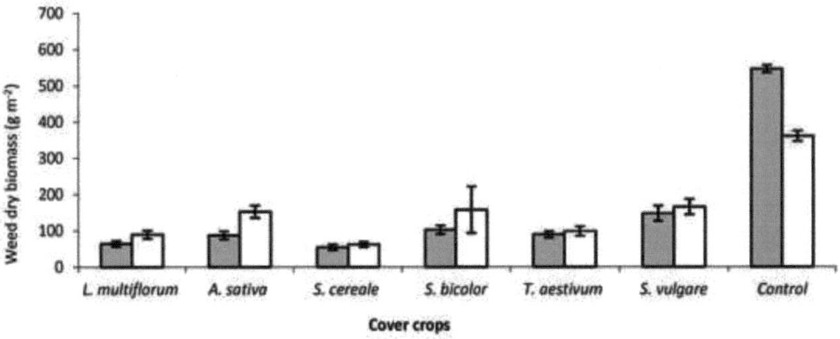

Figure 5.
Effects of various cereal cover crops in different vegetable production systems on the dry biomass production ($g\ m^{-2}$) of weed species at the time of cover crop termination in 2005 (gray bars) and 2006 (white bars). Vertical lines represent standard errors of the means ($p < 0.05$).

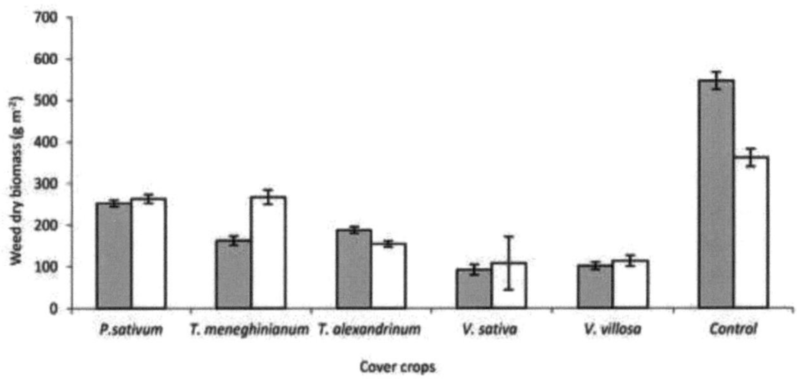

Figure 6.
Effects of various legume cover crops in different vegetable production systems on the dry biomass production ($g\ m^{-2}$) of weed species at the time of cover crop termination in 2005 (gray bars) and 2006 (white bars). Vertical lines represent standard errors of the means ($p < 0.05$).

and herbivore-induced resistance (including PSM production) [75]. The former can be traced back to 120 years ago, while the latter be traced back to 50 years ago. Recently it was realized that both were similar in nature and were controlled by plant hormones, salicylic acid, and jasmonic acid, respectively [75].

Another group of PSMs, allelochemicals, is generally thought of as constitutive, i.e., routinely produced. Compared to PSM induced by pathogens and herbivores, allelochemical induction is a big gap in our knowledge, although the study of allelopathy can be traced back 90 years ago. A primary reason for the difference is that for pathogen/herbivore-induced PSM production, the PSM may (or may not) need activation upon induction (they do not need to be expelled out of the plant body to defend), while in allelopathy, the PSM needs to be expelled out of the plant body to be effective. Sporadic information on the induction of root exudation exists in the literature; for example, Dineli et al. [76] studied the translocation and root exudation of herbicide after foliar treatment of wheat and ryegrass using 14C-labeled diclofop-methyl and triasulfuron. The results showed the presence of untreated plants (wheat or ryegrass) in the same pot as triasulfuron-treated ryegrass or wheat induced the exudation of the herbicide 7 to 32 times more. In the case of diclofop, the induced root exudation of the herbicide was 3 to 6 times more

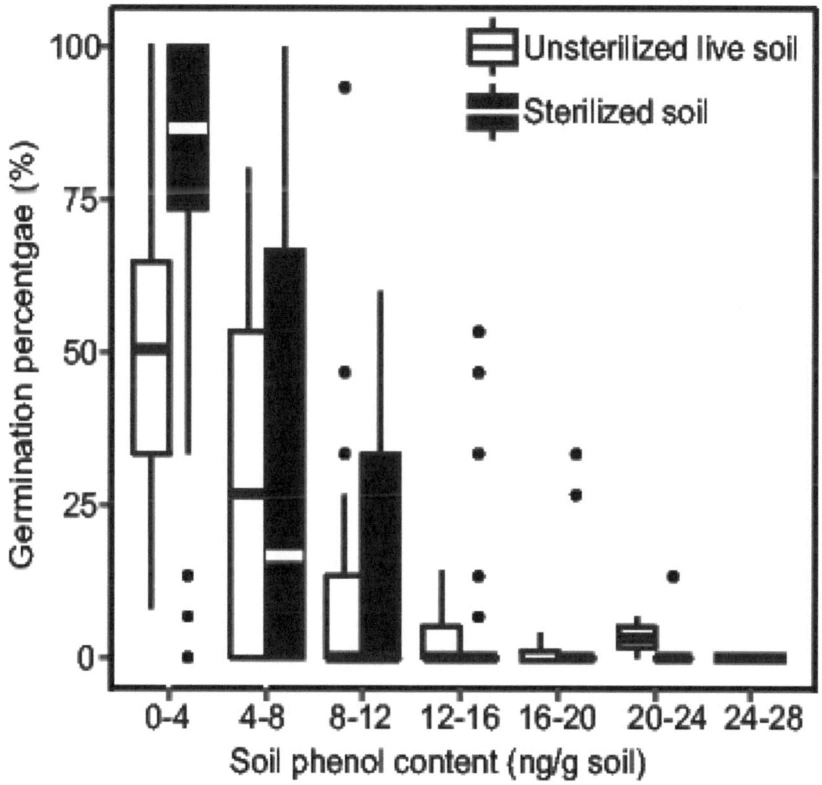

Figure 7.
Germination is inhibited by high concentrations of soil phenols.

in the presence of untreated wheat or ryegrass. The root exudated herbicides suppressed the adjacent plants, indicating a form of allelopathy. This study demonstrated that the presence of adjacent plants induces the release of allelopathic compounds. An immediate question following this case study is—could the biosynthesis of allelopathic compounds (PSM) be induced? If so, how were the signals transmitted during these processes, including the release of the compounds?

As we reviewed previously, PSMs are biopesticides widely used in agriculture. As the PSM are lineage-specific, the selection of a specific crop cultivar or cover crop is similar to selecting what kind of biopesticides to use. Similarly, understanding and application of PSM induction is the dose control of the selected biopesticides. Furthermore, in the pathogen and herbivore-induced resistance (expressed as PSM), the resistance was often called systematic acquired resistance, meaning the resistance was expressed as normal PSM for toxicity and included thickening of cell wall lignin, etc. Hence such systemic acquired resistance is more effective and lasts longer than toxic PSM increase. In this context, filling the knowledge gap of induction of allelopathic compound biosynthesis and release is similar to understanding the dose control of bioherbicide.

3.3.1 Use of plant hormones to regulate secondary metabolite biosynthesis

Generally, it has been accepted that salicylic acid (SA) and jasmonic acid (JA) or methyl jasmonate (MeJA) are recognized plant hormones specialized for defense. These defense hormones have been used to induce PSM production to defend pathogens and herbivores in agricultural studies [77]. This method has not been used in allelopathy.

3.3.2 Use of plant extracts to induce secondary metabolites production

A field study looked at the deer and insect repelling efficacy of coffee senna extract on soybean [12]. After 40 days, the soybean leaf holes were significantly lower than the control or other treatments. This was in contrast to the leaf disc assay results, where soybean loopers were exposed to both coffee senna and sesbania extracts for 24 hours. The soybean looper mortality for sesbania extract was higher than that of coffee senna. A possible explanation for the difference between the field and leaf disc results is that leaf disc experiments used detached leaves. In contrast, field experiments used living soybean plants where the coffee senna extract might have induced defense response in soybean plants. The active ingredient of coffee senna extract may be SA and JA, or a new type of defense response inducer, which is to be determined.

3.3.3 Other chemicals for crop defense activation

Besides plant hormones and plant extracts, some inorganic chemicals have also been used as crop defense activators. Juric et al. [78] reported that Ca^{2+} and Cu^{2+} increased secondary metabolites contents in lettuce. Such chemical crop activator is much less toxic for humans and their defense effects last much longer than insecticides or fungicides, hence they are more preferable to the agriculture community.

3.4 Use of transcriptomic and genomic tools to employ secondary metabolites to reduce crop stresses

During the past ten years, the transcriptome was widely used to study the gene expression of secondary metabolites [79]. While plant secondary metabolites are thought to be the readouts of plant defense activation, usually PSM quantity increase can be detected around 20 days or more after treatment (defense activation). PSM increase can be detected from several hours to 40 hours by transcriptome analysis (qPCR). *Senna tora* is a medicinal plant in Asia, and it is also a close relative to the weed sicklepod (*Senna obtusifolia*) in the US. Both sicklepod and *Senna tora* fruits contain high contents of anthraquinone secondary metabolites. Kang et al. [80] used differential expression analysis and showed that the expression level of genes involved in the anthraquinone biosynthetic pathway regulates differently depending on the degree of tissues and seeds development.

With improvements in sequencing technology, the sequencing cost has plunged during the past decades. Crop or cultivar genome is not far from being available. One discovery with the available genome sequences is that plants devote a significant amount of their genes to secondary metabolites, implying plant ecological functions are equivalent to its growth and development. Kang et al. [80] sequenced the genome of *S. tora,* and found that the CHS-L gene family expanded most notably *in S. tora*. This might explain in part why *S. tora* was rich in anthraquinones.

4. Conclusions

Compared to the estimated number of primary metabolites of 10,000, PSMs are estimated to be more than 200,000 in the plant kingdom. These PSMs function in various ecological roles, including defending pathogens, herbivores, and neighboring plants. Use of these PSM in agriculture includes (1) extraction of the PSM and applying it directly to the crop to reduce biotic stresses, (2) use of PSM in vivo/in situ by screening crop cultivars with desired PSM profiles to achieve better resistance to pests, (3) use of PSM biosynthesis regulation or plant defense activators to achieve defense readiness, (4) filling the knowledge gap on allelochemical induction, biosynthesis, and release, as it will be helpful in improving weed management practices in agriculture, and (5) employing transcriptomic and genomic tools to understand PSM biosynthesis and pathways.

Acknowledgements

The authors appreciate the funding from the Mississippi Soybean Promotion Board (MSPB) and Cotton Incorporated. This work is supported by the National Institute of Food and Agriculture, U.S. Department of Agriculture, Hatch project under accession number 230100, and is a contribution of the Mississippi Agricultural and Forestry Experiment Station.

Conflict of interest

The authors declare that this work was presented in the absence of any commercial or financial relationships that could be construed as a potential conflict of interest.

Author details

Ziming Yue, Varsha Singh, Josiane Argenta, Worlanyo Segbefia, Alyssa Miller and Te Ming Tseng*
Department of Crop and Soil Sciences, Mississippi State University, USA

*Address all correspondence to: t.tseng@msstate.edu

IntechOpen

© 2022 The Author(s). Licensee IntechOpen. This chapter is distributed under the terms of the Creative Commons Attribution License (http://creativecommons.org/licenses/by/3.0), which permits unrestricted use, distribution, and reproduction in any medium, provided the original work is properly cited.

References

[1] Hartmann T. From waste products to ecochemicals: Fifty years research of plant secondary metabolism. Phytochemistry. 2007;**68**:2831-2846

[2] Isah T. Stress and defense responses in plant secondary metabolites production. Biological Research. 2019;**52**:39

[3] Matthias E, Kliebenstein DJ. Plant secondary metabolites as defenses, regulators, and primary metabolites: The blurred functional trichotomy. Plant Physiology. 2020;**184**(1):39-52

[4] Fraenkel GS. The raison d'Etre of secondary plant substances: These odd chemicals arose as a means of protecting plants from insects and now guide insects to food. Science. 1959;**129**:1466-1470

[5] Pichersky E, Lewinsohn E. Convergent evolution in plant specialized metabolism. Annual Review of Plant Biology. 2011;**62**:549-566

[6] Hartmann T. Plant-derived secondary metabolites as defensive chemicals in herbivorous insects: A case study in chemical ecology. Planta. 2004;**219**:1-4

[7] Freeman BC, Beattie GA. An Overview of Plant Defenses against Pathogens and Herbivores. The Plant Health Instructor. 2008. DOI: 10.1094/PHI-I-2008-0226-01

[8] De Luca V, Salim V, Atsumi SM, Yu F. Mining the biodiversity of plants: A revolution in the making. Science. 2012;**336**:1658-1661

[9] Wink M. Evolution of secondary plant metabolism. In: Wiley, Encyclopedia of Life Sciences. Chichester: Wiley; 2016. DOI: 10.1002/9780470015902.a0001922.pub3

[10] Saxena RC. Insecticides from neem. In: Anason JT, Philogene MP, editors. Insecticides of Management. Boca Raton: Lewis Publishers; 1988

[11] Han Y-S, van der Heijden R, Verpoorte R. Biosynthesis of anthraquinones in cell cultures of the rubiaceae. Plant Cell Tissue and Organ Culture. 2001;**67**:201-220

[12] Yue Z, Tseng TM, Lashley M. Characterization and deer-repellent property of chrysophanol and emodin from sicklepod weed. American Journal of Plant Sciences. 2018;**9**:266-280

[13] Constabel CP. A survey of herbivore-inducible defensive proteins and phytochemicals. In: Agrawal AA, Tuzun S, Bent E, editors. Induced Plant Defenses against Pathogens and Herbivores. St. Paul: APS Press; 2000. pp. 137-166

[14] Conover MR. Resolving human-wildlife conflicts: The science of wildlife damage behavior in aquatic environments. ACS Symposium Series. 2002;**991**:3-25

[15] Flyger VF, Thoerig T. Crop damage caused by Maryland deer. In: Proceedings Annual Conference, Southeastern Association of Game and Fish Commissioners. 1962;**16**:45-52

[16] Wagner KK, Nolte DL. Comparison of active ingredients and delivery systems in deer repellents. Wildlife Society Bulletin. 2001;**29**

[17] Nolte DL, Kimball BA, Perry KR, Villalba JJ, Provenza FD. Effects of forage nutritional quality (energy and protein) on deer acceptance of foods containing secondary metabolites. USDA National Wildlife Research Center - Staff Publications; 2004.

p. 374. https://digitalcommons.unl.edu/icwdm_usdanwrc/374

[18] Webster TM et al. Weed survey-southern states. Proceedings, Southern Weed Science Society. 2013;**66**:275-287

[19] Crawford L, McDonald GM, Friedman M. Composition of Sicklepod (*Cassia obtusifolia*) Toxic Weed Seeds. Journal of Agriculture and Food Chemistry. 1990;**38**:2169-2175

[20] Avery ML, Humphrey JS, Decker DG. Feeding deterrence of anthraquinone, anthracene, and anthrone to rice-eating birds. Journal of Wildlife Management. 1997;**61**:1359-1365

[21] Deliberto ST, Werner SJ. Review of anthraquinone applications for pest management and agricultural crop protection. Pest Management Science. 2016;**72**:1813-1825

[22] Qasem JR, Foy CL. Weed allelopathy, its ecological impacts and future prospects. Journal of Crop Production. 2001;**4**(2):43-119

[23] Scavo A, Mauromicale G. Crop allelopathy for sustainable weed management in agroecosystems: Knowing the present with a view to the future. Agronomy. 2021;**11**:2104

[24] Muthayya S, Sugimoto JD, Montgomery S, Maberly GF. An overview of global rice production, supply, trade, and consumption. Annals of the New York Academy of Sciences. 2014;**1324**(1):7-14

[25] Kraehmer H, Jabran K, Mennan H, Chauhan BS. Global distribution of rice weeds–a review. Crop Protection. 2016;**80**:73-86

[26] Oerke EC. Crop losses to pests. The Journal of Agricultural Science. 2006;**144**(1):31-43

[27] Khanh TD, Xuan TD, Chung IM. Rice allelopathy and the possibility for weed management. Annals of Applied Biology. 2007a;**151**(3):325-339

[28] Serra Serra N, Shanmuganathan R, Becker C. Allelopathy in rice: A story of momilactones, kin recognition, and weed management. Journal of Experimental Botany. 2021;**72**(11):4022-4037

[29] Xu M, Galhano R, Wiemann P, Bueno E, Tiernan M, Wu W, et al. Genetic evidence for natural product-mediated plant–plant allelopathy in rice (*Oryza sativa*). New Phytologist. 2012;**193**(3):570-575

[30] Olofsdotter M, Jensen LB, Courtois B. Improving crop competitive ability using allelopathy—an example from rice. Plant Breeding. 2002;**121**(1):1-9

[31] Belz RG. Allelopathy in crop/weed interactions—An update. Pest Management Science: Formerly Pesticide Science. 2007;**63**(4):308-326

[32] Basu C, Halfhill MD, Mueller TC, Stewart CN Jr. Weed genomics: New tools to understand weed biology. Trends in Plant Science. 2004;**9**(8):391-398

[33] Gage KL, Krausz RF, Walters SA. Emerging challenges for weed management in herbicide-resistant crops. Agriculture. 2019;**9**(8):180

[34] Riyaz M, Shah RA, Sivasankaran K. Pesticide Residues: Impacts on Fauna and the Environment. IntechOpen; 2021

[35] Jabran K, Mahajan G, Sardana V, Chauhan BS. Allelopathy for weed control in agricultural systems. Crop Protection. 2015;*72*:57-65

[36] Alsaadawi IS, Sarbout AK, Al-Shamma LM. Differential allelopathic

potential of sunflower (*Helianthus annuus* L.) genotypes on weeds and wheat (*Triticum aestivum* L.) crop. Archives of Agronomy and Soil Science. 2012;**58**(10):1139-1148

[37] Kong CH, Chen XH, Hu F, Zhang SZ. Breeding of commercially acceptable allelopathic rice cultivars in China. Pest Management Science. 2011;**67**(9):1100-1106

[38] Hoffman ML, Weston LA, Snyder JC, Regnier EE. Separating the effects of sorghum (Sorghum bicolor) and rye (Secale cereale) root and shoot residues on weed development. Weed Science. 1996;**1996**:402-407

[39] Chen ZJ, Sreedasyam A, Ando A, et al. Genomic diversifications of five *Gossypium* allopolyploid species and their impact on cotton improvement. Nature Genetics. 2020;**52**:525-533

[40] Fuller GA, Saha S, Stelly DM, Jenkin J, Tseng TM. Assessing the weed-suppressing potential of cotton chromosome substitution lines using the stair-step assay. Plants. 2021;**10**(11):2450

[41] Nwosisi S, Nandwani D, Hui D. Mulch treatment effect on weed biomass and yields of organic sweetpotato cultivars. Agronomy. 2019;**9**(4):190

[42] Shen S, Xu G, Li D, Clements DR, Jin G, Liu S, et al. Allelopathic potential of sweet potato (*Ipomoea batatas*) germplasm resources of Yunnan Province in southwest China. Acta Ecologica Sinica. 2018;**38**(6):444-449

[43] Treadwell DD, Creamer NG, Schultheis JR, Hoyt GD. Cover crop management affects weeds and yield in organically managed sweetpotato systems. Weed Technology. 2007;**21**(4):1039-1048

[44] Norsworthy JK, Ward SM, Shaw DR, Llewellyn RS, Nichols RL, Webster TM, et al. Reducing the risks of herbicide resistance: Best management practices and recommendations. Weed Science. 2012;**60**(SP1):31-62

[45] Harrison HF, Peterson JK. Allelopathic effects of sweet potatoes (*Ipomoea batatas*) on yellow nutsedge (*Cyperus esculentus*) and alfalfa (*Medicago sativa*). Weed Science. 1986;**34**(4):623-627

[46] Varsha V, Werle IS, Shankle MW, Tseng TM. Allelopathic effects of sweetpotato varieties on palmer amaranth growth. Proceedings of Southern Weed Science Society. 2021;**74**:18

[47] Chon SU, Boo HO. Difference in allelopathic potential as influenced by root periderm colour of sweet potato (*Ipomoea batatas*). Journal of Agronomy and Crop Science. 2005;**191**(1):75-80

[48] Soni B, Tseng TMP, Yue Z. Identification and quantification of allelochemicals from selected sweet potato (*Ipomoea batatas* (L.) lam.) cultivars. American Journal of Plant Sciences. 2019;**10**(12):2354

[49] Sène M, Gallet C, Doré T. Phenolic compounds in a Sahelian sorghum (*Sorghum bicolor*) genotype (CE145-66) and associated soils. Journal of Chemical Ecology. 2001;**27**(1):81-92

[50] Głąb L, Sowiński J, Bough R, Dayan FE. Allelopathic potential of sorghum (*Sorghum bicolor* (L.) Moench) in weed control: A comprehensive review. Advances in Agronomy. 2017;**145**:43-95

[51] Sowiński J, Szydełko-Rabska E. Comparison of sorghum type yielding in Polish conditions. *Annales Universitatis*

Mariae Curie-Skłodowska. Agricultura. 2013;**68**(1):30-40

[52] Sowiński J, Liszka-Podkowa A. Fresh and dry matter yield quantity and quality of maize (*Zea mays* L.) and sweet sorghum (*Sorghum bicolor* (L.) Moench.) on sandy soil depending on nitrogen fertilization. Acta Scientiarum Polonorum Agricultura. 2008

[53] Czarnota MA, Paul RN, Weston LA, Duke SO. Anatomy of sorgoleone-secreting root hairs of Sorghum species. International Journal of Plant Sciences. 2003;**164**(6):861-866

[54] Netzly DH, Butler LG. Roots of sorghum exude hydrophobic droplets containing biologically active components 1. Crop Science. 1986;**26**(4):775-778

[55] Dayan FE, Duke SO. Trichomes and root hairs: Natural pesticide factories. Pesticide Outlook. 2003;**14**(4):175-178

[56] Kagan IA, Rimando AM, Dayan FE. Chromatographic separation and in vitro activity of sorgoleone congeners from the roots of *Sorghum bicolor*. Journal of Agricultural and Food Chemistry. 2003;**51**(26):7589-7595

[57] Al-tawaha ARM, Nidal ODAT. Use of sorghum and maize allelopathic properties to inhibit germination and growth of wild barley (*Hordeum spontaneum*). Notulae Botanicae Horti Agrobotanici Cluj-Napoca. 2010;**38**(3):124-127

[58] Cheema ZA, Khaliq A, Abbas M, Farooq M. Allelopathic potential of sorghum (*Sorghum bicolor* L. Moench) cultivars for weed management. Allelopathy Journal. 2007;**20**(1):167

[59] Won OJ, Uddin MR, Park KW, Pyon JY, Park SU. Phenolic compounds in sorghum leaf extracts and their effects on weed control. Allelopathy Journal. 2013;**31**(1):147

[60] Alsaadawi IS, Dayan FE. Potentials and prospects of sorghum allelopathy in agroecosystems. Allelopathy Journal. 2009;**24**(2)

[61] Cheema ZA, Farooq M, Khaliq A. Application of allelopathy in crop production: Success story from Pakistan. In: Allelopathy. Berlin, Heidelberg: Springer; 2013. pp. 113-143

[62] Alsaadawi IS, Khaliq A, Lahmod NR, Matloob A. Weed management in broad bean (*Vicia faba* L.) through allelopathic *Sorghum bicolor* (L.) Moench residues and reduced rate of a pre-plant herbicide. Allelopathy Journal. 2013;*32*(2)

[63] Mahmood A, Khaliq A, Ihsan MZ, Naeem M, Daur I, Matloob A, et al. Estimation of weed dry biomass and grain yield as a function of growth and yield traits under allelopathic weed management in maize. Planta Daninha. 2015;**33**:23-31

[64] Jamil MU, Ata ZAHID, Khaliq ABDUL. Increasing the efficiency of sorghum water extract (sorgaab) by mixing with lower doses of isoproturon to control weeds in wheat. International Journal of Agricultural Biology. 2005;7(8):712-718

[65] Sharif MM, Cheema ZA, Khaliq A. Reducing herbicide dose in combination with sorghum water extract for weed control in wheat (*Triticum aestivum* L.). International Journal of Agricultural Biology. 2005;7(4):560-563

[66] Rehman A, Cheema ZA, Khaliq A, Arshad M, Mohsan S. Application of sorghum, sunflower and rice water extract combinations helps in reducing

herbicide dose for weed management in rice. International Journal of Agricultural Biology. 2010;**12**(6):901-906

[67] Stratton, Anne Elise, Comin, Jucinei José, Siddique, Ilyas, Zak, Donald R., Dambroz Filipini, Letícia, Rodrigues Lucas, Blesh, Jennifer, 2021, Assessing Cover Crop and Intercrop Performance Along a Farm Management Gradient. SSRN Available from: https://ssrn.com/abstract=3964460 or http://dx.doi.org/10.2139/ssrn.3964460

[68] Dias T, Dukes A, Antunes PM. Accounting for soil biotic effects on soil health and crop productivity in the design of crop rotations. Journal of the Science of Food and Agriculture. 2014;**95**:447-454

[69] Ana Luisa A. Allelopathy as a tool in the management of biotic resources in agroecosystems. Critical Reviews in Plant Sciences. 1999;**18**(6):697-739

[70] Li Y, Allen VG, Chen J, Hou F, Brown CP, Green P. Allelopathic influence of a wheat or rye cover crop on growth and yield of no-till cotton. Agronomy Journal. 2013;**105**:1581-1587

[71] Mirsky SB, Ryan MR, Teasdale JR, Curran WS, Reberg-Horton CS, Spargo JT, et al. Overcoming weed management challenges in cover crop–based organic rotational no-till soybean production in the eastern United States. Weed Technology. 2013;**27**:193-203

[72] Mennan H, Jabran K, Zandstra BH, Pala F. Non-chemical weed management in vegetables by using cover crops: A review. Agronomy. 2020;**2**:257

[73] Lou Y, Davis AS, Yannarell AC. Interactions between allelochemicals and the microbial community affect weed suppression following cover crop residue incorporation into soil. Plant Soil. 2016;**399**:357-371

[74] Sturm DJ, Peteinatos G, Gerhards R. Contribution of allelopathic effects to the overall weed suppression by different cover crops. Weed Research. 2018;**58**(5):331-337

[75] Kaeban R, Kuc J. Induced resistance against pathogen and herbivore: An overview. In: Agrawal AA, Tuzun S, Bent E, editors. Induced Plant Defenses against Pathogens and Herbivores: Biochemistry, Ecology and Agriculture. St. Paul, MN, USA: APS Press; 2000

[76] Dinelli G, Bonetti A, Marottii M, Busi S, Catizone P. Root exudation of diclofop-methyl and triasulfuron from foliar-treated durum wheat and ryegrass. Weed Research. 2007;**47**:25-33

[77] Beyer SF, Bel PS, Flors V, et al. Disclosure of salicylic acid and jasmonic acid-responsive genes provides a molecular tool for deciphering stress responses in soybean. Scientific Reports. 2021;**11**:20600

[78] Jurić S, Stracenski SK, Król-Kilińska Ż, Žutić I, Uher SF, Đermić E, et al. The enhancement of plant secondary metabolites content in *Lactuca sativa* L. by encapsulated bioactive agents. Scientific Reports. 2020;**10**:3737. DOI: 10.1038/s41598-020-60690-3

[79] Kang S-H, Lee W-H, Lee C-M, Sim J-S, Won SY, Han S-R, et al. De novo transcriptome sequence of Senna tora provides insights into anthraquinone biosynthesis. PLoS ONE. 2020b;**15**(5):e0225564

[80] Kang SH, Pandey RP, Lee CM, et al. Genome-enabled discovery of anthraquinone biosynthesis in *Senna tora*. Nature Communications. 2020a;**11**:5875

Chapter 9

Secondary Metabolites of Fruits and Vegetables with Antioxidant Potential

Ravneet Kaur, Shubhra Shekhar and Kamlesh Prasad

Abstract

An antioxidant is of great interest among researchers, scientists, nutritionists, and the public because of its ability to prevent oxidative damage, as indicated by various studies. This chapter mainly focuses on the free radicals and their types; antioxidants and their mode of action against free radicals; fruits, vegetables, and their byproducts as a source of antioxidants; and various analytical methods employed for assessing antioxidant activity. Antioxidants discussed in this chapter are ascorbic acid, Vitamin E, carotenoids and polyphenols, and their mechanism of action. Different antioxidant activity assay techniques have been reported. Fruits and vegetables are abundant sources of these secondary metabolites. The waste generated during processing has many bioactive materials, which possibly be used in value-added by-products.

Keywords: antioxidant, free radical, oxidative stress, secondary metabolite, ascorbic acid, carotenoids, polyphenol, degenerative diseases

1. Introduction

The word antioxidant is commonly heard nowadays, especially whenever there comes a topic of health concern. People consume antioxidants as a symbol of a healthy lifestyle to fight against various health problems, better skin, and anti-aging benefits. What makes antioxidants so important? The trait responsible for such importance of antioxidants is their ability to stop free radical reactions that can have potentially deleterious effects [1]. This gives rise to various questions, such as What are the free radicals? What are the sources of free radicals? What are their harmful effects? What are antioxidants? What are the common sources of antioxidants? How do they work against free radicals? Answers to these questions are discussed in the present chapter.

2. Free radicals

Free radicals are those atoms or molecules with an unpaired electron in their outer orbit [2]. Any electron present alone in an orbital is referred to as an unpaired electron, and it is accountable for the reactive and unstable state of the free radical.

IntechOpen

The vital class of free radicals generated in a living system is usually derived from oxygen and reactive oxygen species (ROS) [3]. Hydroperoxyl (HO_2^{o}), alkoxy (RO^{o}), peroxyl (RO_2^{o}), hydroxyl (OH^{o}), and superoxide radical (HO_2^{o}) are common among oxygen free radicals. Nitrosative stress is the condition that occurs due to the overproduction of reactive nitrogen species (RNS) [3, 4]. Nitric oxide (NO^{o}) and nitrogen dioxide (NO_2^{o}), the nitrogen-free radicals can also be converted into other nonreactive species under the antioxidant-dependent reactions. Thus, ROS and RNS include radicals and nonradical species, such as hydrogen peroxide, singlet oxygen, ozone, organic peroxide, peroxynitrite, nitrosyl cation, nitroxyl cation, dinitrogen trioxide, and nitrous acid [5]. When reactive oxygen species (ROS) react with thiols, they give rise to reactive sulfur species (RSS) [6].

The most reactive hydroxyl free radical is formed by exposure to ionizing radiations. These radiations lead to the formation of H^{o} and OH^{o} by causing the fission of OH bonds in water.

$$H_2O \rightarrow H^{o} + OH^{o} \tag{1}$$

Harmful effects are initiated when the hydroxyl radical reacts with macronutrients such as carbohydrates, protein, and lipids along with DNA, the genetic material [7].

Molecular oxygen receives one electron and is converted to superoxide anion, a reduced form [8]. Superoxide anion is formed in the mitochondria during the initial step of the electron transport system [9]. Oxygen is reduced to water during the electron chain reaction. The electrons escape a chain reaction and react directly with oxygen in its formation [8].

$$\underset{\text{Oxygen}}{O_2} \xrightarrow[H^+]{e^-} \underset{\text{Superoxide radical}}{HO_2^{o}} \xrightarrow[H^+]{e^-} \underset{\text{Hydrogen peroxide}}{H_2O_2} \xrightarrow[H^+]{e^-} \underset{\text{Hydroxyl radical}}{OH^{o}} \xrightarrow[H^+]{e^-} \underset{\text{Water}}{H_2O} \tag{2}$$

Many other reactive oxygen species are also formed in the living system by the formed superoxide anions. These include hydrogen peroxide, hydroxyl radicals, or singlet oxygen [10].

Hydrogen peroxide (H_2O_2) is a nonradical that is formed by the superoxide radical when it undergoes nonenzymatic or enzyme-catalyzed (superoxide dismutase, SOD) dismutation reaction. It is very diffusible within and between the cells [11].

$$\underset{\text{Superoxide radical}}{2HO_2^{o}} + \underset{\text{Hydrogen ion}}{2H^+} \longrightarrow \underset{\text{Hydrogen peroxide}}{H_2O_2} + \underset{\text{Oxygen}}{O_2} \tag{3}$$

In the presence of metal ions and superoxide anion, hydrogen peroxide generates hydroxyl radical.

$$\underset{\text{Superoxide radical}}{HO_2^{o}} + \underset{\text{Hydrogen peroxide}}{H_2O_2} \longrightarrow \underset{\text{Hydroxyl radical}}{OH^{o}} + \underset{\text{Hydroxyl ion}}{OH^-} + \underset{\text{Oxygen}}{O_2} \tag{4}$$

Nitric oxide is formed during the metabolization of arginine to citrulline by the enzyme nitric oxide synthase (NOSs) via five electron oxidative reactions [12]. Nitric oxide readily diffuses through cytoplasm and plasma membranes due to its solubility in both liquid and liquid media [13].

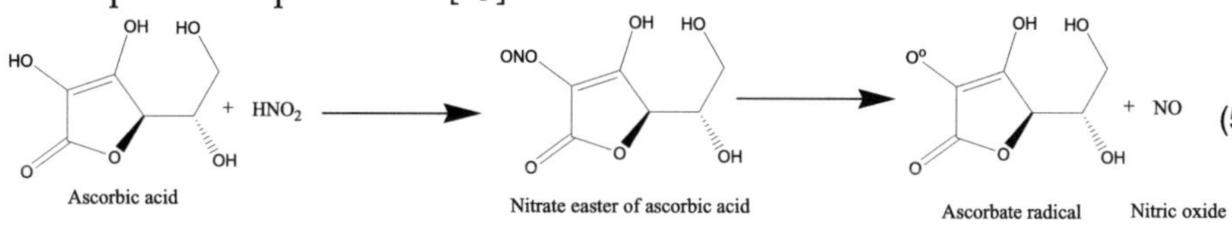

(5)

3. Sources of free radicals and harmful effects

Oxygen, an essential element of life, also has harmful effects on the human body by forming reactive oxygen species [14]. Free radicals are produced internally as well as due to external factors.

Internally

- Normal metabolism within mitochondria during electron transport reactions and another mechanism [15]
- Xanthine oxides
- Inflammation processes – by neutrophils and macrophages
- Phagocytosis
- Ischaemia
- Peroxisomes [14]

External factors

- Radiation
- UV rays, X rays, γ rays
- Environmental pollutants
- Certain drugs, pesticides, anesthetics
- Ozone
- Cigarette smoke [16]

Reactive oxygen species mediate damage to cells structures, including lipids and membrane protein and the nucleic acid under the presence of its higher concentration. This condition is termed oxidative stress [17]. These free radicals' processes are also associated with various food products. The rancidity of fatty foods, such as potato chips and butter, is due to free radical chain oxidation. Oxidation of polyunsaturated fatty acid (PUFA) is also associated with free-radical processes [18]. The importance of antioxidants is because of their property to stop the free radical chain reaction.

4. Antioxidants

An antioxidant is a chemical compound that has free radical scavenging properties, can delay or inhibit cellular damage and neutralize the effect of free radical by donating an electron [19]. Antioxidants thus counteract oxidative stress. A series of defense mechanisms have been developed to combat the exposure to free radicals from various sources [20]. Antioxidants further contribute to disease prevention and protect cells from the toxic effects of free radicals by neutralizing their excess. Antioxidants can be endogenous, generated in situ or exogenous, supplied through food [21].

To prevent condition like oxidative stress, it is essential to maintain a balance between the production of free radicals and antioxidants defense [22]. Fruits and vegetables are consumed by people as a source of antioxidants, as they are rich in flavonoids and antioxidants. It contributes by protecting the human being from cancer and cardiovascular problems, the ill effects of free radicals [23].

Antioxidants remove free radical intermediates and prevent or slow down the oxidation of other molecules by being oxidized themselves and terminate the chain reactions [24].

Antioxidants can act as

- Scavenging the peroxidation initiating species
- Decomposition of lipid peroxide
- prevent the generation of reactive species by chelating metal ion
- Preventing the formation of peroxides by quenching activity
- Reducing localized O_2 concentrations [25]

Antioxidants also play an essential role in food products by preventing oxidation reactions, browning in fruits and vegetables, and rancidity in fats and oil [23].

Antioxidants may be of natural or synthetic origin. Natural antioxidants are the important secondary metabolites of plant origins mainly explored in preparing some functional foods. In food systems, during storage, the use of nutritional antioxidants and the micro-nutrient, such as Vitamin E, helps maintain the color, texture, and flavor of the food product by preventing or retarding lipid peroxidation and reducing lipid peroxidation protein oxidation [26].

4.1 Vitamin C

Ascorbic acid is a water-soluble vitamin commonly known as vitamin C and was reckoned as L-ascorbic acid in 1965 by the IUPAC-IUB commission on biochemical nomenclature. Ascorbic acid has a 2,3-enediol group responsible for its antioxidant activity [27]. It is a 6-carbon lactone and cannot be synthesized in the human body, and is water-soluble, it must be regularly supplied through external means.

It plays an essential role in the biosynthesis of collagen, carnitine, and neurotransmitters [27]. The normal metabolic respiration process of the body produces potentially damaging free radicals. These free radicals can be efficiently quenched by ascorbic acid due to its reducing nature [28].

Ascorbic acid, after oxidation, leads to the formation of a dimer called dehydroascorbic acid (DHA). DHA is an oxidized form of ascorbic acid and can be reduced back to ascorbic acid by the action of glutathione (GSH) [29]. In aqueous solutions, dehydroascorbic acid exists as hydrated hemiketal [30].

The formation of dehydroascorbic acid from ascorbic acid is a two-step reversible oxidation process, during which ascorbyl radical is formed as an intermediate [31]. Ascorbyl radical is involved in the termination of free radical reactions, due to the delocalized nature of unpaired electrons present in it, it reacts with free radicals [32].

Dehydro-ascorbate is irreversibly converted to 2,3-diketo-L-gluconic acid with the hydrolysis of lactone ring [33, 34]. Diketo-L-gluconic acid is unstable and does not have biological activity [35].

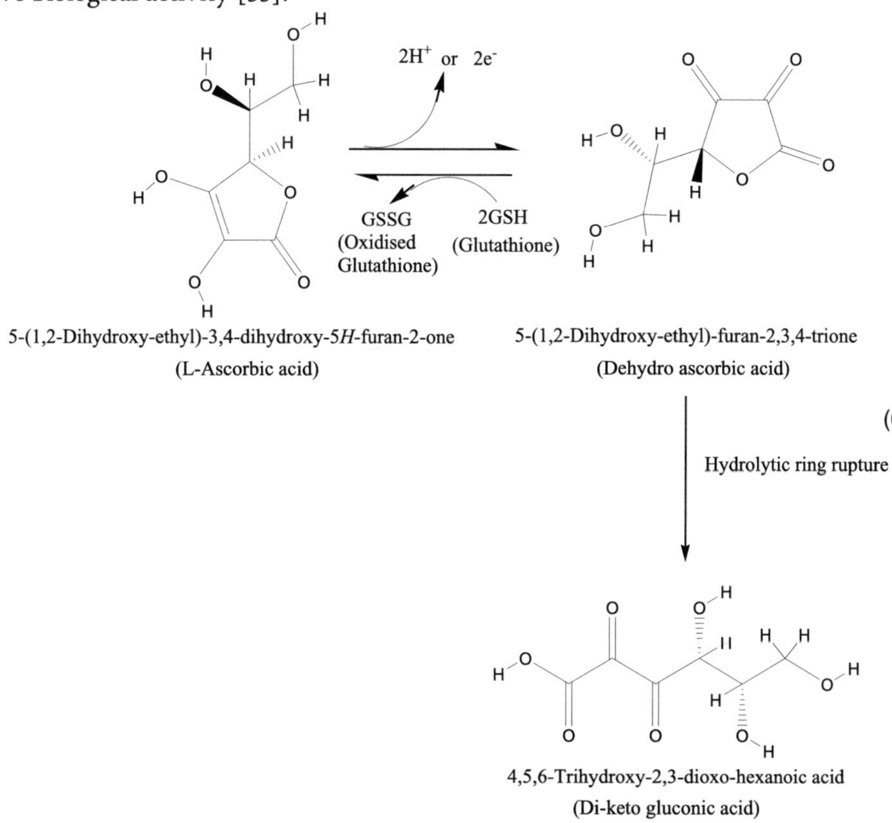

(6)

In fruits and vegetables with low levels of antioxidant (Vitamin C), on cutting, there is the exposure of the phenolic group to oxygen, and the cresolase and catecholase activity act and form quinone, which converts further to dopachrome before its polymerization into brown melanin pigment. Ascorbic acid can reverse this reaction, which converts quinones back to phenolic form [36].

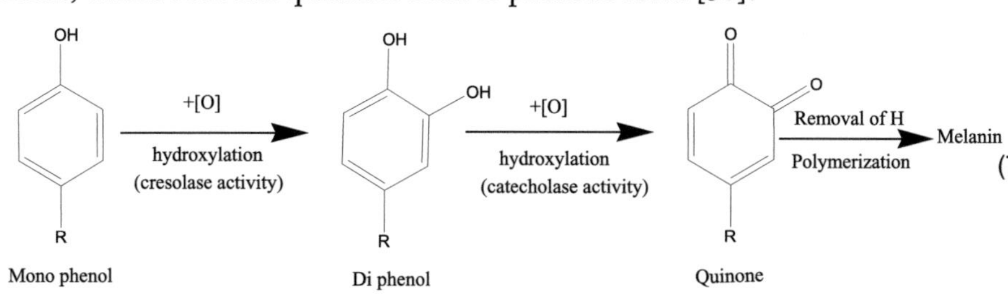

(7)

Termination of lipid peroxidation chain reaction is carried by ascorbic acid by donating an electron to lipid radical, which gets converted to ascorbate radical. These ascorbate radicals further react with each other to form ascorbate and dehydroascorbate molecules. Dehydroascorbate molecule on the addition of two electrons is converted back to ascorbate molecule because DHA does not have the antioxidant capacity, and this process is carried out by oxidoreductase [37].

Ascorbic acid prevents the formation of N-nitrosamines in nitrate-cured meats. It results in NO's formation, which is desirable for cured meats color [36]. L-ascorbic acid protects against oxidation of low-density lipoprotein implicated in the development of atherosclerosis by scavenging reactive oxygen species, which prevent oxidative stress [38].

4.2 Vitamin E

Vitamin E is a fat-soluble vitamin found in tocopherol and tocotrienol structures that exists in eight different isomeric forms equal configurations for both forms [39]. All eight forms are lipophilic. Chromanol group is responsible for antioxidant activities, and its methylation differs among all the members of the Vitamin E group [40].

The amount of methyl groups attached to phenol ring and pattern of methylation are responsible for reactive antioxidant activities for these isomers, which is found to be $\alpha > \beta > \gamma > \delta$. The highest activity of α-tocopherol is due to the presence of 3-methyl substituents [41]. The Food and Nutrition Board defines vitamin E requirements in the human body are fulfilled only by α-tocopherol.

Vitamin E repairs the oxidizing radicals during lipid auto-oxidation and halts the propagation step, thus acting as a chain-breaking antioxidant [42].

$$\underset{\text{Peroxyl radical}}{\text{ROO}^{\text{o}}} + \underset{\alpha\text{-tocopherol}}{\text{Toc-OH}} \longrightarrow \underset{\text{Lipid hydroperoxide}}{\text{ROOH}} + \underset{\alpha\text{-tocopherol radical}}{\text{Toc-O}^{\text{o}}} \quad (8)$$

Ascorbic acid is responsible for the regeneration of α-tocopherol from α-tocopherol radical. Thus, there is a synergistic effect between α-tocopherol and ascorbic acid [43].

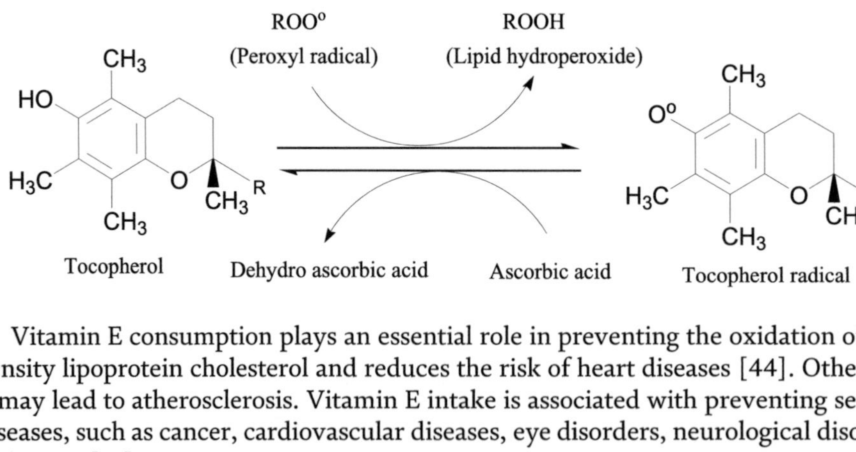

(9)

Vitamin E consumption plays an essential role in preventing the oxidation of low-density lipoprotein cholesterol and reduces the risk of heart diseases [44]. Otherwise, it may lead to atherosclerosis. Vitamin E intake is associated with preventing several diseases, such as cancer, cardiovascular diseases, eye disorders, neurological disorders, and aging [22].

4.3 Carotenoids

Carotenoids are yellow-red pigments synthesized naturally by plants and some microorganisms [45]. They have an isoprenoid polyene structure [46]. These are a group of tetra terpenoids that contain eight isoprene units with 40 carbon atoms.

Carotenoids can be categorized into two groups, which are as follows:

1. Carotenoid hydrocarbons (carotenes) contains specific end group as in β-carotene or lycopene.

2. Oxygen carotenoids (xanthophylls) as zeaxanthin and lutein [45].

Consumption of foods that are a rich source of carotenoids is related to a decrease in age-related diseases. Coronary heart diseases associated with oxidation of LDL cholesterol can be prevented by lycopene and β carotene [47].

Antioxidant activities of carotenoids are due to their structure that contains conjugated double bond, and their ability to delocalize unpaired electrons [48]. Singlet molecular oxygen $^1O_2$ and peroxyl radicals are among the two reactive oxygen species that are most likely to be scavenged by carotenoids [49]. At a low concentration of oxygen, the antioxidant activity of carotenoids increases, and at higher concentrations, it acts as a pro-oxidant (Ruth [50]).

Scavenging of superoxide anions ($\bullet O_2^-$) by β-carotene occurs as follows (R. [51]).

$$\underset{\text{Superoxide anion r+adical}}{{}^{o}O_2^-} + \underset{\text{Carotenoid}}{CAR} + 2H^+ \rightleftharpoons \underset{\text{Carotenoid}}{CAR^+} + \underset{\text{Hydrogen peroxide}}{H_2O_2} \quad (10)$$

Carotenoids can hinder free radical chain reactions that occur during lipid peroxidation due to their antioxidant activity. Free radical reactions proceed in the following manner [52].

Initiation

$$\text{Initiator} + RH \rightarrow R^o \quad (11)$$

Propagation

$$R^o + O_2 \rightarrow ROO^o \quad (12)$$

Termination

$$ROO^o + ROO^o \rightarrow \text{Product} \quad (13)$$

This chain reaction can be inhibited by carotenoids in three ways [53].

i. Electron transfer:

$$ROO^o + CAR \rightarrow ROO^- + CAR^{o+} \quad (14)$$

ii. Hydrogen abstraction:

$$ROO^o + CAR \rightarrow ROOH + CAR^o \quad (15)$$

iii. Addition of radical species:

$$ROO^{o} + CAR \rightarrow (ROO - CAR)^{o} \quad (16)$$

The photooxidative process leads to eye and skin diseases on exposure to light. The light filtering effect and antioxidant activity of carotenoids can protect against the ill effects of these processes [54]. β carotene acts as a provitamin and precursor for the formation of Vitamin A in the human body.

4.4 Polyphenols

Polyphenols are chemical compounds having phenolic structures and are obtained from plant sources [55]. These have several bioactive properties, such as they may act as attracting agents for pollinators, contribute to pigmentation of plants, as an antioxidant, and protection from UV light [56].

The chemical structure of these compounds comprises an aromatic ring with one or more hydroxyl groups. These can be simple phenolics or in polymeric form having high molecular mass [57]. The most important group of polyphenols is flavonoids (glycosides with benzopyrone nucleus). Flavonoids include flavones, flavonols, flavanone, flavonols, and anthocyanins [58]. Flavonoids consist of 15 carbon atoms having an arrangement, as shown below in the figure. These are compounds having a low molecular weight [59].

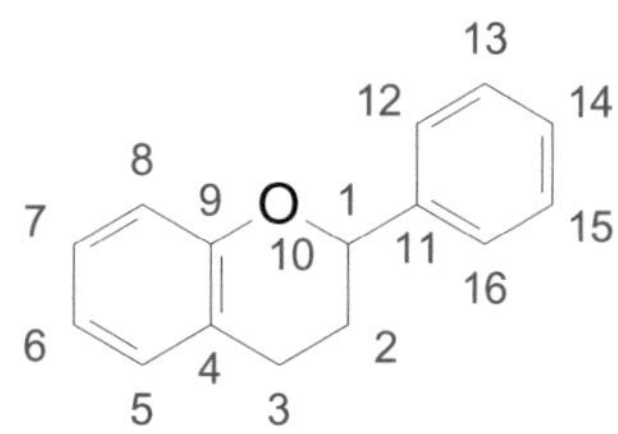

The antioxidant activity of these compounds is due to their ability to donate hydrogen and metal ion chelation [60]. Phenolic radicals formed after presenting hydrogen atoms do not readily participate in other radical reactions, as they become resonance stabilized [61]. Flavonoids can form a complex with metals and thus prevents metal-initiated lipid oxidation [62].

The difference in structure and glycosylation patterns of these compounds are responsible for their different antioxidant activity. Glycosides of anthocyanidins are called anthocyanins. These are the most extensive water-soluble pigments, commonly present in flowers and fruits [60].

Tannins are an important group of polyphenolic compounds, having high molecular weight. These are categorized as hydrolyzable and condensed tannins [63]. Hydrolyzable tannins are derived from the esterification of gallic acid (3,4,5-trihydroxy benzoic acid). Galloyl group of core polyol (formed from esterification of gallic acid) is further esterified to obtain hydrolyzable tannins [64]. Condensed tannins are the polymeric compounds obtained from polyhydroxy flavan-3-ol. These are also known as pro-anthocyanidins [63]. Tannins have metal ion chelating properties, act as an agent for protein precipitation, and possess antioxidant activity [64].

Polyphenolic compounds act as antioxidants to inactivate free radicals by two mechanisms, which are as follows:

i. Hydrogen atom transfer mechanism.

ii. single electron transfer mechanism.

It is supposed that an antioxidant ArOH transfers its hydrogen atom to react with free radical in the hydrogen atom transfer mechanism.

$$ArOH + R^o \rightarrow ArO^o + RH \tag{17}$$

In the single electron transfer mechanism, it is supposed that an oxidant donates an electron to the antioxidant molecule [65]:

$$ArOH + R^o \rightarrow ArOH^+ + R^- \tag{18}$$

5. Methods for antioxidant assessment

Antioxidants play an essential role in problems related to oxidative stress, such as neurodegenerative and cardiovascular diseases. People nowadays are more focused on antioxidant-rich foods, so it is vital to assess these components' antioxidant activity or free radical scavenging capacity. There are various ways of measuring antioxidant activity (**Figure 1**). Different methods follow different reaction mechanisms. These can be classified according to the reaction mechanism as:

Hydrogen atom transfer (HAT) method is based on the determination of free radical scavenging activity of antioxidants by donating a hydrogen atom. These are rapid reactions and do not depend on pH and solvent but are affected by the existence of reducing agents [66]. In contrast, the single electron transfer (SET) method is based on the ability of an antioxidant component to reduce the compounds such as carbonyls, radicals, or metal ions by transferring a single electron [67]. The most commonly used method is the oxygen radical absorbance capacity (ORAC) assay. This method is based on the principle of decrease in intensity of fluorescent compounds, such as β-phycoerythrin or fluorescein, due to the oxidative degradation by radicals (which leads to the formation of non-fluorescent compound) generated from thermal decomposition of AAPH (2, 2′-azobis (2-amidino propane) dihydrochloride) that is used as a free radical generator. The antioxidant activity is measured as a decrease in the amount and rate of formation of non-fluorescent products [68, 69]. This method provides an advantage that by altering the solvent and source of free radicals, it is possible to determine the hydrophilic and hydrophobic antioxidants. In this method, a controlled source of radicals is provided that simulates the reactions between lipids and antioxidants in food [70, 71].

The total radical-trapping antioxidant parameter (TRAP) assay is based on the same principle as ORAC. The antioxidant activity is measured as the moles of peroxyl radicals that are trapped by 1 L of antioxidant solution. Like the ORAC method, the loss of fluorescence is monitored. Trolox is used as a standard to compare the plasma-induced lag phase to that induced by antioxidant sample solution in the same plasma sample. It determines the activity of non-enzymatic antioxidants, such as ascorbic acid and glutathione, but this method is time-consuming and requires expertise [67].

Ferric reducing antioxidant power (FRAP) assay is based on the formation of the blue-colored ferrous complex by the antioxidants by reducing ferric 2,4,6-tripyridyl-s-triazine complex $[Fe^{3+}\text{-}(TPTZ)_2]^{3+}$ in an acidic medium [72]. Reactions were carried out under acidic conditions (pH 3.6) to maintain the solubility of iron. Reducing 1 M ferric ions to ferrous ions is known as one FRAP unit [73].

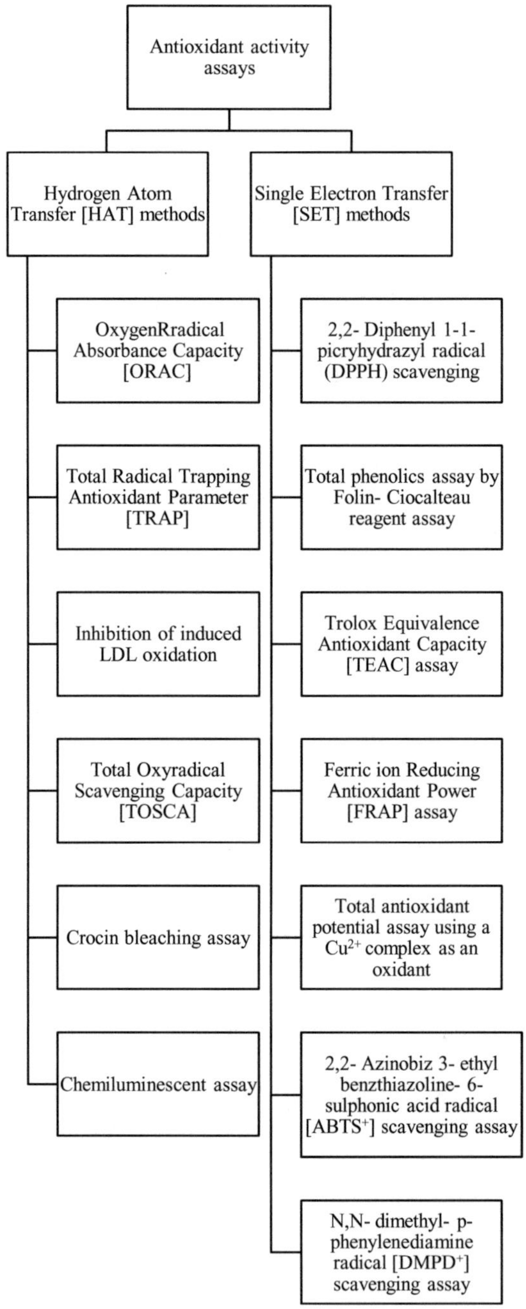

Figure 1.
Antioxidant assay techniques.

In the method of DPPH (2,2-Diphenyl-1-picryl hydrazyl) assay, the ability of an antioxidant to scavenge DPPH radical (purple color) and reduce it to diphenyl picryl hydrazine (yellow color) is measured. The reaction is carried out in an alcoholic solution [74]. Generally, the results are described as efficient concentration (EC_{50}). To bring about a 50% decrease in the concentration of DPPH, the amount of antioxidant required is reported as EC_{50} value [75].

6. Sources of antioxidants

Besides providing essential nutrients, fruits and vegetables also contain substantial amounts of biologically active secondary metabolites [76]. The secondary metabolites of plants that provide numerous health benefits are covered elsewhere [77, 78]. The principal dietary components found in the antioxidant properties of fruits and vegetables are polyphenols, flavonoids, carotenoids, Vitamin C, Vitamin E, glutathione, selenium indoles, and protease inhibitors (**Table 1**) [79].

The varying amounts of waste material are generated during the preparation of cut or processed fruits and vegetables [80]. Peels and seeds are the byproducts generated in large amounts during minimal processing of fruits and vegetables and comprise of large quantities of phytochemical components with antimicrobial and antioxidant properties [81–83]. All of these can be effectively utilized as a source of antioxidants. The fruits and vegetable tissues are rich in bioactive compounds, such as phenolics, vitamins, and carotenoids. These are even present in higher amounts in byproducts compared to the final product [84].

6.1 Grapes

Fresh grapes, grape juice, and grape wine are excellent sources of phenolic antioxidants (**Figure 2**). Flavonoids and other phenolic compounds present in grapes have anticarcinogenic, anti-allergic, anti-inflammatory, hepatotoxic, and antioxidative effects [85–87]. The majority of phenolics present in grapes are 60–70% in seeds and 28–35% in the skin, whereas pulp contains utmost to 10%. These phenolics can act as free radical scavengers and act as antioxidants. The grape seed oil also offers various health benefits, such as improving vision, protection of skin from sun damage, improved blood circulation, reduced oxidation of low-density lipoproteins, and reduced risk of coronary heart disease [88]. The antioxidant activity of grape juice is highest among the commercial juices, followed by grapefruit juice, tomato, orange, and apple [89]. Phenolic antioxidants obtained from grape pomace were found to exhibit the property to retard oxidation of human low-density lipoprotein (LDL) cholesterol [90].

6.2 Apple

"An apple a day keeps the doctor away," can be attributed to the number of phytochemicals present in apples. Apple is a rich source of polyphenols, vitamins, and carotenoids that prevent free radical damage due to their high antioxidant activity. Antioxidant compounds in apples are quercitin-3-glucoside, quercitin-3-galactoside, catechin, epicatechin, procyanidin, cyanidin-3-galactoside, chlorogenic acid, coumaric acid, and gallic acid [91]. The amount of these compounds varies with the cultivars and between the flesh and peel of an apple. These phytochemicals are rice in peels as compared to flesh. Peels contain a high amount of quercetin conjugates whereas, chlorogenic acid is present in higher concentrations in the flesh [92]. Phloridzin, an antioxidant compound, mainly present in apple seeds [93], is a derivative of chalcone, also having anti-diabetic activity because of its ability to inhibit sodium-linked glucose transport, thus limiting the absorption of glucose in the intestine and kidney [94, 95].

Antioxidant	Chemical structure	Antioxidant activity (mM)	Sources
Vitamins			
Vitamin C	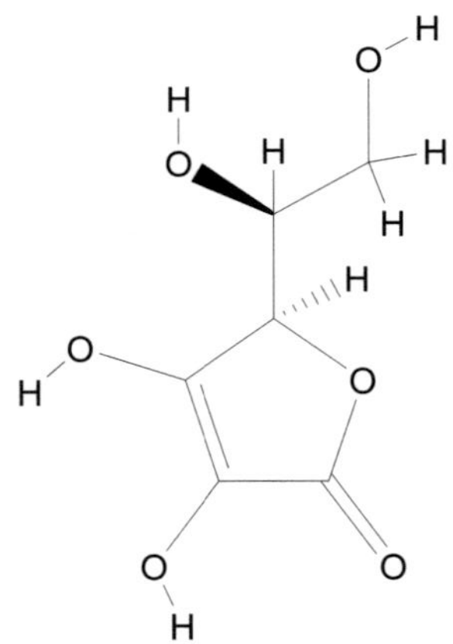	1.0 ± 0.02	Citrus fruits, gooseberry, broccoli, spinach
Vitamin E	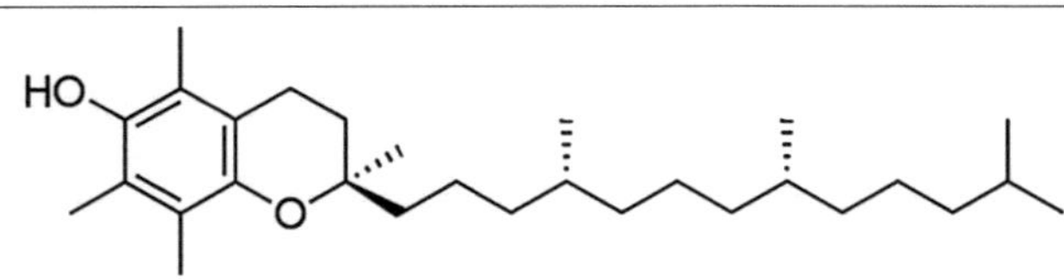	1.0 ± 0.03	Green leafy vegetables, nuts
Carotenoids			
β-Carotene	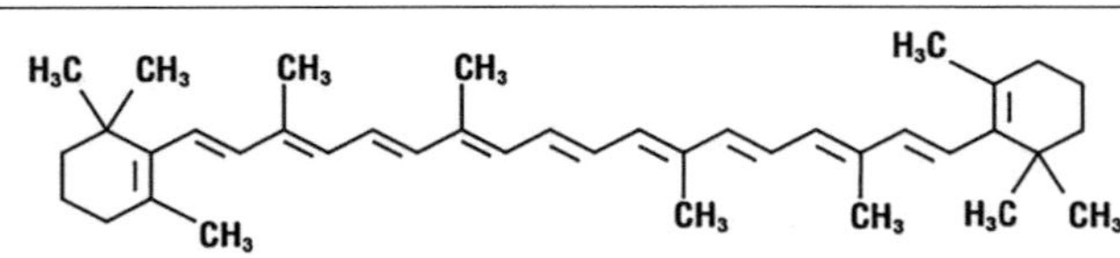	1.9 ± 0.1	Beetroot, apricots, carrots, tomatoes, mango, papaya, oranges

DOI: http://dx.doi.org/10.5772/intechopen.103707

Antioxidant	Chemical structure	Antioxidant activity (mM)	Sources
α-Carotene	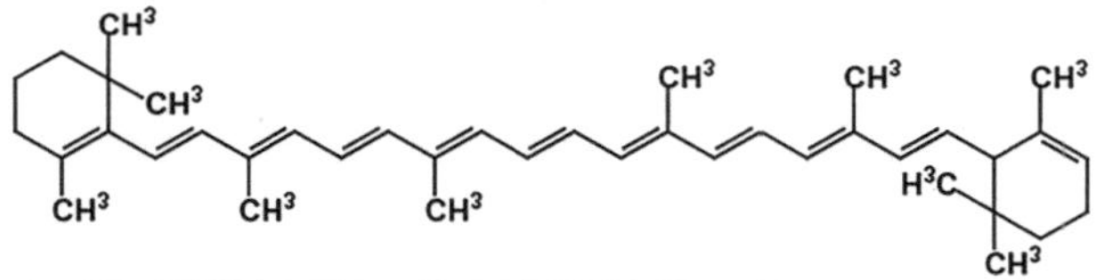	1.3 ± 0.04	Carrots, leafy vegetables
Lycopene	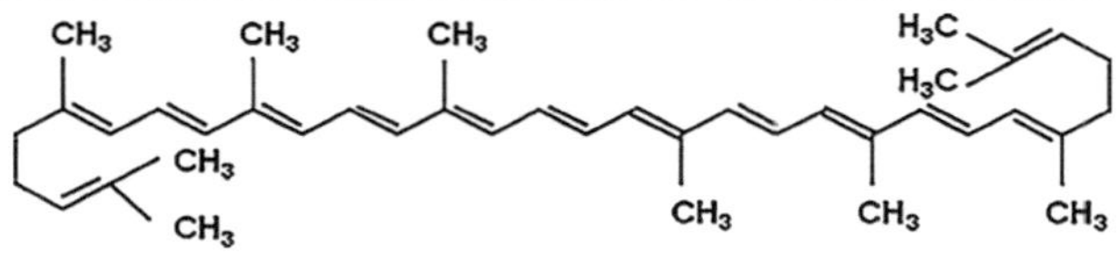	2.9 ± 0.15	Apricots, grapefruit, guava, watermelon, papaya, carrots, tomato
Lutein	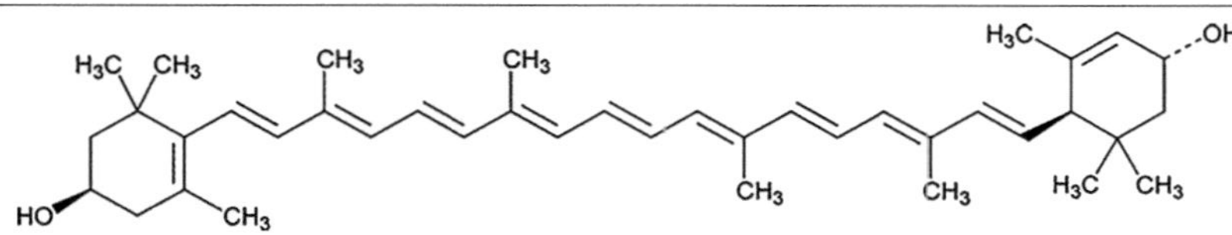	1.5 ± 0.1	Banana, satsuma peel, egg yɔlk, green vegetables
Flavonoids *Flavanones*			
Naringenin	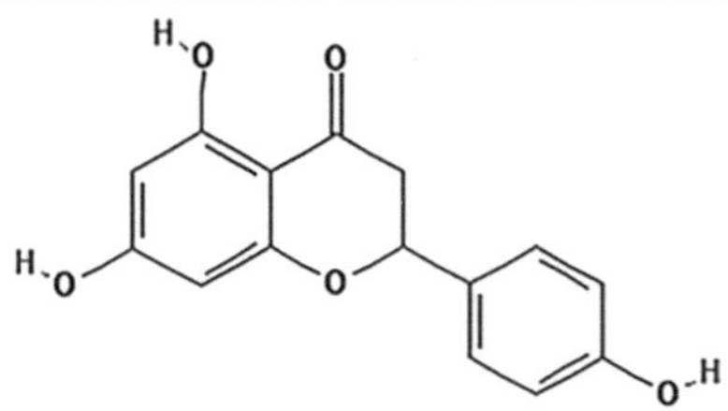	1.5 ± 0.05	Citrus fruits

Antioxidant	Chemical structure	Antioxidant activity (mM)	Sources
Taxifolin	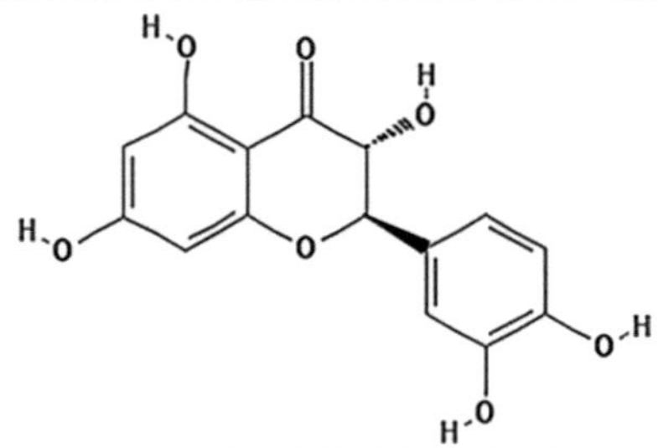	1.9 ± 0.03	Citrus fruits
Hesperidin	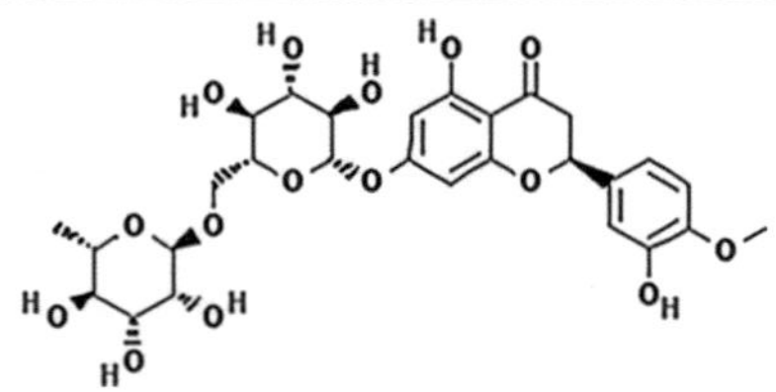	1.0 ± 0.03	Orange
Flavones			
Luteolin	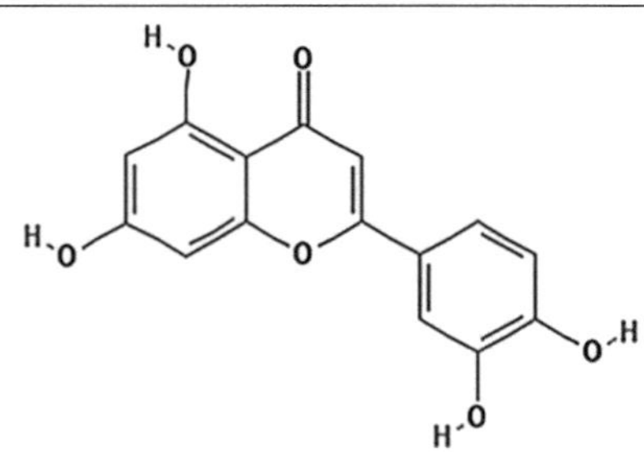	2.1 ± 0.05	Red pepper, celery, olive

DOI: http://dx.doi.org/10.5772/intechopen.103707

Antioxidant	Chemical structure	Antioxidant activity (mM)	Sources
Apigenin	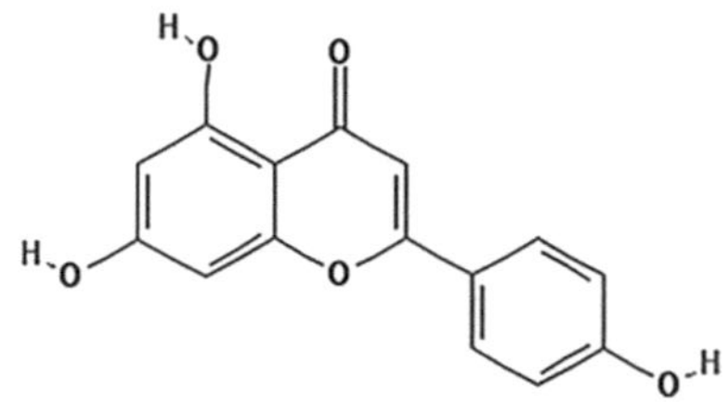	1.5 ± 0.08	Celery, parsley
Flavonols			
Quercitin	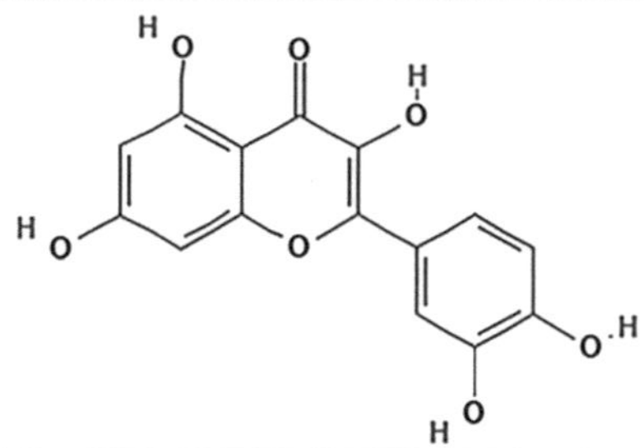	4.7 ± 0.10	Onion, lettuce, broccoli, tomato, tea, red wine, apple
Kaempferol	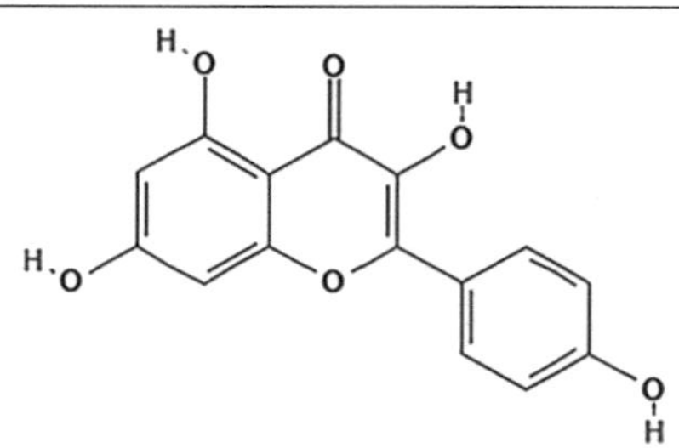	1.3 ± 0.08	Leek, broccoli, grapefruit, tea

Antioxidant	Chemical structure	Antioxidant activity (mM)	Sources
Flavanals			
Epicatechin	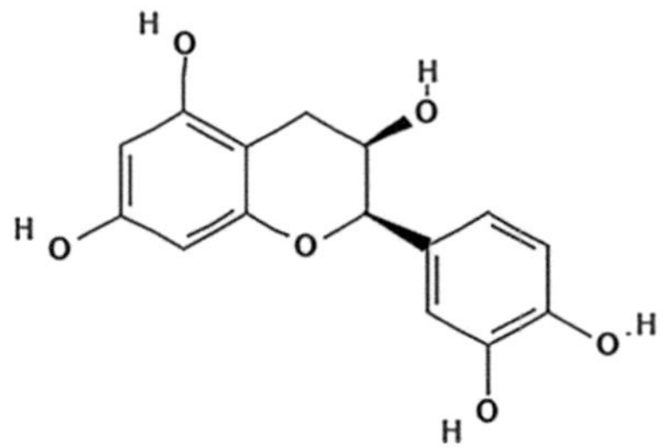	2.4 ± 0.02	Apricot, cherry, grape, peach, blackberry, apple, tea
Epigallocatechin	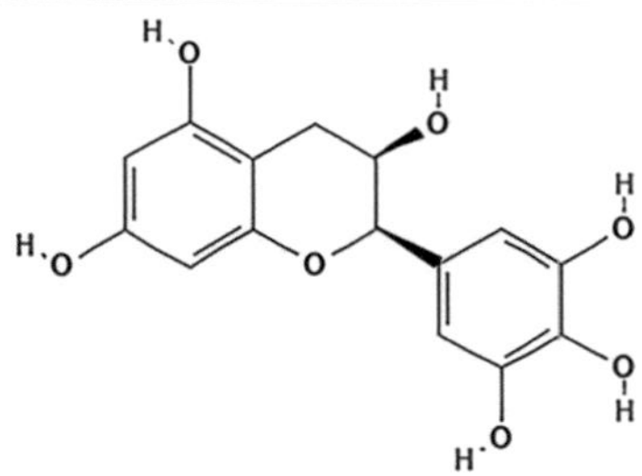	3.8 ± 0.06	tea
Epi catechin gallate	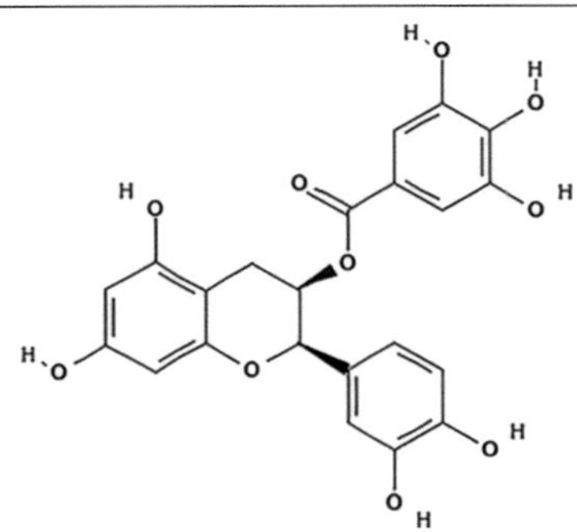	4.9 ± 0.02	tea

Antioxidant	Chemical structure	Antioxidant activity (mM)	Sources
AnthocyAnidins			
Cyanidin	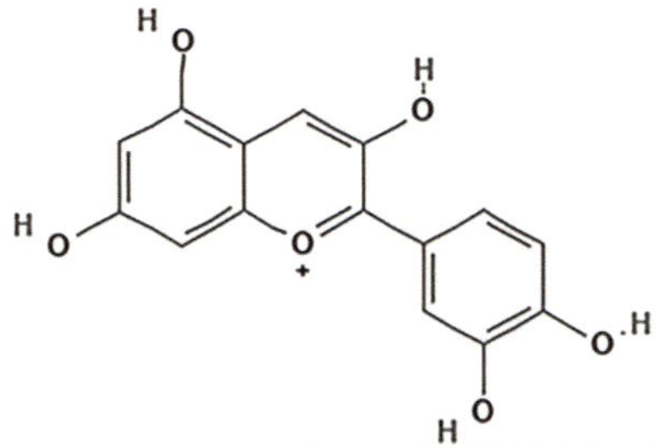	4.4 ± 0.12	Cherry, raspberry, strawberry
Delphinidin	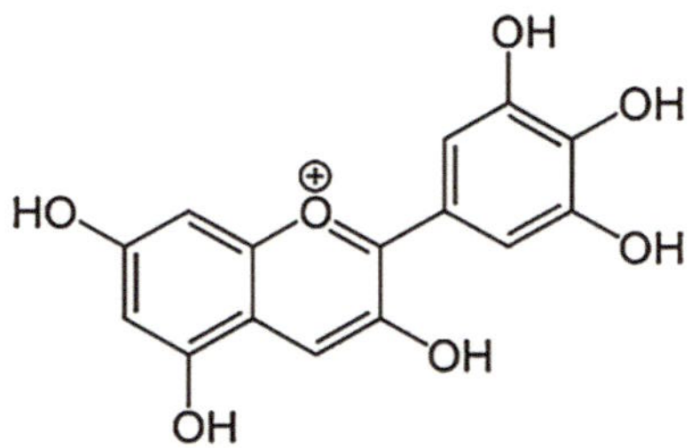	4.4 ± 0.11	Aubergine skin
Hydroxycinnamic acid			
Caffeic acid	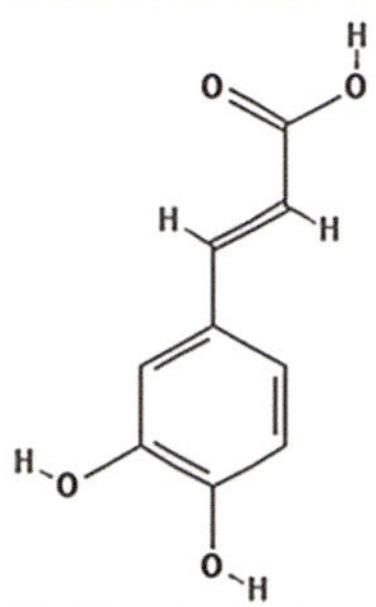	1.3 ± 0.01	White grapes, cabbage, asparagus, olive

Antioxidant	Chemical structure	Antioxidant activity (mM)	Sources
Ferulic acid	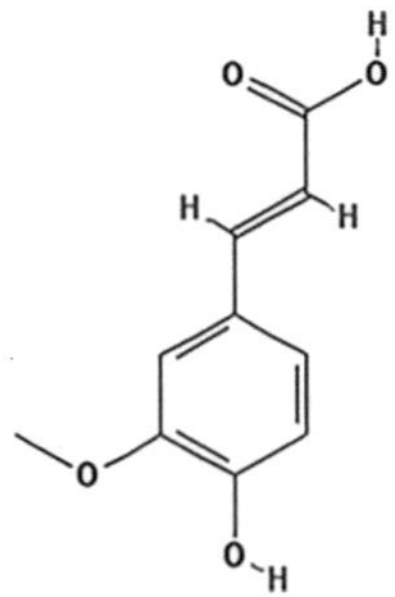	1.9 ± 0.02	Cabbage, tomato, asparagus
p-coumaric acid	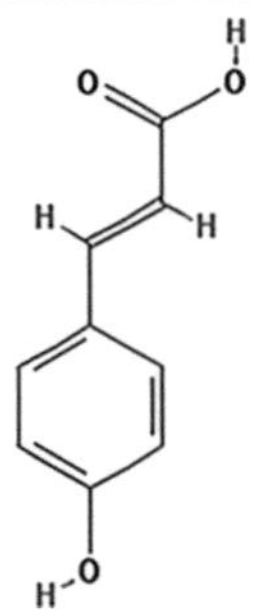	2.2 ± 0.06	White grapes, cabbage, asparagus, tomato
Chlorogenic acid	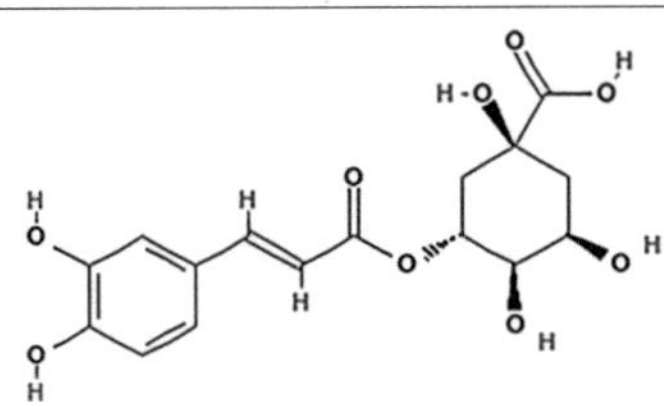	1.3 ± 0.02	Apple, peach, pear, cherry, tomato

Table 1.
Antioxidant, chemical structure, antioxidant activity, and sources.

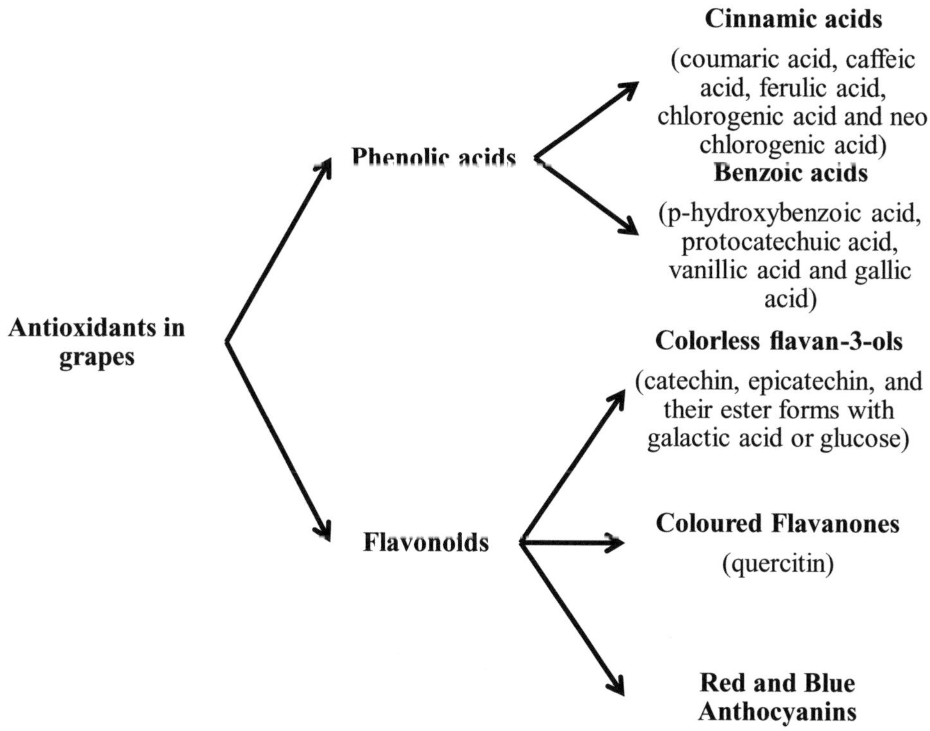

Figure 2.
Antioxidants present in grapes.

6.3 Berries

Berries are highly perishable, soft fruits, including strawberries, raspberries, blueberry, blackberry, blackcurrant, bilberry, and cranberry are rich sources of bioactive compounds, mainly phenolics [96]. Blackcurrant, bilberry, and chokeberry contain a higher amount of phenolic content as compared to other berries [97]. Phenolic acids present in berry fruits include *p*-hydroxybenzoic acid, gallic acid, salicylic acid, ellagic acid (benzoic acid derivatives), and *p*-coumaric acid ferulic acid, caffeic acid (cinnamic acid derivatives) [98]. Hydrocinnamic acids can inhibit LDL oxidation [86]. There is a decrease in the phenolic content of strawberries during the development from the unripe to the ripened stage [99]. Strawberries have a similar total antioxidant capacity as that of blackberries and raspberries but are lower than blueberries [100].

Other than cut fruits, most of the berries are used as raw material for the preparation of various processed products, such as jams, jellies, and juices. During the processing, a large amount of waste is generated. This waste can be used to recover highly valuable bioactive compounds. Blackberry and raspberry seeds can be used for the extraction of oil that is rich in antioxidant compounds such as phenols, carotenoids, and tocopherols along with linoleic acid (omega −6) and α-linoleic acid (omega-3) in 2 to 4:1 ratio [101]. Leaves and pomace from cranberry juice processing have more antioxidant activity and contain a higher amount of polyphenols than the juice [102].

6.4 Pomegranate

Pomegranate (*Punica granatum*) arils have high antioxidant activity [103] due to phenolic compounds, such as anthocyanins (cyanidin, delphinidin, pelargonidin 3-glucoside, and 3,5-diglucoside), punicalagin isomers, and ellagic acid derivatives. These compounds inhibit lipid oxidation due to their free radical scavenging activity [104].

Pomegranate peels and seeds that are the byproducts of juice processing are wasted or used as animal feed. But, it has been found that the amount of bioactive compounds or the antioxidant activity of the extracts of peel is higher than that of juice [105]. Pomegranate seeds can be used for oil extraction, that contain bioactive components. The oil extracted from pomegranate seeds has a fatty acid called punicic acid (conjugated linoleic acid isomer) [106] that constitutes about 70–76% of the seed oil and has high phytosterol content [107]. There are various health benefits of this pomegranate seed oil due to its unique chemical composition. Some of these benefits include modifying blood lipid profile in people suffering from hyperlipidemia [108].

6.5 Orange

Orange segments are a rich source of carotenoids (a class of natural pigments), such as zeaxanthin, β-cryptoxanthin, antheraxanthin, violaxanthin, and mutatoxanthin. Consumption of carotenoids is linked with reducing the risk of degenerative diseases in the body [109]. Oranges are rich in various antioxidant compounds, mainly ascorbic acid and phenolic compounds [110].

During the processing of oranges for juice manufacturing, a large amount of waste comprises peels and seeds. These are an abundant source of phytochemicals that are associated with a reduction of free radical damages. Various flavonoids have been identified in the orange peel, including hydroxylated poly ethoxy flavones and methylated flavonoids. These bioactive compounds are found to have protective action against oxidative stress [111].

6.6 Banana

Banana is a global food that belongs to the genus *Musa* [112]. Major producers of bananas in the world are India, China, the Philippines, and Ecuador. The largest importer and exporter of bananas globally are the USA and Ecuador, respectively [113]. There is a distinctive arrangement of secondary metabolites in a banana that is responsible for its antioxidant properties. Dessert banana is a rich source of various polyphenolic compounds and flavonols. The major polyphenolic compounds present in the edible part are catechins, epicatechins, gallic acid, tannins, and anthocyanins [114]. Bananas are also an abundant source of carotenoids, mainly present in peels. The major carotenoids include lutein, violaxanthin, neoxanthin, isoleucine, α- and β-carotene [115]. Seratonin is a biogenic amine found in bananas that imparts the feeling of happiness and wellbeing. The antioxidant potential of banana peels is more potent than that of pulp (Sulaiman et al., 2011), inhibits lipid peroxidation, and has high free radical scavenging activity [116].

Banana offers several health benefits, such as retardation of the aging process, reducing the risk of degenerative diseases like heart problems, atherosclerosis, brain dysfunction, and inflammation. It also provides resistance against oxidative changes in low-density lipoprotein and reduces oxidative stress due to bioactive compounds, such as dopamine and ascorbic acid. Serotonin stimulates the intestinal smooth muscles and thus inhibiting gastric secretion [117]. Banana peel can be utilized as a potential source of antioxidant compounds instead of discarding it.

6.7 Mango

Mango comes second, after the banana, regarding production. India is the largest producer of mango. The edible slices of mongo consist of significant amounts of antioxidant compounds. Xanthones are found in high concentrations comprising mainly of mangiferin (1,3,6,7-tetrahydroxyxanthone 2-glucopyranoside) and c-glucoside xanthone [118].

Mango byproducts, mainly peels, have shown high antioxidant activity. The phenolics and flavonoid content of mango peels is responsible for its anti-proliferative potential against cancer cells [119]. Mangiferin content of peels is about three times higher than that of pulp [120]. Gallo-tannins are found in higher amounts in mango kernels (15.5 mg/g dry matter) followed by peel (1.4 mg/g dry matter) and lowest in pulp (0.2 mg/g) [121]. Mango peel extracts can scavenge singlet oxygen ($^1O_2$), hydroxyl radical, and superoxide anion due to the presence of compounds, such as ethyl gallate and penta-o-galloyl glucoside [122].

6.8 Tomato

Tomato is an important and widely consumed vegetable (M. W. [123]). It is considered beneficial for health as it provides carotenoids, flavonoids, and phenolic acids [124]. During the production of tomato juice, about 3–7% of raw material is wasted, which comprises skin and seed.

Tomatoes are a rich source of carotenoid, the lycopene responsible for their characteristic red color [125]. The lycopene content of tomato peel is five times higher than pulp [126]. Thus, the hot break method is preferred in tomato juice extraction to get the tomato product of intense redness due to higher lycopene concentration.

6.9 Carrot

Carrot is a significant and widely consumed root vegetable that is a rich source of dietary fiber and secondary metabolites, mainly carotenoids and phenolics [77, 127, 128]. Carrots provide substantial health benefits [129] due to compounds like tocopherol, ascorbic acid, and β-carotene and hence is also called vitaminized food [130].

Carotenoids acts as a precursor of Vitamin A, especially the β-carotene, which are the major bioactive components present in carrots [131]. The most prevalent phenolic acid present in carrots is caffeic acid and thiamin, folic acid, riboflavin, and Vitamin C, which are in considerable amounts of carrot roots [132]. Carrot peel, the by-product of the processing industry, accounts for about 11% of the fresh carrot and can provide 54.1% of the total phenolic content of carrot. Therefore, these peels can be utilized for the value addition of various food products [133].

6.10 Garlic

Garlic is widely consumed as a spice and flavoring agent. Because of its preventive and curative action against various ailments, it is widely utilized for dietary and medicinal values [134, 135]. Garlic consists of a high content of γ-glutamylcysteine, which is believed to be responsible for various health benefits provided by garlic, along with other sulfur-containing compounds [136]. The chief bioactive component of garlic is allicin (diallyl this sulphonate). Raw garlic homogenate also consists of other significant sulfur-containing compounds, including allyl methyl sulphonate, γ-glutamyl cysteine, and 1-propenyl allyl thio sulphonate [137].

Garlic has been found to increase the resistance against LDL oxidation and thus is beneficial for heart and blood vessels because oxidative modification of LDL can lead to the formation of plaque in blood vessels by deposition of fatty streaks [138]. Garlic in the form of 10% homogenate in a salt solution and its supernatant fraction was found to be capable of reducing the free radicals generated from the Fenton reaction, and it was also effective in reducing the free radicals in cigarette smoke [139].

Garlic shows protective action against oxidative damage of tissues induced by nicotine. It was also found to be effective against carbon tetrachloride damage. Rats intoxicated with carbon tetrachloride were given an oral dosage of garlic oil, and it was found to prevent liver damage by peroxidation of lipids, alkaline phosphatase, and serum transaminase. These results are similar to that of Vitamin E [140].

6.11 Onion

Onion consists of substantial bioactive compounds, mainly flavonoids [141]. Flavonols are the significant flavonoids present in onions, quercetin derivatives being the most important ones [142]. Quercitin 3,4′-diglucoside and quercitin 4′-glucoside accounts for around 80–95% of total flavonols [143].

Some varieties of red onions also consist of anthocyanins. Anthocyanins are mainly concentrated in the outer skin of the onion (63%), and flavonoids in the skin are present mainly in aglycone forms [144]. So, the onion skin, which is generated as a waste, can be used to extract bioactive components.

6.12 Potato

Potato is considered the king of vegetables. It is the most widely consumed vegetable and the significant raw material for processed products, such as chips, fingers, and fries, during the processing of potatoes; peels are a considerable waste. Potato waste consists of various antioxidant compounds: caffeic acid, chlorogenic acid, protocatechuic acid, gallic acid, and para-hydroxybenzoic acid [145]. The antioxidant capacity of polyphenolic extracts obtained from potato peels is found to be analogous to that of synthetic antioxidants [BHA (butylated hydroxyanisole) and BHT (butylated hydroxytoluene)]. It was found that soybean and sunflower oil thermal degradation was suppressed when potato peel extract was incorporated in these oils, which may be attributed to chlorogenic and gallic acid in the extract [146].

6.13 Beetroot

Beetroots are an abundant source of valuable bioactive components such as carotenoids [147], betacyanins [148], betanin, flavonoids, and polyphenols [149]. The antioxidant activity of beetroots is primarily attributed to the total phenolic content of about 50–60 μmol/gm of dry weight [150]. The entire phenolic content in different portions of beetroot is found to be in the following order: Flesh (13%)

< crown (37%) < peel (50%) [151]. Beetroot peel's major phenolic compounds are p-coumaric acid, ferulic acid, and cyclodopa glucoside derivatives [152].

Betacyanins found in red beets have antioxidant activity and free-radical scavenging properties [153]. These are responsible for the inhibition of cervical and ovarian cancer cells [154]. Betalains improve the antioxidant profile of humans by reducing the oxidative degradation of lipids by scavenging the free radicals [155].

A large amount of horticultural waste is generated when preparing either cut fruits and vegetables or processed products. So, it can be a better option to utilize these wastes into valuable byproducts by extracting various phytochemicals to utilize in pharmaceuticals, cosmetics, and food products as functional ingredients. These bioactive compounds can be used in vegetable oils to prevent oxidation and edible coatings to increase shelf life.

7. Conclusion

Antioxidants prevent oxidative damage in food products and protect the human body from damage caused by reactive species, such as ROS, RNS, RSS, and free radicals. Antioxidants prevent the damage induced by free radicals acting through different mechanisms, such as free radical scavenging, prevention of free radical formation, or decomposition of reactive species. Antioxidants such as ascorbic acid, Vitamin E, carotenoids, and polyphenols can be obtained from plant sources, mainly fruits and vegetables fresh and processed products. The by-products obtained during the processing of fruits and vegetables can be utilized as a potential source for the extraction of antioxidants as these consist of high amounts of bioactive compounds. Secondary metabolites in peels and seeds of some fruits and vegetables, such as grapes, berries, pomegranate, garlic, and onion, can be higher than their pulp and juice. Such horticultural by-products can be utilized as a source of bioactive compounds in pharmaceutical, cosmetic, and food products.

Author details

Ravneet Kaur[1], Shubhra Shekhar[2] and Kamlesh Prasad[1]*

1 Department of Food Engineering and Technology, SLIET, Longowal, Punjab, India

2 Department of Food Process Engineering, NIT, Rourkela, India

*Address all correspondence to: profkprasad@gmail.com

IntechOpen

© 2022 The Author(s). Licensee IntechOpen. This chapter is distributed under the terms of the Creative Commons Attribution License (http://creativecommons.org/licenses/by/3.0), which permits unrestricted use, distribution, and reproduction in any medium, provided the original work is properly cited.

References

[1] Thomas MJ. The role of free radicals and antioxidants. Nutrition. 2000;**16**(7):716-718

[2] Gilbert DL. Fifty years of radical ideas. Annals of the New York Academy of Sciences. 2000;**899**(1):1-14

[3] Klatt P, Lamas S. Regulation of protein function by S-glutathiolation in response to oxidative and nitrosative stress. European Journal of Biochemistry. 2000;**267**(16):4928-4944

[4] Ridnour LA, Thomas DD, Mancardi D, Espey MG, Miranda KM, Paolocci N, et al. The chemistry of nitrosative stress induced by nitric oxide and reactive nitrogen oxide species. Putting perspective on stressful biological situations. Biological Chemistry. 2004;**385**(1):1-10

[5] Evans P, Halliwell B. Micronutrients: Oxidant/antioxidant status. British Journal of Nutrition. 2001;**85**(S2):S67-S74

[6] Lü JM, Lin PH, Yao Q, Chen C. Chemical and molecular mechanisms of antioxidants: Experimental approaches and model systems. Journal of Cellular and Molecular Medicine. 2010;**14**(4):840-860

[7] von Sonntag C. The Chemical Basis of Radiation Biology. London: Taylor & Francis; 1987

[8] Harman D. Aging: Overview. Annals of the New York Academy of Sciences. 2001;**928**(1):1-21

[9] Fridovich I. Fundamental aspects of reactive oxygen species, or what's the matter with oxygen? Annals of the New York Academy of Sciences. 1999;**893**(1):13-18

[10] Stief TW. The physiology and pharmacology of singlet oxygen. Medical Hypotheses. 2003;**60**(4):567-572

[11] Chance B, Sies H, Boveris A. Hydroperoxide metabolism in mammalian organs. Physiological Reviews. 1979;**59**(3):527-605

[12] Ghafourifar P, Cadenas E. Mitochondrial nitric oxide synthase. Trends in Pharmacological Sciences. 2005;**26**(4):190-195

[13] Chiueh CC. Neuroprotective properties of nitric oxide. Annals of the New York Academy of Sciences. 1999;**890**(1):301-311

[14] Lobo V, Patil A, Phatak A, Chandra N. Free radicals, antioxidants and functional foods: Impact on human health. Pharmacognosy Reviews. 2010;**4**(8):118

[15] Cadenas E. Biochemistry of oxygen toxicity. Annual Review of Biochemistry. 1989;**58**(1):79-110

[16] Langseth L. Oxidants, Antioxidants, and Disease Prevention. Brussels: ILSI Europe; 1995

[17] Poli G, Leonarduzzi G, Biasi F, Chiarpotto E. Oxidative stress and cell signalling. Current Medicinal Chemistry. 2004;**11**(9):1163-1182

[18] Porter NA, Weber BA, Weenen H, Khan JA. Autoxidation of polyunsaturated lipids. Factors controlling the stereochemistry of product hydroperoxides. Journal of the American Chemical Society. 1980;**102**(17):5597-5601

[19] Halliwell B. How to characterize an antioxidant: An update. Biochemical Society Symposia. 1995;**61**:73-101

[20] Cadenas E, Packer L. Handbook of Antioxidants. Vol. 712. New York: Marcel Dekker; 2002

[21] Pham-Huy LA, He H, Pham-Huy C. Free radicals, antioxidants in disease and health. International Journal of Biomedical Science: IJBS. 2008;**4**(2):89

[22] Bagchi K, Puri S. Free radicals and antioxidants in health and disease: A review. EMHJ-Eastern Mediterranean Health Journal. 1998;**4**(2):350-360

[23] Hamid A, Aiyelaagbe O, Usman L, Ameen O, Lawal A. Antioxidants: Its medicinal and pharmacological applications. African Journal of Pure and Applied Chemistry. 2010;**4**(8):142-151

[24] Sies H. Oxidative stress: Oxidants and antioxidants. Experimental Physiology: Translation and Integration. 1997;**82**(2):291-295

[25] Nawar W. In: Fennema OR, editor. Lipids in Food Chemistry. New York: Marcel Dekker; 1996

[26] Elias RJ, Kellerby SS, Decker EA. Antioxidant activity of proteins and peptides. Critical Reviews in Food Science and Nutrition. 2008;**48**(5): 430-441

[27] Li Y, Schellhorn HE. New developments and novel therapeutic perspectives for vitamin C. The Journal of Nutrition. 2007;**137**(10):2171-2184

[28] Arrigoni O, De Tullio MC. Ascorbic acid: Much more than just an antioxidant. Biochimica et Biophysica Acta (BBA)-General Subjects. 2002;**1569** (1–3):1-9

[29] Aguirre R, May JM. Inflammation in the vascular bed: Importance of Vitamin C. Pharmacology & Therapeutics. 2008;**119**(1):96-103

[30] Seib PA, Tolbert BM. Ascorbic Acid: Chemistry, Metabolism, and Uses: Based on a Symposium Sponsored by the Division of Carbohydrate Chemistry at the Second Chemical Congress of the North American Continent (180th ACS National Meeting), Las Vegas, Nevada, August 26-27, 1980. Vol. 200. Washington, United States: Amer Chemical Society; 1982

[31] Laroff GP, Fessenden RW, Schuler RH. Electron spin resonance spectra of radical intermediates in the oxidation of ascorbic acid and related substances. Journal of the American Chemical Society. 1972;**94**(26): 9062-9073

[32] Bielski BH. Chemistry of ascorbic acid radicals. In: Advances in Chemistry. Vol. 200. Washington, DC, United States: ACS Publications; 1982

[33] Bode A, Cunningham L, Rose R. Spontaneous decay of oxidized ascorbic acid (dehydro-L-ascorbic acid) evaluated by high-pressure liquid chromatography. Clinical Chemistry. 1990;**36**(10):1807-1809

[34] Chatterjee I. Biosynthesis Ofl-Ascorbate in Animals Methods in Enzymology. Vol. 18. Amsterdam, Netherlands: Elsevier; 1970. pp. 28-34

[35] DeMan JM. Principles of Food Chemistry. New York: Springer; 2007

[36] Liao M-L, Seib PA. Chemistry of L-ascorbic acid related to foods. Food Chemistry. 1988;**30**(4):289-312

[37] Nimse SB, Pal D. Free radicals, natural antioxidants, and their reaction mechanisms. RSC Advances. 2015;**5**(35): 27986-28006

[38] Jialal I, Vega GL, Grundy SM. Physiologic levels of ascorbate inhibit the

oxidative modification of low density lipoprotein. Atherosclerosis. 1990;**82**(3): 185-191

[39] Evans HM, Bishop KS. On the existence of a hitherto unrecognized dietary factor essential for reproduction. Science. 1922;**56**(1458):650-651

[40] Niki E. Antioxidants in relation to lipid peroxidation. Chemistry and Physics of Lipids. 1987;**44**(2–4):227-253

[41] Kamal-Eldin A, Appelqvist LÅ. The chemistry and antioxidant properties of tocopherols and tocotrienols. Lipids. 1996;**31**(7):671-701

[42] Serbinova EA, Packer L. Antioxidant Properties of α-tocopherol and α-tocotrienol Methods in Enzymology. Vol. 234. Amsterdam, Netherlands: Elsevier; 1994. pp. 354-366

[43] Yeum K-J, Beretta G, Krinsky NI, Russell RM, Aldini G. Synergistic interactions of antioxidant nutrients in a biological model system. Nutrition. 2009;**25**(7–8):839-846

[44] Hathcock J. Vitamins and minerals: Efficacy and safety. The American Journal of Clinical Nutrition. 1997;**66**(2): 427-437

[45] Paiva SA, Russell RM. β-Carotene and other carotenoids as antioxidants. Journal of the American College of Nutrition. 1999;**18**(5):426-433

[46] Britton G. Structure and properties of carotenoids in relation to function. The FASEB Journal. 1995;**9**(15): 1551-1558

[47] Weisburger J. Mechanisms of action of antioxidants as exemplified in vegetables, tomatoes and tea1. Food and Chemical Toxicology. 1999;**37**(9–10): 943-948

[48] Mortensen A, Skibsted LH, Truscott T. The interaction of dietary carotenoids with radical species. Archives of Biochemistry and Biophysics. 2001;**385**(1):13-19

[49] Young AJ, Lowe GM. Antioxidant and prooxidant properties of carotenoids. Archives of Biochemistry and Niophysics. 2001; **385**(1):20-27

[50] Edge R, Land EJ, McGarvey DJ, Burke M, Truscott TG. The reduction potential of the β-carotene+/β-carotene couple in an aqueous micro-heterogeneous environment. FEBS Letters. 2000;**471**(2–3):125-127

[51] Edge R, McGarvey DJ, Truscott TG. The carotenoids as anti-oxidants—A review. Journal of Photochemistry and Photobiology B: Biology. 1997;**41**(3): 189-200. DOI: 10.1016/S1011-1344(97) 00092-4

[52] Krinsky NI, Mayne ST, Sies H. Carotenoids in Health and Disease. Boca Raton, Florida, United States: CRC Press; 2004

[53] Krinsky NI. Carotenoids as antioxidants. Nutrition. 2001;**17**(10): 815-817

[54] Stahl W, Sies H. Antioxidant activity of carotenoids. Molecular Aspects of Medicine. 2003;**24**(6):345-351

[55] Tsao R. Chemistry and biochemistry of dietary polyphenols. Nutrients. 2010; **2**(12):1231-1246. DOI: 10.3390/ nu2121231

[56] Naczk M, Shahidi F. Phenolics in cereals, fruits and vegetables: Occurrence, extraction and analysis. Journal of Pharmaceutical and Biomedical Analysis. 2006;**41**(5): 1523-1542

[57] Balasundram N, Sundram K, Samman S. Phenolic compounds in plants and Agri-industrial by-products: Antioxidant activity, occurrence, and potential uses. Food Chemistry. 2006; **99**(1):191-203

[58] Sisein EA. Biochemistry of free radicals and antioxidants. Scholars Academic Journal of Biosciences. 2014; **2**(2):110-118

[59] Merken HM, Beecher GR. Measurement of food flavonoids by high-performance liquid chromatography: A review. Journal of Agricultural and Food Chemistry. 2000; **48**(3):577-599

[60] Bravo L. Polyphenols: Chemistry, dietary sources, metabolism, and nutritional significance. Nutrition Reviews. 1998;**56**(11):317-333

[61] Lee J, Koo N, Min DB. Reactive oxygen species, aging, and antioxidative nutraceuticals. Comprehensive Reviews in Food Science and Food Safety. 2004; **3**(1):21-33

[62] Hendrich S, Wang G, Lin H, Xu X, Tew B, Wang H, et al. Isoflavone metabolism and bioavailability. Antioxidant Status, Diet, Nutrition and Health. 1999:211-230

[63] Porter L. Condensed Tannins Natural Products of Woody Plants. Berlin/Heidelberg, Germany: Springer; 1989. pp. 651-690

[64] Hagerman AE. Hydrolyzable tannin structural chemistry. In: Tannin Handbook. Cambridge, Massachusetts, United States: Academic Press; 2002. pp. 1-8

[65] Leopoldini M, Russo N, Toscano M. The molecular basis of working mechanism of natural polyphenolic antioxidants. Food Chemistry. 2011; **125**(2):288-306

[66] Prior RL, Wu X, Schaich K. Standardized methods for the determination of antioxidant capacity and phenolics in foods and dietary supplements. Journal of Agricultural and Food Chemistry. 2005;**53**(10):4290-4302

[67] Huang D, Ou B, Prior RL. The chemistry behind antioxidant capacity assays. Journal of Agricultural and Food Chemistry. 2005;**53**(6):1841-1856

[68] Cao G, Alessio HM, Cutler RG. Oxygen-radical absorbance capacity assay for antioxidants. Free Radical Biology and Medicine. 1993;**14**(3): 303-311

[69] Ou B, Hampsch-Woodill M, Prior RL. Development and validation of an improved oxygen radical absorbance capacity assay using fluorescein as the fluorescent probe. Journal of Agricultural and Food Chemistry. 2001; **49**(10):4619-4626

[70] Ou B, Hampsch-Woodill M, Flanagan J, Deemer EK, Prior RL, Huang D. Novel fluorometric assay for hydroxyl radical prevention capacity using fluorescein as the probe. Journal of Agricultural and Food Chemistry. 2002; **50**(10):2772-2777

[71] Prior RL, Hoang H, Gu L, Wu X, Bacchiocca M, Howard L, et al. Assays for hydrophilic and lipophilic antioxidant capacity (oxygen radical absorbance capacity (ORACFL)) of plasma and other biological and food samples. Journal of Agricultural and Food Chemistry. 2003;**51**(11):3273-3279

[72] Benzie IF, Strain J. Ferric Reducing/ Antioxidant Power Assay: Direct Measure of Total Antioxidant Activity of Biological Fluids and Modified Version

for Simultaneous Measurement of Total Antioxidant Power and Ascorbic Acid Concentration Methods in Enzymology. Vol. 299. Amsterdam, Netherlands: Elsevier; 1999. pp. 15-27

[73] MacDonald-Wicks LK, Wood LG, Garg ML. Methodology for the determination of biological antioxidant capacity in vitro: A review. Journal of the Science of Food and Agriculture. 2006; **86**(13):2046-2056

[74] Roginsky V, Lissi EA. Review of methods to determine chain-breaking antioxidant activity in food. Food Chemistry. 2005;**92**(2):235-254

[75] Brand-Williams W, Cuvelier M-E, Berset C. Use of a free radical method to evaluate antioxidant activity. LWT-Food Science and Technology. 1995;**28**(1):25-30

[76] Siddiqui M, Prasad K. Plant Secondary Metabolites, Volume One: Biological and Therapeutic Significance. OAkville, ON, USA: Apple Academic Press; 2017

[77] Haq R-U, Prasad K. Antioxidant activity, phenolic, carotenoid and color changes in packaged fresh carrots stored under refrigeration temperature. Journal of Food Measurement and Characterization. 2017;**11**(4):1542-1549

[78] Singh R, Prasad K, Siddiqui MW, Prasad K. Medicinal Plants in Preventive and Curative Role for Various Ailments Plant Secondary Metabolites, 3 Volume Set. Apple Academic Press; 2016. pp. 63-100

[79] Karakaya S, Kavas A. Antimutagenic activities of some foods. Journal of the Science of Food and Agriculture. 1999; **79**(2):237-242

[80] Joshi V, Kumar A, Kumar V. Antimicrobial, antioxidant and phyto-chemicals from fruit and vegetable wastes: A review. International Journal of Food and Fermentation Technology. 2012;**2**(2):123

[81] Muthuswamy S, Rupasinghe H, Stratton G. Antimicrobial effect of cinnamon bark extract on Escherichia coli O157: H7, listeria innocua and fresh-cut apple slices. Journal of Food Safety. 2008;**28**(4):534-549

[82] Shrikhande AJ. Wine by-products with health benefits. Food Research International. 2000;**33**(6):469-474

[83] Tuchila C, Jianu I, Rujescu CI, Butur M, Ahmadi-Khoie M, Negrea I. Evaluation of the antimicrobial activity of some plant extracts used as food additives. Journal of Food, Agriculture and Environment. 2008;**6**(3–4):68-70

[84] Ayala-Zavala J, Rosas-Domínguez C, Vega-Vega V, González-Aguilar G. Antioxidant enrichment and antimicrobial protection of fresh-cut fruits using their own byproducts: Looking for integral exploitation. Journal of Food Science. 2010;**75**(8):R175-R181

[85] de Gaulejac NS-C, Glories Y, Vivas N. Free radical scavenging effect of anthocyanins in red wines. Food Research International. 1999;**32**(5): 327-333

[86] Meyer AS, Donovan JL, Pearson DA, Waterhouse AL, Frankel EN. Fruit hydroxycinnamic acids inhibit human low-density lipoprotein oxidation in vitro. Journal of Agricultural and Food Chemistry. 1998a;**46**(5):1783-1787

[87] Meyer AS, Jepsen SM, Sørensen NS. Enzymatic release of antioxidants for human low-density lipoprotein from grape pomace. Journal of Agricultural and Food Chemistry. 1998b;**46**(7): 2439-2446

[88] Shi J, Yu J, Pohorly JE, Kakuda Y. Polyphenolics in grape seeds—Biochemistry and functionality. Journal of Medicinal Food. 2003;**6**(4):291-299

[89] Wang H, Cao G, Prior RL. Total antioxidant capacity of fruits. Journal of Agricultural and Food Chemistry. 1996; **44**(3):701-705

[90] Meyer AS, Yi O-S, Pearson DA, Waterhouse AL, Frankel EN. Inhibition of human low-density lipoprotein oxidation in relation to composition of phenolic antioxidants in grapes (*Vitis vinifera*). Journal of Agricultural and Food Chemistry. 1997;**45**(5):1638-1643

[91] Boyer J, Liu RH. Antioxidants of apples. New York Fruit Quarterly. 2003; **11**(4):11-15

[92] Escarpa A, Gonzalez M. High-performance liquid chromatography with diode-array detection for the determination of phenolic compounds in peel and pulp from different apple varieties. Journal of Chromatography A. 1998;**823**(1–2):331-337

[93] Ehrenkranz JRL, Lewis NG, Ronald Kahn C, Roth J. Phlorizin: A review. Diabetes/Metabolism Research and Reviews. 2005;**21**(1):31-38. DOI: 10.1002/dmrr.532

[94] Dudash J, Zhang X, Zeck RE, Johnson SG, Cox GG, Conway BR, et al. Glycosylated dihydrochalcones as potent and selective sodium glucose co-transporter 2 (SGLT2) inhibitors. Bioorganic & Medicinal Chemistry Letters. 2004;**14**(20):5121-5125. DOI: 10.1016/j.bmcl.2004.07.082

[95] Manzano S, Williamson G. Polyphenols and phenolic acids from strawberry and apple decrease glucose uptake and transport by human intestinal Caco-2 cells. Molecular Nutrition & Food Research. 2010; **54**(12):1773-1780. DOI: 10.1002/mnfr.201000019

[96] Szajdek A, Borowska E. Bioactive compounds and health-promoting properties of berry fruits: A review. Plant Foods for Human Nutrition. 2008; **63**(4):147-156

[97] Kähkönen MP, Hopia AI, Heinonen M. Berry phenolics and their antioxidant activity. Journal of Agricultural and Food Chemistry. 2001; **49**(8):4076-4082

[98] Häkkinen SH, Törrönen AR. Content of flavonols and selected phenolic acids in strawberries and vaccinium species: Influence of cultivar, cultivation site and technique. Food Research International. 2000;**33**(6):517-524

[99] Reganold JP, Andrews PK, Reeve JR, Carpenter-Boggs L, Schadt CW, Alldredge JR, et al. Fruit and soil quality of organic and conventional strawberry agroecosystems. PLoS One. 2010;**5**(9): e12346

[100] Kalt W, Forney CF, Martin A, Prior RL. Antioxidant capacity, vitamin C, phenolics, and anthocyanins after fresh storage of small fruits. Journal of Agricultural and Food Chemistry. 1999; **47**(11):4638-4644

[101] Radočaj O, Vujasinović V, Dimić E, Basić Z. Blackberry (*Rubus fruticosus* L.) and raspberry (*Rubus idaeus* L.) seed oils extracted from dried press pomace after longterm frozen storage of berries can be used as functional food ingredients. European Journal of Lipid Science and Technology. 2014;**116**(8):1015-1024

[102] Oszmiański J, Wojdyło A, Lachowicz S, Gorzelany J, Matłok N. Comparison of bioactive potential of cranberry fruit and fruit-based products

versus leaves. Journal of Functional Foods. 2016;**22**:232-242

[103] Gil MI, Tomás-Barberán FA, Hess-Pierce B, Holcroft DM, Kader AA. Antioxidant activity of pomegranate juice and its relationship with phenolic composition and processing. Journal of Agricultural and Food Chemistry. 2000; **48**(10):4581-4589

[104] Noda Y, Kaneyuki T, Mori A, Packer L. Antioxidant activities of pomegranate fruit extract and its anthocyanidins: Delphinidin, cyanidin, and pelargonidin. Journal of Agricultural and Food Chemistry. 2002;**50**(1):166-171

[105] Zhang L-H, Li L-L, Li Y-X, Zhang Y-H. In vitro antioxidant activities of fruits and leaves of pomegranate. Acta Horticulturae. 2008; **765**:31-34

[106] Kaufman M, Wiesman Z. Pomegranate oil analysis with emphasis on MALDI-TOF/MS triacylglycerol fingerprinting. Journal of Agricultural and Food Chemistry. 2007;**55**(25): 10405-10413

[107] Kýralan M, Gölükcü M, Tokgöz H. Oil and conjugated linolenic acid contents of seeds from important pomegranate cultivars (*Punica granatum* L.) grown in Turkey. Journal of the American Oil Chemists' Society. 2009; **86**(10):985-990

[108] Johanningsmeier SD, Harris GK. Pomegranate as a functional food and nutraceutical source. Annual Review of Food Science and Technology. 2011;**2**: 181-201

[109] Okino Delgado CH, Fleuri LF. Orange and mango by-products: Agro-industrial waste as source of bioactive compounds and botanical versus commercial description—A review. Food Reviews International. 2016;**32**(1):1-14

[110] Galaverna G, Di Silvestro G, Cassano A, Sforza S, Dossena A, Drioli E, et al. A new integrated membrane process for the production of concentrated blood orange juice: Effect on bioactive compounds and antioxidant activity. Food Chemistry. 2008;**106**(3): 1021-1030

[111] Manthey JA, Grohmann K. Phenols in citrus peel byproducts. Concentrations of hydroxycinnamates and polymethoxylated flavones in citrus peel molasses. Journal of Agricultural and Food Chemistry. 2001;**49**(7): 3268-3273

[112] Lassoudière A. Le bananier et sa culture. Versailles, France: Editions Quae; 2007

[113] FAOSTAT. 2017. Available from: http://faostat.fao.org/beta/en/#data/QC [Accessed: November 04, 2016]

[114] Mattila P, Hellström J, Törrönen R. Phenolic acids in berries, fruits, and beverages. Journal of Agricultural and Food Chemistry. 2006;**54**(19): 7193-7199

[115] Subagio A, Morita N, Sawada S. Carotenoids and their fatty-acid esters in banana peel. Journal of Nutritional Science and Vitaminology. 1996;**42**(6): 553-566

[116] González-Montelongo R, Lobo MG, González M. Antioxidant activity in banana peel extracts: Testing extraction conditions and related bioactive compounds. Food Chemistry. 2010; **119**(3):1030-1039

[117] Kumar KS, Bhowmik D, Duraivel S, Umadevi M. Traditional and medicinal uses of banana. Journal of Pharmacognosy and Phytochemistry. 2012;**1**(3):51-63

[118] Berardini N, Knödler M, Schieber A, Carle R. Utilization of mango peels as a source of pectin and polyphenolics. Innovative Food Science & Emerging Technologies. 2005;**6**(4): 442-452

[119] Mohan C, Viswanatha G, Savinay G, Rajendra C, Halemani PD. 1, 2, 3, 4, 6 Penta-O-galloyl-β-d-glucose, a bioactivity guided isolated compound from Mangifera indica inhibits 11β-HSD-1 and ameliorates high fat diet-induced diabetes in C57BL/6 mice. Phytomedicine. 2013;**20**(5):417-426

[120] Ruiz-Montañez G, Ragazzo-Sánchez J, Calderón-Santoyo M, Velazquez-De La Cruz G, De León JR, Navarro-Ocaña A. Evaluation of extraction methods for preparative scale obtention of mangiferin and lupeol from mango peels (*Mangifera indica* L.). Food Chemistry. 2014;**159**:267-272

[121] Berardini N, Carle R, Schieber A. Characterization of gallotannins and benzophenone derivatives from mango (*Mangifera indica* L. cv.'Tommy Atkins') peels, pulp and kernels by high-performance liquid chromatography/ electrospray ionization mass spectrometry. Rapid Communications in Mass Spectrometry. 2004;**18**(19): 2208-2216

[122] Jiang LY, He S, Pan YJ, Sun CR. Bioassay-guided isolation and EPR-assisted antioxidant evaluation of two valuable compounds from mango peels. Food Chemistry. 2010;**119**(4):1285-1292. DOI: 10.1016/j.foodchem.2009.09.005

[123] Siddiqui MW, Lara I, Ilahy R, Tlili I, Ali A, Homa F, et al. Dynamic changes in health-promoting properties and eating quality during off-vine ripening of tomatoes. Comprehensive Reviews in Food Science and Food Safety. 2018; **17**(6):1540-1560

[124] Slimestad R, Verheul M. Review of flavonoids and other phenolics from fruits of different tomato (*Lycopersicon esculentum* Mill.) cultivars. Journal of the Science of Food and Agriculture. 2009;**89**(8):1255-1270. DOI: 10.1002/ jsfa.3605

[125] Sharma S, Le Maguer M. Lycopene in tomatoes and tomato pulp fractions. Italian Journal of Food Science: IJFS= Rivista italiana di scienza degli alimenti. 1996;**8**(2):107-113

[126] Martins N, Ferreira IC. Wastes and by-products: Upcoming sources of carotenoids for biotechnological purposes and health-related applications. Trends in Food Science & Technology. 2017;**62**:33-48

[127] Haq R-U, Prasad K. Physico-chemical, antioxidant and bioactive changes in cortex core sections of carrot (*Daucus carota* var. *Pusa rudhira*). Journal of Food Measurement and Characterization. 2016;**10**(3):701-708. DOI: 10.1007/s11694-016-9354-8

[128] Raees-ul H, Prasad K. Nutritional and processing aspects of carrot (*Daucus carota*)—A review. South Asian Journal of Food Technology and Environment. 2015;**1**(1):1-14

[129] Hager T, Howard L. Processing effects on carrot phytonutrients. HortScience. 2006;**41**(1):74-79

[130] Hashimoto T, Nagayama T. Chemical composition of ready to eat fresh carrot. Journal of the Food Hygienic Society of Japan. 2004;**39**: 324-328

[131] Nicolle C, Cardinault N, Aprikian O, Busserolles J, Grolier P, Rock E, et al. Effect of carrot intake on cholesterol metabolism and on antioxidant status in cholesterol-fed rat.

European Journal of Nutrition. 2003; **42**(5):254-261

[132] Howard FD, MacGillivray JH, Yamaguchi M. Nutrient composition of fresh California-grown vegetables. Bulletin of the California Agricultural Experiment Station. 1962;**788**

[133] Zhang D, Hamauzu Y. Phenolic compounds and their antioxidant properties in different tissues of carrots (*Daucus carota* L.). Journal of Food Agriculture and Environment. 2004;**2**: 95-100

[134] Prasad K, Kaushik V, Prasad KK. Garlic: A potential source of herbal medicine. Beverage and Food World. 2002;**29**(12):24-26

[135] Rivlin RS. Historical perspective on the use of garlic. The Journal of Nutrition. 2001;**131**(3):951S-954S

[136] Banerjee S, Mukherjee PK, Maulik S. Garlic as an antioxidant: The good, the bad and the ugly. Phytotherapy Research: An International Journal Devoted to Pharmacological and Toxicological Evaluation of Natural Product Derivatives. 2003;**17**(2): 97-106

[137] Lawson L. Human Medicinal Agents from Plants in Bioactive Organosulfur Compounds of Garlic and Garlic Products. Washington DC: American Chemical Society; 1994

[138] Lau BHS. Suppression of LDL oxidation by garlic. The Journal of Nutrition. 2001;**131**(3):985S-988S. DOI: 10.1093/jn/131.3.985S

[139] Török B, Belagyi J, Rietz B, Jacob R. Effectiveness of garlic on the radical activity in radical generating systems. Arzneimittel-Forschung. 1994;**44**(5): 608-611

[140] Augusti K. Therapeutic values of onion (*Allium cepa* L.) and garlic (*Allium sativum* L.). Indian Journal of Experimental Biology. 1996;**34**(7): 634-640

[141] Sampson L, Rimm E, Hollman PC, de VRIES JH, Katan MB. Flavonol and flavone intakes in US health professionals. Journal of the American Dietetic Association. 2002; **102**(10):1414-1420

[142] Slimestad R, Fossen T, Vågen IM. Onions: A source of unique dietary flavonoids. Journal of Agricultural and Food Chemistry. 2007;**55**(25): 10067-10080

[143] Grzelak K, Milala J, Król B, Adamicki F, Badełek E. Content of quercetin glycosides and fructooligosaccharides in onion stored in a cold room. European Food Research and Technology. 2009;**228**(6):1001-1007

[144] Gennaro L, Leonardi C, Esposito F, Salucci M, Maiani G, Quaglia G, et al. Flavonoid and carbohydrate contents in Tropea red onions: Effects of homelike peeling and storage. Journal of Agricultural and Food Chemistry. 2002; **50**(7):1904-1910

[145] Mohdaly AAA, Sarhan MA, Mahmoud A, Ramadan MF, Smetanska I. Antioxidant efficacy of potato peels and sugar beet pulp extracts in vegetable oils protection. Food Chemistry. 2010; **123**(4):1019-1026

[146] Zeyada NN, Zeitoum M, Barbary O. Utilization of some vegetables and fruit waste as natural antioxidants. Alexandria Journal of Food Science and Technology (AJFS). 2008;**5**:1-11

[147] Dias MG, Camões MFG, Oliveira L. Carotenoids in traditional Portuguese fruits and vegetables. Food Chemistry. 2009;**113**(3):808-815

[148] Patkai G, Barta J, Varsanyi I. Decomposition of anticarcinogen factors of the beetroot during juice and nectar production. Cancer Letters. 1997;**114**(1): 105-106

[149] Váli L, Stefanovits-Bányai É, Szentmihályi K, Fébel H, Sárdi É, Lugasi A, et al. Liver-protecting effects of table beet (*Beta vulgaris* var. rubra) during ischemia-reperfusion. Nutrition. 2007;**23**(2):172-178

[150] Cao G, Sofic E, Prior RL. Antioxidant capacity of tea and common vegetables. Journal of Agricultural and Food Chemistry. 1996;**44**(11):3426-3431

[151] Kujala TS, Loponen JM, Klika KD, Pihlaja K. Phenolics and betacyanins in red beetroot (*Beta vulgaris*) root: Distribution and effect of cold storage on the content of total phenolics and three individual compounds. Journal of Agricultural and Food Chemistry. 2000; **48**(11):5338-5342

[152] Kujala T, Loponen J, Pihlaja K. Betalains and phenolics in red beetroot (*Beta vulgaris*) peel extracts: Extraction and characterisation. Zeitschrift für Naturforschung C. 2001;**56**(5–6): 343-348

[153] Pedreño MA, Escribano J. Correlation between antiradical activity and stability of betanine from *Beta vulgaris* L roots under different pH, temperature and light conditions. Journal of the Science of Food and Agriculture. 2001;**81**(7):627-631

[154] Zou D-M, Brewer M, Garcia F, Feugang JM, Wang J, Zang R, et al. Cactus pear: A natural product in cancer chemoprevention. Nutrition Journal. 2005;**4**(1):25

[155] Delgado-Vargas F, Jiménez A, Paredes-López O. Natural pigments: Carotenoids, anthocyanins, and betalains—Characteristics, biosynthesis, processing, and stability. Critical Reviews in Food Science and Nutrition. 2000;**40**(3):173-289

Chapter 10

Cytotoxicity and Antitumor Action of Lignans and Neolignans

Ana Laura Esquivel-Campos, Salud Pérez-Gutiérrez, Leonor Sánchez-Pérez, Nimsi Campos-Xolalpa and Julia Pérez-Ramos

Abstract

Lignans and neolignans are plant's secondary metabolites, widely distributed in the plant kingdom, and have been identified in more than 70 plant families. These compounds are mainly localized in lignified tissues, seeds, and roots. Lignans and neolignans present a great variety of biological activities, such as antioxidant, anti-inflammatory, antineurodegenerative, antiviral, antimicrobial, and antitumor. By 2040, it is estimated that the number of new cancer cases per year will rise to 29.5 million; therefore, the development of new anticancer agents and adjuvants is essential. Lignans and neolignans have also indicated a reduction in the risk of cancer at different stages. The objective of this review is to search and analyze the cytotoxic and antitumor activity of lignans and neolignans that can be an important source of new antitumor drugs. We have made a comprehensive summary of 113 lignans and neolignans, obtained from 44 plants and divided between 34 families, which demonstrated cytotoxic activity in several human cancer cell lines evaluated through various in vitro studies and other in vivo models, by inducing mitochondrial apoptosis and cell cycle arrest, inhibiting NF-κβ activity and activation of metalloproteinases (MMPs), among other processes. Overall, 13 compounds, methoxypinoresinol, arctigenin, trachelogenin, 4-O-methylhonokiol, honokiol, bifidenone, (−)-trachelogeninit, deoxypodophyllotoxin, matairesinol, bejolghotin G, H, and I, and hedyotol-B, showed the best anticancer activity.

Keywords: Neolignans, cytotoxic activity, cancer, natural products

1. Introduction

Cancer produces uncontrolled cell proliferation, and one of the treatments used to stop it is chemotherapy. However, although these therapies have advanced over the years, they not only destroy cancer cells but also healthy cells, causing adverse effects in people suffering from this disease. A great variety of tumors are the cause of death in the population; the World Health Organization (WHO) reports that cancer causes approximately 10 million deaths each year, with one out of every six deaths

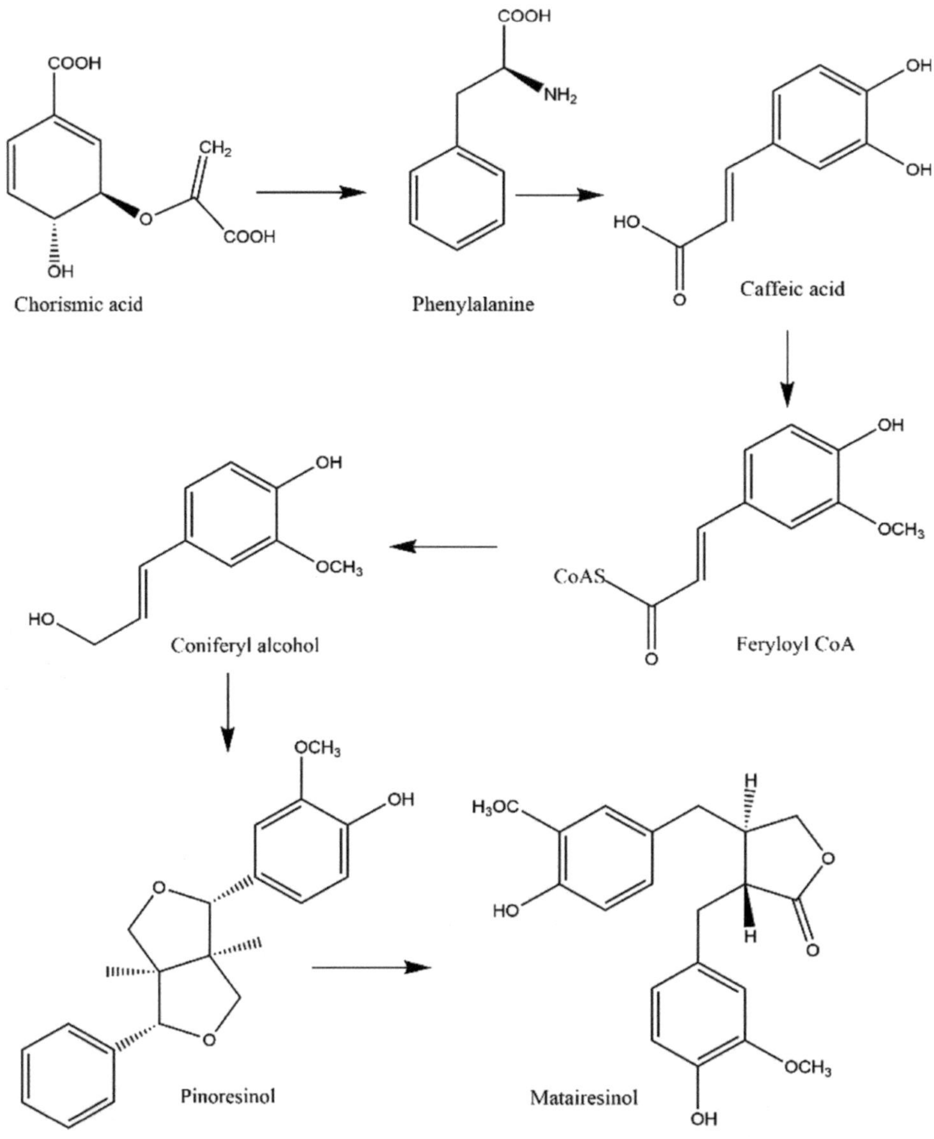

Figure 1.
Shikimic acid pathway for lignan and neolignan biosynthesis.

worldwide due to some type of cancer [1]. The main problem of this disease is that it is often detected at an advanced stage, and the lack of access to health services and the high cost of treatment are common, particularly in developing countries. The WHO suggests that 90% of the population in developed countries has access to treatment for this disease, while only 15% of the population in developing countries has access to treatment [2].

At present, the search for new chemotherapy drugs continues, with the purpose of having a wide range of compounds that help improve the quality of life of people with cancer. For many years, plants have played a very important role, as a source of compounds with biological activity. As a treatment alternative, multiple plant genera and species have demonstrated their cytotoxic potential in cancer cells and have been

Compound	Method	Results		Reference
1. 3-(1, 3-benzodioxol-5- yl methyl)-4-[(3, 4-dimethoxyphenyl)methyl] dihydro-, (3S-cis)-2(3H)- furanone **2.** 4-[(R)-1, 3-benzodioxol-5-ylhydroxymethyl]-3-(1, 3-benzodioxol-5-ylmethyl)dihydro-, (3S, 4R)-2(3H)-furanone **3.** (−)-Dihydrosesamin **4.** Phenol, 4, 4′-(2R, 3S, 4S)-tetrahydro2-methoxy-3, 4-furandiyl]bis(methylene)] bis[2-methoxy **5.** 4, 4′-dihydroxy-3, 3′, 9-trimethoxy-9, 9′-epoxylignan **6.** (+)-1-hydroxypinoresinol	MTT assay HL-60 SMMC-7721 A549 MCF-7 SW480	IC_{50} µM > 40		[11]
7. (+)-Nortrachelogenin **8.** -(3″-methoxy-4″-hydroxybenzyl)-3-(3′-methoxy-4′- hydroxylbenzyl)-γ-butyrolactone	MTT assay	IC_{50} µM		[12]
		(7)	(8)	
	A549	19.6	17.0	
	HepG2	17.6	15.1	
	U251	39.1	23.9	
	Bcap-37	51.6	50.3	
	MCF-7	45.6	25.3	
9. Sesamin (SE)	MTT assay	Cytotoxicity %		[13]
	MCF-7	23		
	Caco-2	15		
	CCK-8 assay in EL4 Cell apoptosis assay in EL4 lymphoma (EL4) induced in BALB/c mice	% Viability (40 µM) 50 to 80 (48, 72 y 96 h) SE Induced apoptosis by increased expression levels of apoptotic markers (Bax/Bcl-2) and cleaved-Caspase 3 SE decreased the size of tumor (10 mg/kg for 21 days)		[14]
10. Methoxypinoresinol	MTT assay PANC-1	IC_{50} µM 3.7		[15]
11. Erythro-austrobailignan-6 (EA6)	MTT assay 4 T-1 MCF-7 Western blot	IC_{50} µM (24 h) 4.3 12.6 EA6 increased the levels of p38 MAPK and caspase-3, in 4 T-1 and MCF-7		[16]
12. Mappiodoinin A **13.** Mappiodoinin B **14.** Mappiodoinin C **15.** Conocarpan **16.** Odoratisol A **17.** Trichobenzolignan **18.** Prunustosanan AI **19.** Simulanol **20.** Woorenogenin	MTT assay	IC_{50} µM		[9]
	HL-60	0.8–5.8		
	SMMC-7721	1.8–8.8		
	A-549	2.2–16.2		
	MCF-7	1.3–15.9		
	SW480	0.2–12.5		

Compound	Method	Results	Reference
21. Noralashinol B **22.** Noralashinol C	MTT assay HepG2	IC_{50} μM 21 / 22 31.7 / 15.8	[17]
23. Arctigenin (ATN)	MTT assay	IC_{50}μM	[18] [19]
	MCF-7	40.8	
	MCF-10A	24.1	
	SK-BR-3	20.7	
	MDA-MB-435S	3.8	
	MDA-MB-453	2.9	
	MDA-MB-231	0.8	
	MDA-MB-468	0.3	
	SRB assay in MCF-7 Colony formation assay. Cell cycle analysis by flow cytometry	At 200 μM arctigenin inhibited cell viability around 50%. ATN induced autophagy in MCF-7cells. The lignan might inhibit downstream effector molecules of the TOR resulting in a decreased expression of Erα in ER-positive MCF-7	
	Cell Count Reagent Western blot. JC-1 mitochondrial membrane potential	CC_{50} μM BC3 / BCBL1 2.8 / 2.3 ATN induced the caspase-9-mediated apoptosis of glucose-starved PEL cells (BC3). ATN induced mitochondrial disruption in glucose-starved BC3 cells by decreasing ATP levels and disrupting the mitochondrial membrane, and suppressed ERK and p38 MAPK signaling	[20]
24. Honokiol (HNK)	CCK-8 assay OC2 OCSL Apoptosis by annexin Xenograft nude mice model	GI_{50} μM at 48 h 22 13 This compound induced apoptosis cell death HNK had antitumour activity	[21]
	MTT assay	IC_{50} μg/mL	[22]

Compound	Method	Results	Reference
	KKU-213 L5 Apoptosis by Muse™ Cell Analyzer Western blot Flow cytometer analysis	24 (h)	48 (h) 50.0	26.3 % apoptosis 50 μM	70 μM 30.4	52.0 HNK increased apoptosis by decrease of intact caspase-3, whereas cleaved caspase-3 increased The antitumor activity of dendritic cells (DC) is increased using a lysate derived from a cell line (KKU-2113 L5) treated with HNK HNK increased antitumor activity of DCs stimulated with cell lysates derived from KKU-213 L5	
25.1-(2′,6′-dimethoxy-7′,8′-peroxyphenylpropyl)-2,10-dimethoxybibenzyl-6,9′-diol **26.** Aloifol I **27.** Moscatilin **28.** Moniliformine **29.** Balanophonin	MTT assay HL-60	IC_{50} μM 25	26	27	28	29 4.5	4.5	5.1	10.7	11.0	[23]
30. (−)-Trachelogenin (TA)	MTT assay HL-60 OVCAR-8 HCT-116 HCT-8 PC-3 SF-295 Membrane integrity and viability by the exclusion of propidium iodide	IC_{50} μM 32.4 3.5 1.9 5.2 15.0 0.8 TA did not induce apoptosis, but it was induced by autophagic death mediated by the increase of LC3 activation. Also promoted changes in the expression of Beclin-1 levels	[24]
31. 4-O-methylhonokiol (MH)	MTT assay OSCC PE/CA-PJ41	IC_{50} μM 1.3	[25]
32. Bifidenone (BF)	Sequoia Sciences Assay NCI-H460 Caspase-Glo 3/7 assay LDH assay Tubulin Polymerization assay Tubulin	IC_{50} μM 0.26 BF increased the levels of caspase (2.5-fold) BF increased the level of LDH released BF inhibits tubulin polymerization in a dose-dependent manner BF interfered with mitosis by	[26]

Compound	Method	Results	Reference
	competition assay PC-3 SF-295 ACHN	disrupting the microtubule dynamics necessary for cell division IC_{50} μM 0.49 0.25 0.36	
	M14 A375 UACC-62 SKMEL-2 HCC-2998	0.064 0.075 0.044 0.095 1.41	
33. (+)-Hinokinin	WST-8 Assay PANC-1 MIA-PaCa2 CAPAN-1 SN-1 KLM-1	PC_{50} μM 64.1 21.3 50.1 60.1 92.5	[27]
34. (−)-Deoxypodophyllotoxin (DPT)	MTT assay U2OS Annexin-V/ propidium iodide (PI) assay Acridine orange assay	IC_{50} nM 40 DPT induced apoptosis related with proteins Annexin-V positive cells were increased in DPT-treated cells, compared with control group. Formation of acidic vesicular organelles (AVOs) was significantly increased in DPT-treated cells in a dose-dependent manner	[28]
35. Lariciresinol (LA)	CCK-8 assay HepG2 Flow cytometry Immuno-fluorescence staining Annexin V/PI double-staining assay Mitochondrial membrane potential (ΔΨm)	IC_{50} μg/mL 208 after 48 h LA exhibited an apoptosis-inducing effect LA decreased Ki-67 expression and induced apoptosis LA was a concentration- and time-dependent manner resulted in an increasing percentage of apoptosis, which might result in the cytotoxic activity of LA on HepG2 cells LA might induce HepG2 cell apoptosis through the mitochondrial-mediated apoptosis pathway	[29]
36. Burserain **37.** Picropolygamain	MTT assay HeLa	IC_{50} μM 36: 21.7 37: 9.1	[30]

Compound	Method	Results			Reference
38. Heilaohulignan C **39.** Kadsuralignan I **40.** Longipedunin B	MTT assay	IC_{50} μM			[31]
		38 39 40			
	HepG2	9.9	21.7	18.7	
	BGC-823	16.6	—	—	
	HCT-116	16.7	—	—	
41. (−)-(7′S,8S,8′R)- 4,4′-dihydroxy-3,3′,5,5′-tetramethoxy-7′,9-epoxylignan-9′-ol-7-one **42.** Burseneolignan **43.** (8R)-3,5′-dimethoxy-8,3′-neoligna-4,4′,9,9′-tetraol	MMP-9 assay	IC_{50} μM			[32]
		41	42	43	
		16.5	18.8	8.7	
44. Oryzativol C	Ez-Cytox cell kit MDA -MB -231	IC_{50} μM 24.8			[33]
45. (−)-Asarinin	MTT assay A2780 SKoV3 Annexin V-FITC/PI Double Staining	IC_{50} μM 38.4 60.9 This compound might induce apoptotic cell death in human ovarian cancer cells			[34]
46. Balanophonin **47.** Dehydrodiconiferyl (DDI) **48.** Methoxyl-balanophonin	MTT assay	IC_{50} μM			[35]
		46	47	48	
	HepG2	36.5	78.6	80.5	
	Hep3B	29.3	65.5	76.8	
	Flow cytometry				
		DDI induced apoptosis			
49. Dehydrodieugenol B **50.** Methyldehydrodieugenol B (MEB)	MTT assay	IC_{50} μg/mL			[36]
		50	51		
	SKMEL-147	4.4	43.6		
	Comet Assay CBMN on SKMEL-29	100% of apoptosis	25% of apoptosis		
		MEB increased DNA damage by cytokinesis			
51. (−)-Rabdosiin	MTT assay	IC_{50} μg/mL			[37].
	MCF-7 SKBR3 HCT-116	75 83.0 84.0			
	Flow Cytometry	% of apoptosis			
	MCF-7 SKBR3 HCT-116	44.9 40.1 43.1			
52. Kalshiolin A	SRB assay A549 MDA-MB-231 MCF-7 KB KB-VIN	IC_{50} μg/mL 35.9 to 43.3			[38]

Compound	Method	Results			Reference
34. (−)-Deoxy podophyllotoxin **53.** (−)-Mataireslnol	SRB assay NB	IC_{50}			[39]
		34	53		
		1.7 ng/mL	3.7 μg/mL		
54. Phengustifols A	CCK-8 assay A375	IC_{50} μM 12.1			[40]
55. Hedyotol-B	MTT assay SGC7901 A549 MDA-MB-231 HepG2	IC_{50} μM 1.7 6.1 24.0 26.0			[41]
56. Heilaohusus C	MTT assay HepG2	IC_{50} μM			[42]
		13.0			
57. Zijusesquilignan A **58.** Zijusesquilignan B **59.** Zijusesquilignan C	MTT assay	IC_{50} μM			[43]
		57	58	59	
	MCF-7	9.8	8.8	8.4	
	HL-60	11	—	—	
60, 61. Crataegifin B (enantiomers) **62.** CrataegifinC	MTT assay	IC_{50} μM			[44]
		60	61	62	
	Hep3B	25.5	59.4		
	HepG2	—	—	34.3	
	Flow cytometry	Compound 61 at 25 μM induced apoptosis in Hep3B cell in 10.76%			
63. Bejolghotin A **64.** Bejolghotin B **65.** Bejolghotin C **66.** Bejolghotin G **67.** Bejolghotin H **68.** Bejolghotin I	MTT assay	IC_{50} μM			[45]
	HCT-116	0.8–39.9			
	A549	0.9–39.9			
	MDA-MB-231	0.8–45.6			
54. (−)-Matairesinol **23.** Arctigenin **34.** (−)-Deoxypodophyllotoxin	MTT assay	IC_{50} μg/mL			[46]
		54	23	34	
	MDA-MB-231b	—	1.1	0.07	
	A549	—	0.8	0.004	
	HepG2	15.1	2.8	—	
69. Niranthin **70.** 7-hydroxy- hinokinin	MTT assay HepG2	IC_{50} μM			[47]
		69	70		
		7.2	8.5		
71. Cleistonkinin A **72.** Cleistonkinin B **73.** Cleistonkinin C **74.** Cleistonkinin D **75.** Cleistonkinin E	MTT assay	IC_{50} μM			[48]
	A549	>20			
	PANC-1	>20			
	HeLa	>20			

Compound	Method	Results			Reference
76. Cleistonkiside A **77.** Cleistonkiside B	Hep3B	>20			
	MCF-7	>20			
78. Crataegusal A **79.** Crataegusal A	MTT assay Hep3B	IC_{50} μM			[49]
		78	79		
		34.97	17.42		
80. Miliusin A **81.** Miliusin B **82.** Miliusin 7R,8S **83.** Miliusin C **84.** Miliusin D **85.** Miliusin E **86.** Miliusin F	MTT assay	IC_{50} (μM)			[50]
	HeLa	0.2–18			
	HN22	0.2–43.1			
	HepG2	2.9–88.5			
	HCT116	4.5–107.5			
87. Pleiocarpumlignan B	MTS assay MCF-7	IC_{50} μM 18.2			[51]
88. Officinalioside (OFD)	MTT assay HepG2	OFD showed cytotoxic effect at 50 μmol/L and 100 μmol/L			[52]
89. 5-((E)-2-carboxyvinyl)-7-methoxy-2-(3′,4′-methylenedioxyphenyl) Benzofuran **90.** Egonol **91.** (−)-Machicendiol	MTT assay	IC_{50} μM			[53]
		89	90	91	
	KB	96.0	22.1	33.5	
	HepG2	86.6	18.1	31.5	
	Lu	106.9	21.5	22.2	
92. Schisphenlignan M **93.** Schisphenlignan N **94.** Gomisin G **95.** Schisantherin D **96.** Schisantherin A **97.** Epigomisin O **98.** (+)-omisin K3 (Schisanhenol) **99.** Schisanhenol B **100.** Gomisin A	MTT assay A549 HCT116 SW620	IC_{50} μM 13.5 to >50			[54]
101. Glalignin B **102.** Glalignin C **103.** Glalignin E **104.** Glaneolignin A **105.** Dihydrodehydro diconiferyl alcohol **106.** Tribulusamide A	MTT assay A549	IC_{50} μM			[55]
		13.5–100			
	HeLa	20.1–79.9			
	MCF-7	11.4–100			
107. Pinoresinol monomethyl ether-β-D-glucoside (PMG)	MTT assay HeLa MDA-MB-231	IC_{50} μg/mL 10.1 (24 h) and 3.54(48 h) >250 (24 and 48 h)			[56]
108. Methylcubebin (MB) **109.** Cubebin (CB) **110.** Dyhydrocubebin (DB) **111.** Ethylcubebin (EB)	MTT assay HEp-2 SCC-25 Transwell cell migration assay	MB and CB decreased cell proliferation at concentrations of 10 and 50 μg/mL DB, EB, and MB decreased cell migration			[57]
112. (1S,2S)-1-(4-hydroxy-3-methoxyphenyl)-2-[2-methoxy-4-[(2S,3R,	MTT assay HL-60 A549	IC_{50} μM 8.2 15.1			[58]

Compound	Method	Results	Reference
4R)-tetrahydro-4-[(4-hydroxy-3-methoxyphenyl)methyl]-3-(hydroxymethyl)-2-furanyl] phenoxy]-1,3-propanediol (MFP)	SMMC-7721 MCF-7 SW480 Flow cytometry	10.6 4.4 16.1 MFP induced dose-dependent apoptosis in MCF-7 cells	

Abbreviations: PC50: Preferential cytotoxicity mean Concentration; IC50 Inhibitory mean Concentration; CC50: cytotoxic effects; GI50:Growth inhibition; LDH deshidrogenase lac tate; MTT: 3-(4,5-dimethylthiazol-2-yl)-2,5-diphenyltetrazolium bromide; SRB: Sulforhodamine B; CCK-8: The Cell Counting Kit 8 assay; CBMNCyt: cytokinesis block micronucleus; MMP-9: Matrix metalloproteinase 9; LC3: a process that involved the bulk degradation of cytoplasmic components (positive structures are prominent in autophagy-deficient); MAPK: protein kinase; ERK: extracellular signal-regulated kinase; MYCN: proto-oncogene; MYCN2: human neuroblastoma cell with MYCN amplification; pCNA nuclear antigen of cell proliferation; STATs: Signal transducers and activators of transcription; JC-1: mitochondrial membrane assay.
Human cancer cell lines: A2780, SKOV3, OVCAR-8: ovarian; A549, NCI-H460: lung; BGC-823, SGC7901: gastric cancer; Caco-2, HCC-2998, HCT-16, HCT-116, HCT-8, SW480, SW620: colon cancer; HeLa: human cervical uterine cancer; KB, KBVIN: papillomavirus; Bcap-37: endocervical adenocarcinoma; Hep3B, HepG2, SMMC 7721: hepatocellular carcinoma; KKU-213 L5: cholangiocarcinoma; HEp-2: laryngeal cancer; HL-60: promyelocytic leukemia; SN-1: leukemia; HN22: head and neck squamous cell carcinoma; TNBC, MCF-10A, MCF-7, MDA-MB-468, MDA-MB-453, MDAMB-231, SK-BR-3: breast cancer; NB: neuroblastoma; SKMEL-147: wild-type human melanoma; SKMEL-29: human melanoma carrying the B-Raf mutation-V600E; SKMEL-2, A375: malignant melanoma skin; M14, UACC-62: melanoma; OC2, SCC-25, OSCC: squamous cell carcinoma; Lu carcinoma; MIA-PaCa2, CAPAN-1, KLM-1 PANC-1: pancreatic cancer; PC-3: prostate cancer; SF-295, U251: glioblastoma; ACHN: renal cancer; U2OS: osteosarcoma; BCBL1: lymphoma cells; muscular cancer cell lines 4 T-1.

Table 1.
Anticancer activity of lignans and neolignan isolated of different plants.

used in traditional medicine in many countries as anti-inflammatory and antirheumatic agents, among others, as well as antirhythmic and antitumor agents, since they inhibit cell proliferation and induce cytotoxicity in a large number of cell lines, as demonstrated through research [3].

Lignans are a group of secondary metabolites found in different plant and animal species. Lignans are biologically synthesized from the shikimic acid pathway [4] and through different reactions (**Figure 1**). Despite their structural variety, lignans are dimers of phenylpropanoid units that are linked via their β-carbon atoms [5]. Dimers of phenylpropanoid units that are coupled via other linkages are named neolignans [6]. The lignan family is classified into the following eight classes, based on how oxygen is incorporated into the skeleton and the cyclization pattern: furofuran, furan, dibenzylbutane, dibenzylbutyrolactone, aryltetralin, arylnaphthalene, dibenzocyclooctadiene, and dibenzylbutyrolactol. The neolignans have structural variety and are divided into more than 15 groups, some of them are: benzofuran, dihydrobenzofuran, diarylethane, benzodioxine, alkyl aryl ether, and bicycloctane derivatives, among others [7]. These metabolites present different biological activities, such as cytotoxicity; as an example, podophyllotoxin is used in cancer treatments today [8].

In this sense, Jiang and col. [9] have suggested that this behavior is not the same with all cell lines, where tested, and that it depends on the type of lignan for its cytotoxicity. Multiple lignans are being studied, particularly for their effectiveness against breast cancer. Because they bind to cells where there are estrogen deposits, they have been shown to be effective against breast cancer [10]. The cytotoxic activity of various lignans has also been studied on colon, pancreatic, throat, and oral cancers,

Wikstroemia scytophylla		
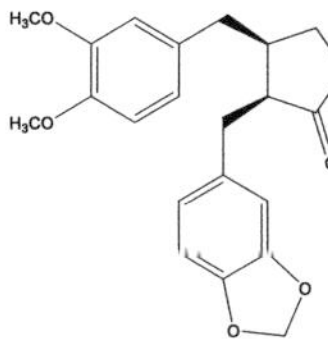 **1.** 3-(1, 3-benzodioxol-5-ylmethyl)-4-[(3, 4-dimethoxyphenyl)methyl] dihydro-, (3S-cis)-2(3H)- furanone	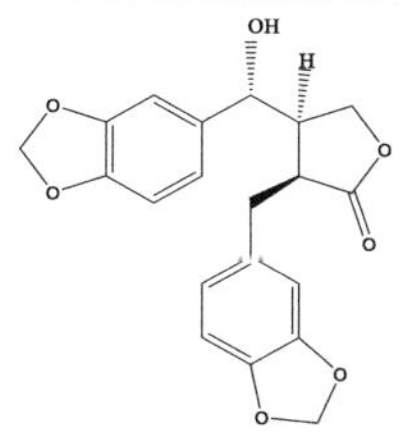 **2.** 4-[(R)-1, 3-benzodioxol-5-ylhydroxymethyl]-3-(1, 3-benzodioxol-5-ylmethyl)dihydro-, (3S, 4R)-2(3H)-furanone	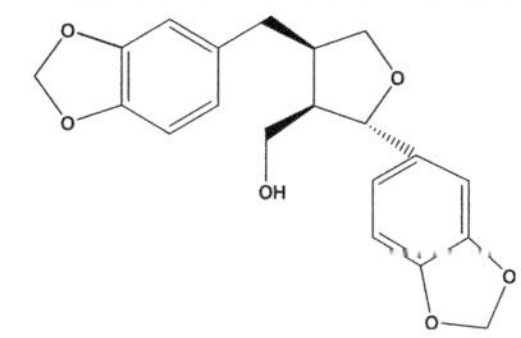 **3.** (−)-Dihydrosesamin
Wikstroemia scytophylla		
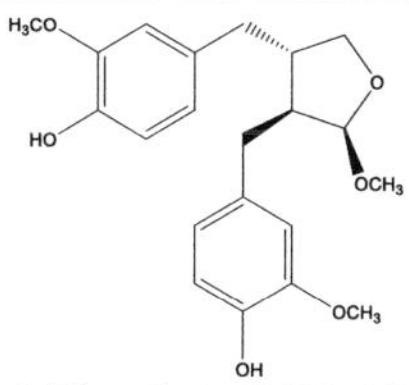 **4.** Phenol, 4, 4′-(2R, 3S, 4S)-tetrahydro2-methoxy-3, 4-furandiyl]bis(methylene)]bis[2-methoxy	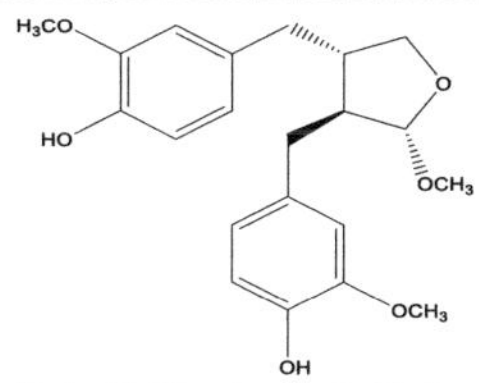 **5.** 4, 4′-dihydroxy-3, 3′, 9-trimethoxy-9, 9′-epoxylignan	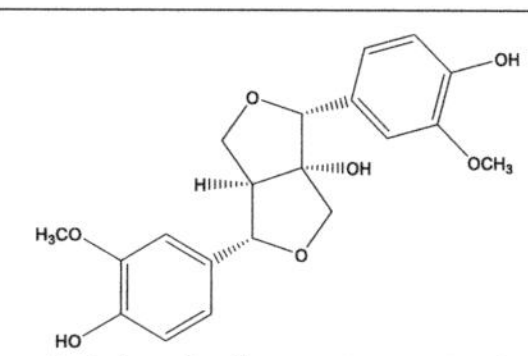 **6.** (+)-1-hydroxypinoresinol
Bupleurum chinense		*Zanthoxylum capense, Sesamun Virola; Piper sp, Camellia sp, Magnolia sp*
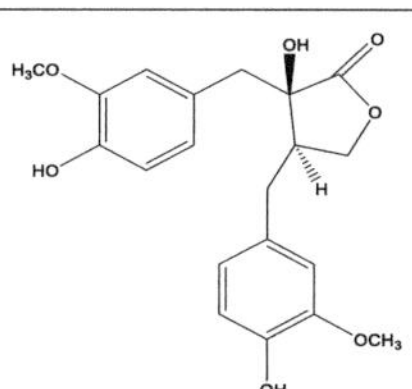 **7.** (+)-Nortrachelogenin	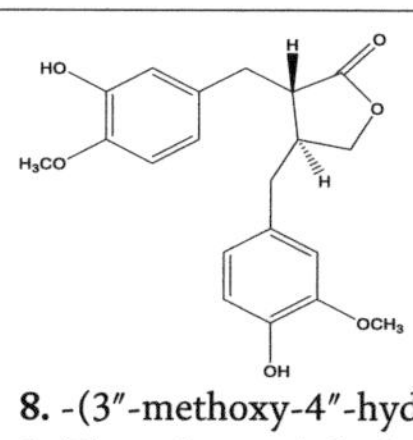 **8.** -(3″-methoxy-4″-hydroxybenzyl)-3-(3′-methoxy-4′- hydroxylbenzyl)-γ-butyrolactone	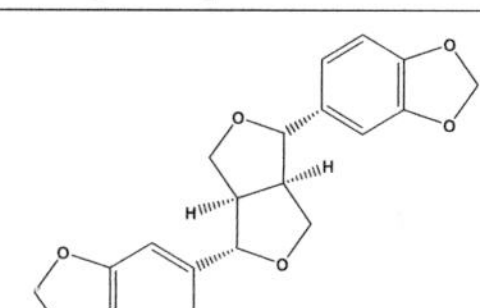 **9.** Sesamin (SE)
Calotropis gigantea	*Saururus chinensis*	*Mappianthus iodoizes*
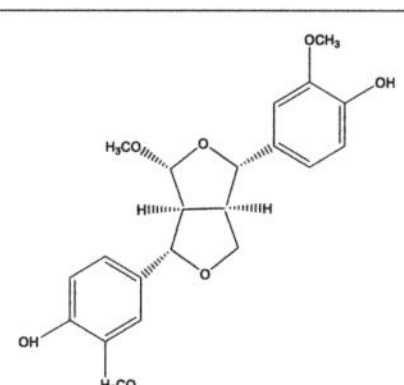 **10.** Methoxypinoresinol	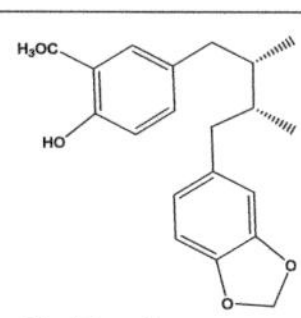 **11.** Erythro-austrobailignan-6 (EA6)	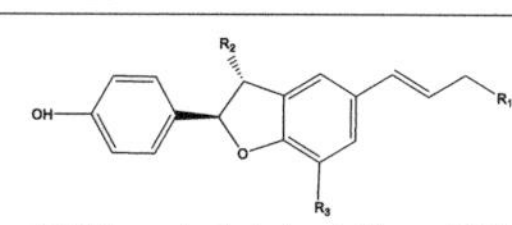 **12.**Mappiodoinin A R_1 = OH, R_2 = H, R_3 = H **13.** Mappiodoinin B R_1 = OCH_3, R_2 = H, R_3 = H **14.** Mappiodoinin C R_1 = COH, R_2 = OH, R_3 = OCH_3 **15.** Conocarpan R_1 = H, R_2 = H, R_3 = H **16.** Odoratisol A R_1 = OH, R_2 = OH, R_3 = H

Wikstroemia scytophylla		
Mappianthus iodoizes		*Syringa pinnatifolia*
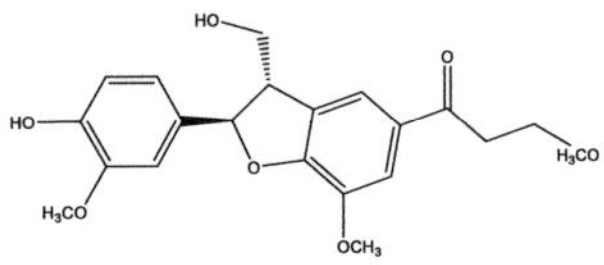 **17.** Trichobenzolignan	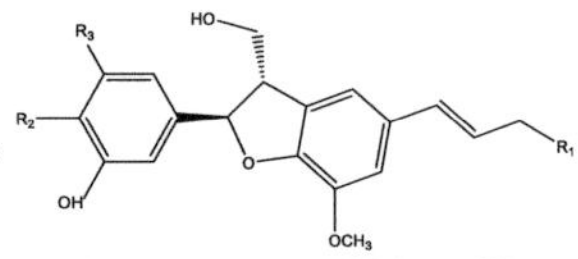 **18.** Prunustosanan AI R_1 = OH, R_2 = OCH_3, R_3 = OCH_3 **19.** Simulanol R_1 = OCH, R_2 = OH, R_3 = H **20.** Woorenogenin R_1 = OCH, R_2 = OCH_3, R_3 = H	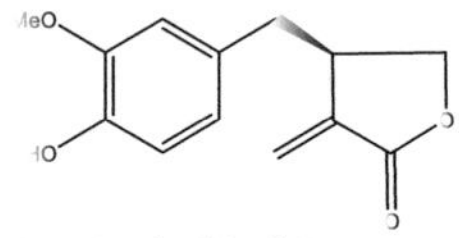 **21.** Noralashinol B
Syringa pinnatifolia	*Arctium lappa, Cupressus macrocarpa*	*Magnolia officinalis*
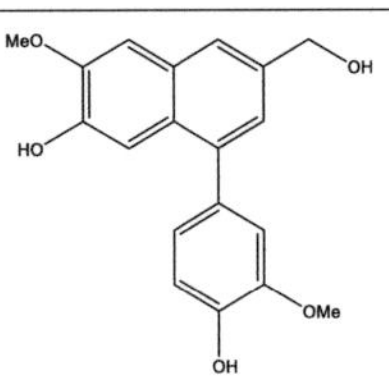 **22.** Noralashinol C	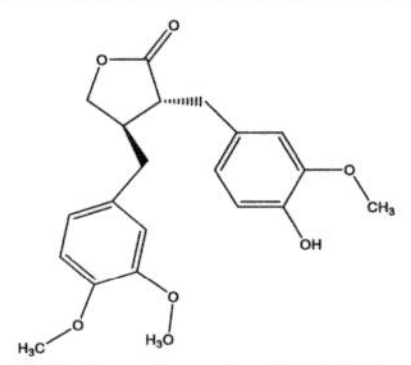 **23.** Arctigenin (ATN)	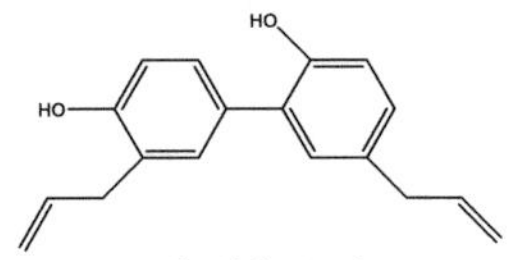 **24.** Honokiol (HNK)
Dendrobium williamsonii		
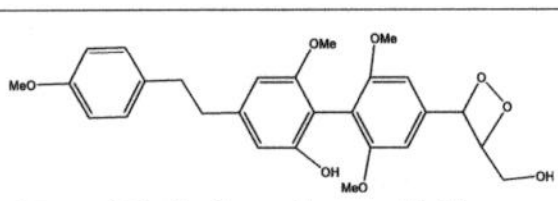 **25.** 1-(2′,6′-dimethoxy-7′,8′-peroxyphenylpropyl)-2,10-dimethoxybibenzyl-6,9′-diol	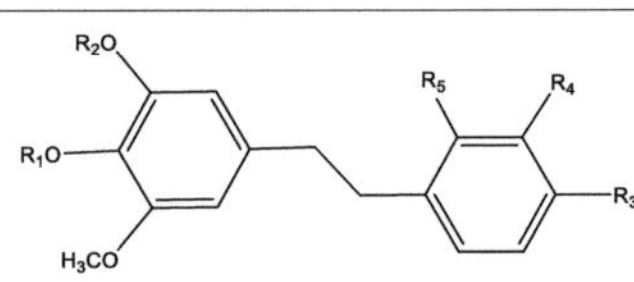 **26.** Aloifol I,R_1 = R_3 = R_4 = H, R_2==CH_3, R_5 = OH **27.** Moscatilin,. R_1 = R_5 = H, R_2 = CH_3, R_3 = OH **28.** Moniliformine R_1 = R_2 = R_4 = R_5 = H, R_3 = OCH_3	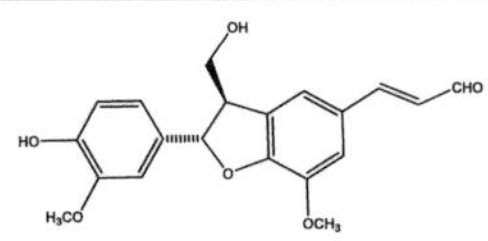 **29.** Balanophonin
Combretum fruticosum	*Magnolia officinalis*	*Beilschmiedia sp*
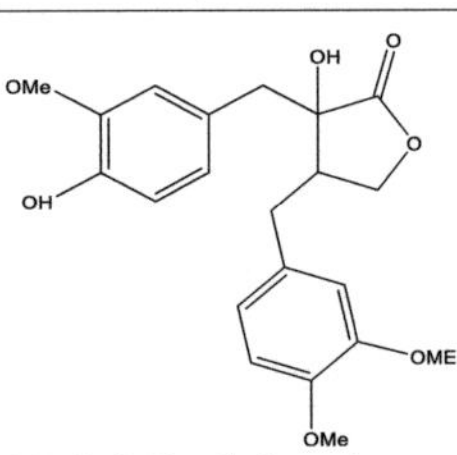 **30.** (−)-Trachelogenin	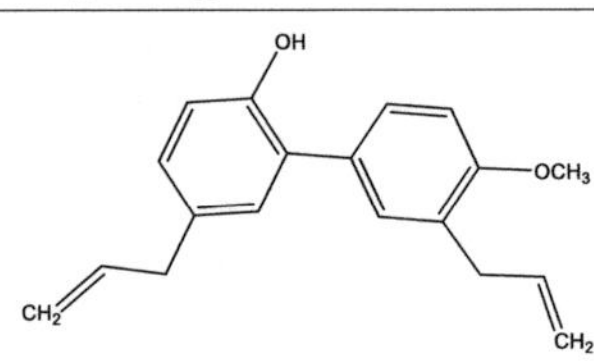 **31.** 4-O-methylhonokiol (MH)	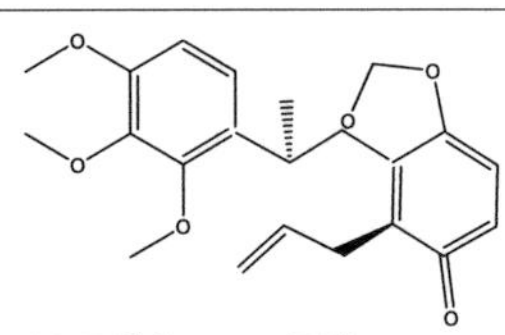 **32.** Bifidenone (BF)
Chamaecyparis obtusa	*Anthriscus sylvestris, C. macrocarpa*	*Patrinia scabra*
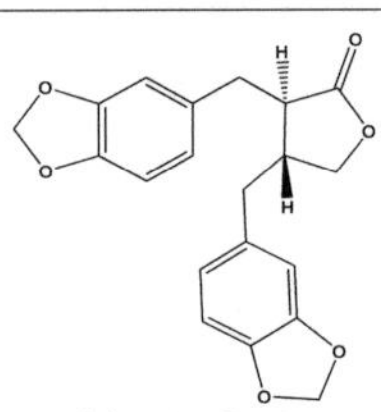 **33.** (+)-Hinokinin	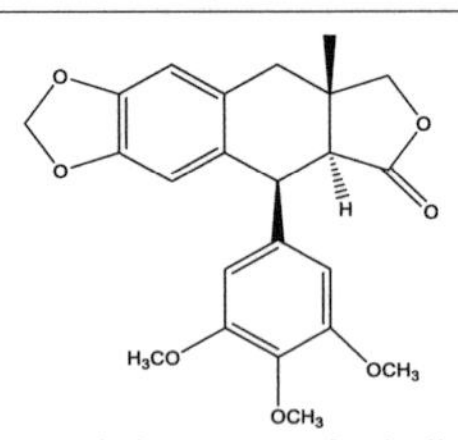 **34.** (−)-Deoxypodophyllotoxin (DPT)	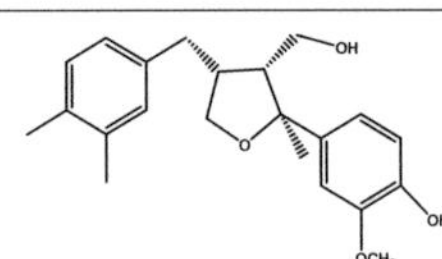 **35.** Lariciresinol (LA)

Wikstroemia scytophylla		
Bursera microphylla		*Kadsura coccinea*
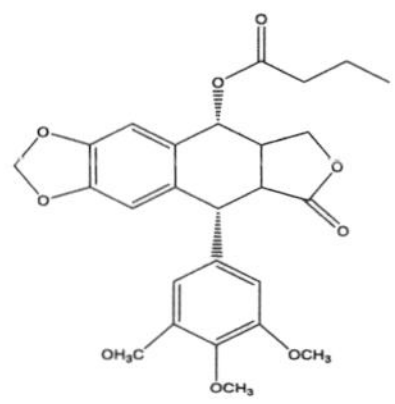 **36.** Burserain	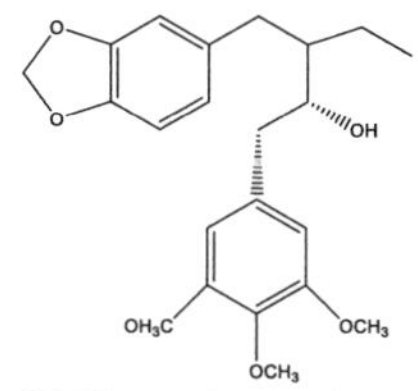 **37.** Picropolygamain	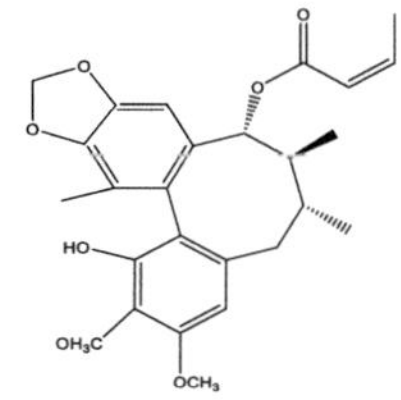 **38.** Heilaohulignan C
Kadsura coccinea	*Selaginella moellendorffii*	
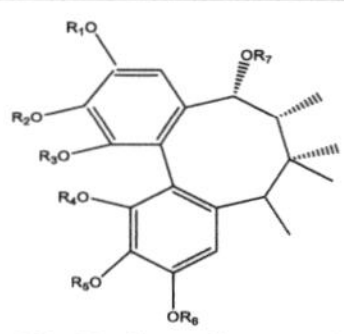 **39.** Kadsuralignan I $R_1 + R_2 = CH_2$; $R_3 = R_5 = R_6 = CH_3$; $R_7 = OH$; $R_4 = OAng$ **40.** Longipedunin B$R_1 + R_2 = CH_2$; $R_3 = R_5 = R_6 = CH_3$; $R_4 = OH$; $R_7 = OProp$	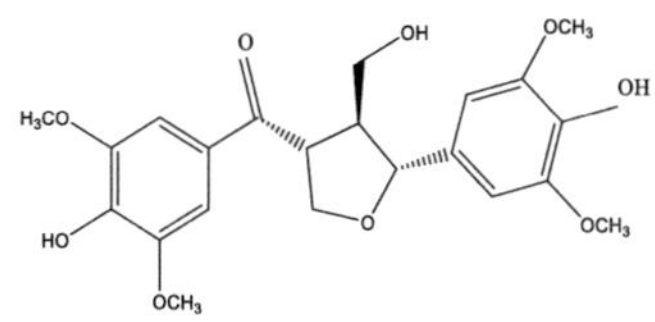 **41.** (−)-(7′S,8S,8′R)- 4,4′-dihydroxy-3,3′,5,5′-tetramethoxy-7′,9-epoxylignan-9′-ol-7-one	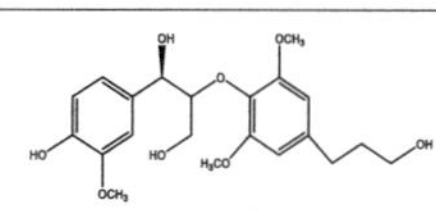 **42.** Burseneolignan
Selaginella moellendorffii	*Oryza sativa*	*Asarum sieboldii*
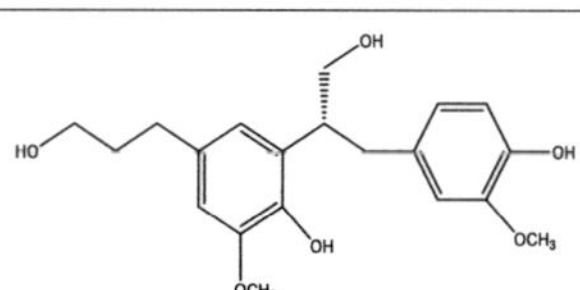 **43.** (8R)-3,5′-dimethoxy-8,3′-neoligna-4,4′,9,9′-tetraol	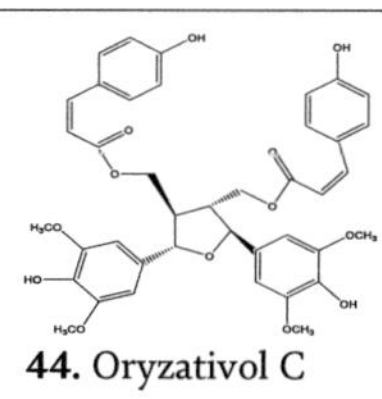 **44.** Oryzativol C	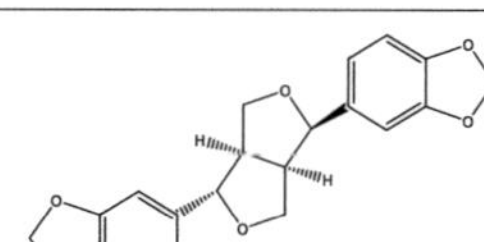 **45.** (−)-Asarinin
Picrasma quassioides	*Nectandra leucantha*	*Ocimum sanctum*
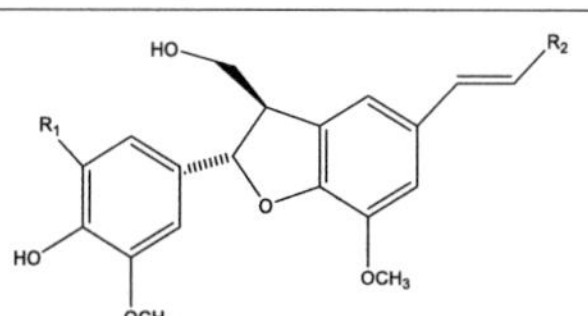 **46.** $R_1 = H$; $R_2 = CHO$ **47.** $R_1 = H$; $R_2 = OH$ **48.** $R_1 = OCH_3$; $R_2 = CHO$	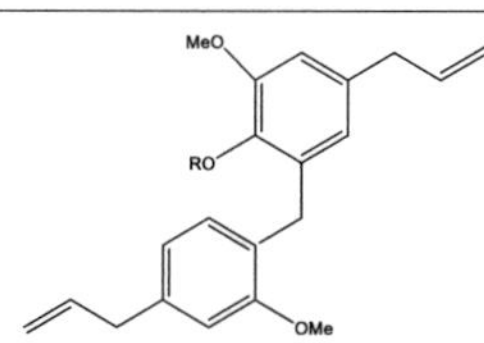 **49.** R = H Dehydroeugenol **50.** R = Me methyl	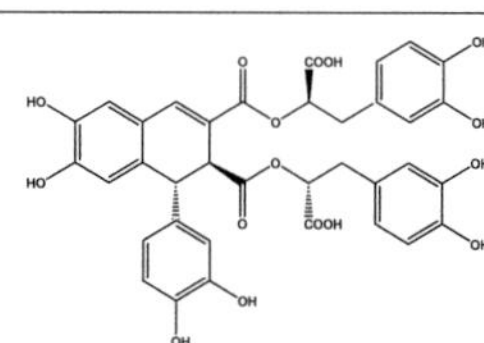
46. Balanophonin, **47.** Dehydrodiconiferyl (DDI). **48.** Methoxyl-balanophonin	**49.** Dehydrodieugenol B **50.** Methyldehydrodieugenol B (MEB)	**51.** (−)-Rabdosiin

Wikstroemia scytophylla		
Kalimeris shimadaies	*C. macrocarpa*	*Elaeagnus angustifolia*
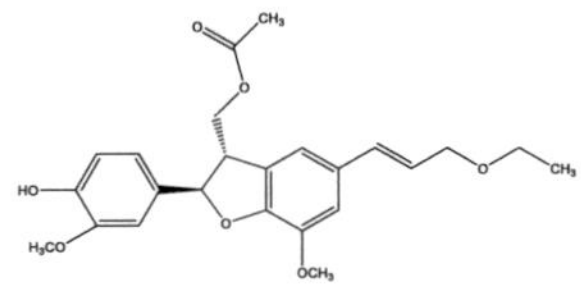 **52.** Kalshiolin A	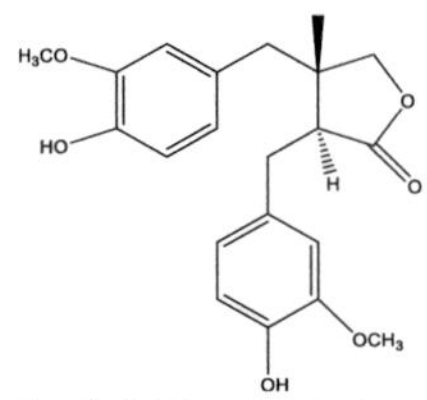 **53.** (−)-Matairesinol	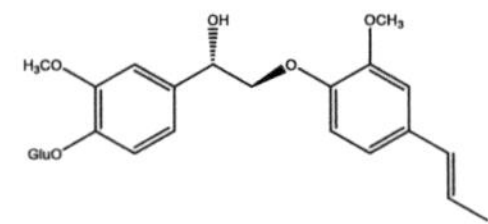 **54.** Phengustifols A
Herpetospermum pedunculosum	*Kadsura coccinea*	*Ziziphus jujuba*
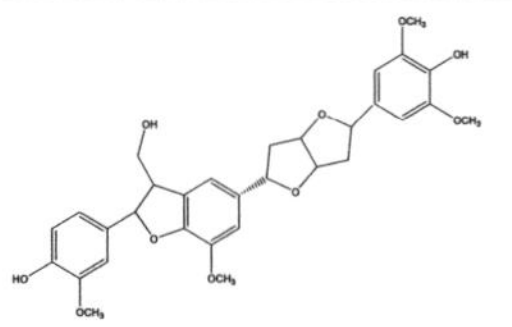 **55.** Hedyotol-B	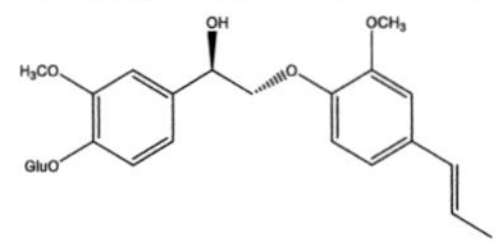 **56.** Heilaohusus C	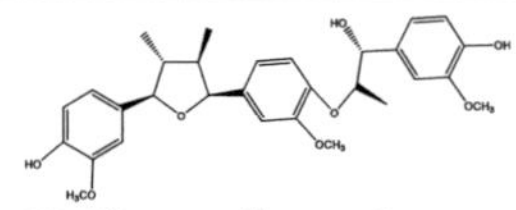 **57.** Zijusesquilignan A
Z. jujuba	*Crataegus pinnatifida*	
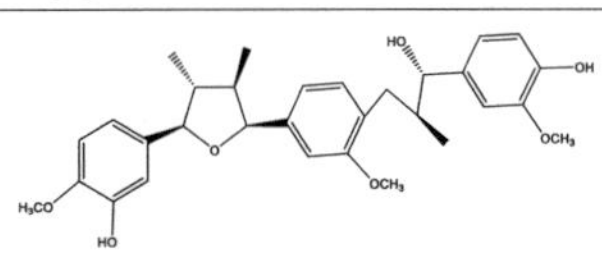 **58.** Zijusesquilignan B	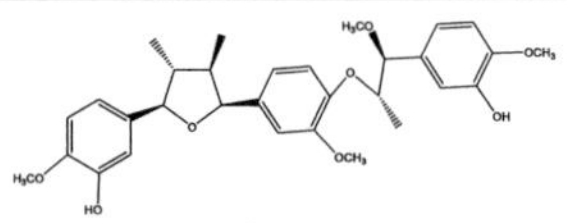 **59.** Zijusesquilignan C	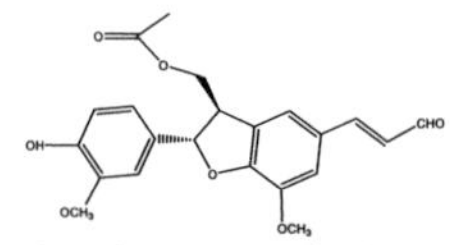 **60–61.** Crataegifin B
C. pinnatifida	*Cinnamomum bejolghota*	
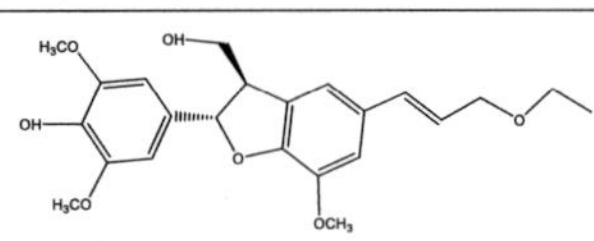 **62.** Crataegifin C	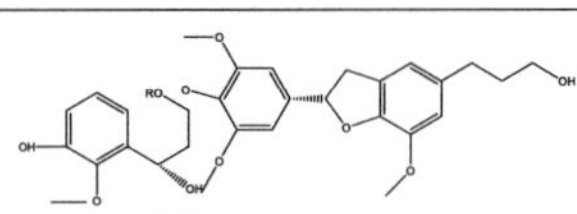 **63.** Bejolghotin A	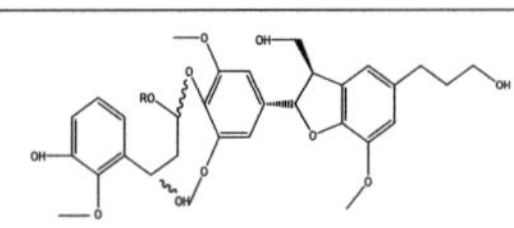 **64.** Bejolghotin B 7”S, 8”R R = E-Feruloyl **65.** Bejolghotin C 7”R, 8”R R = E-Feruloyl
Cinnamomum bejolghota		*Euphorbia hirta*
R = E-Feruloyl		
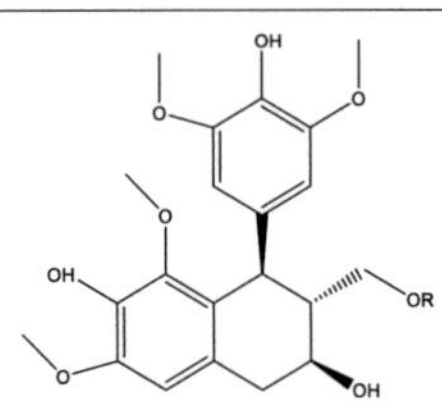 **66.** Bejolghotin G	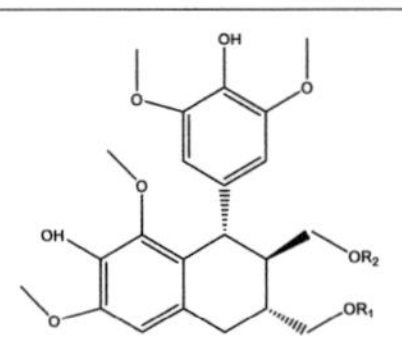 **67.** Bejolghotin H R_1 = E-Feruloyl R_2 = H **68.** Bejolghotin I R_1 = R_2 = E-Feruloyl	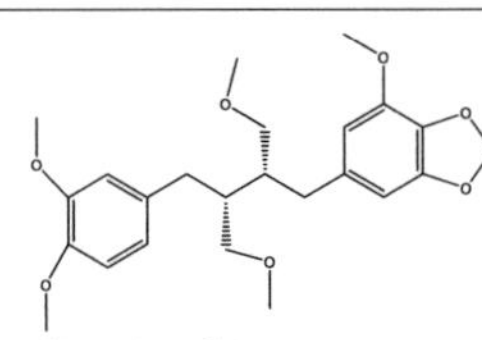 **69,** Niranthin

Wikstroemia scytophylla		
E. hirta	*Cleistanthus tonkinensis*	
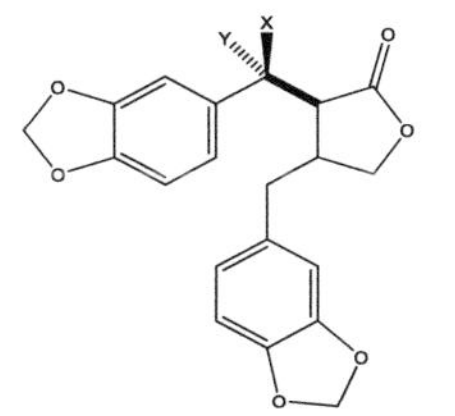 **70.** 7-hydroxy- hinokinin	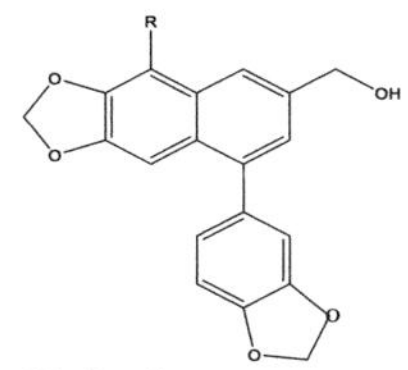 **71.** *R = H* **72.** *R = OCH_3* **71.** Cleistonkinin A **72.** Cleistonkinin B	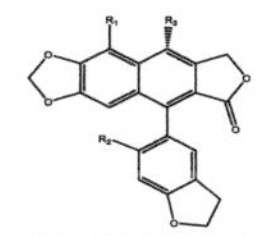 **73.** Cleistonkinin C R_1 = OCH3, R_2 = H, R_3 = H **74.** leistonkinin D R_1 = OCH3, R_2 = H, R_3 = OH **75.** Cleistonkinin E R_1 = H, R_2 = OH, R_3 = OH
Cleistanthus tonkinensis		*C. pinnatifida*
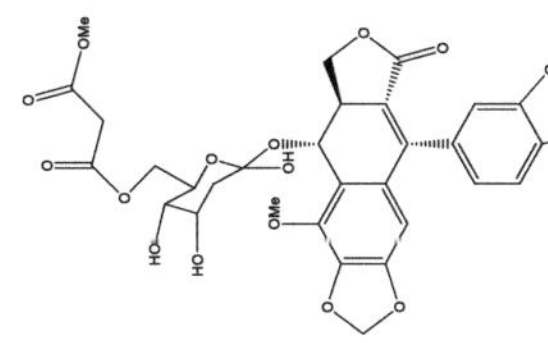 **76.** Cleistonkiside A	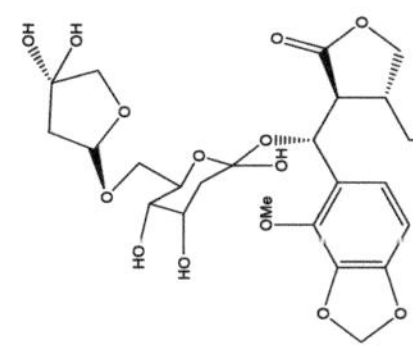 **77.** Cleistonkiside B	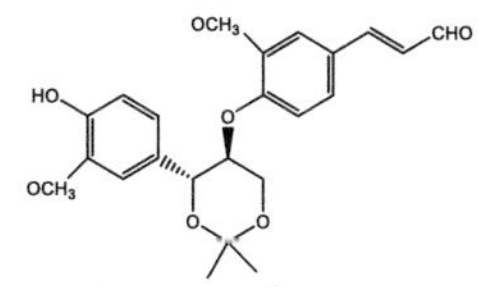 **78.** Crataegusal A
C. pinnatifida	*Miliusa sessilis*	
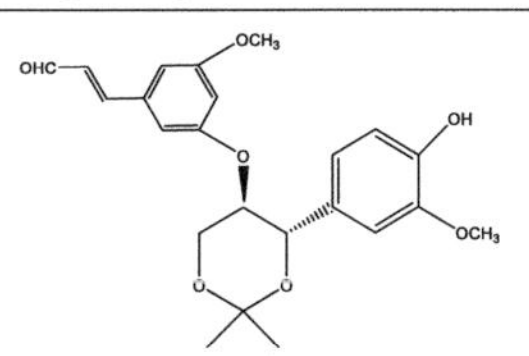 **79.** Crataegusal A	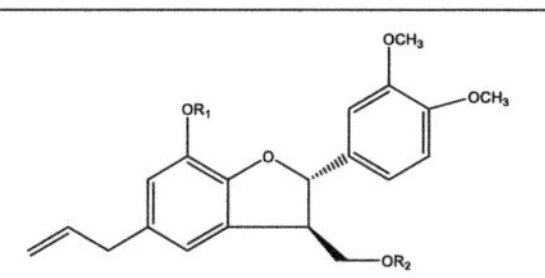 **80.** Miliusin AR_1 = H, R_2 = AC **81.** Miliusin BR_1 = H, R_2 = H **82.** Miliusin 7R,8SR_1 = CH_3, R_2 = H	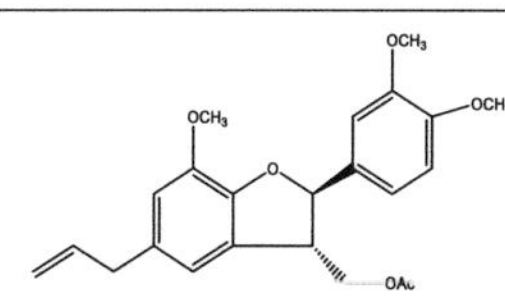 **83.** Miliusin C
Miliusa sessilis		*Piper pleiocarpum*
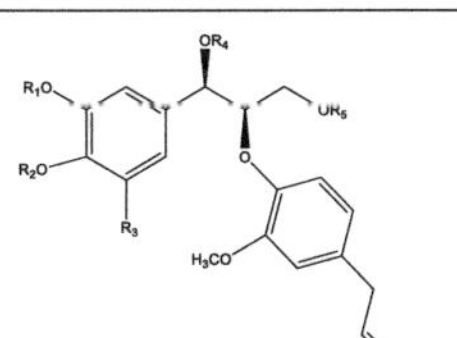 **84.** Miliusin D. *R_1 = R_2 = CH_3, R_3 = R_4 = H, R5 = Ac* **85.** Miliusin E *R_1 = R_2 = CH_3, R_3 = R_4 = R_5 = H*	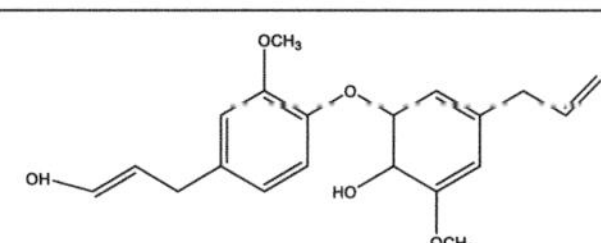 **86.** Miliusin F	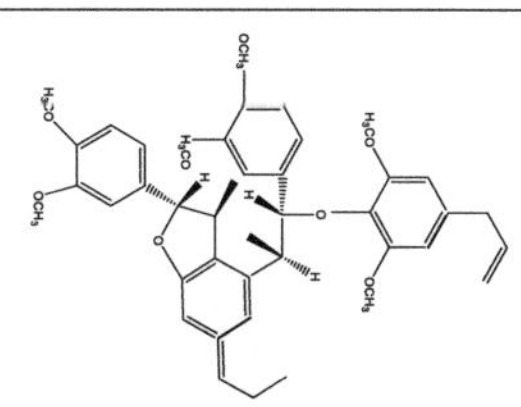 **87.** Pleiocarpumlignan B
Solanum lyratum	*Styrax argentifolius*	
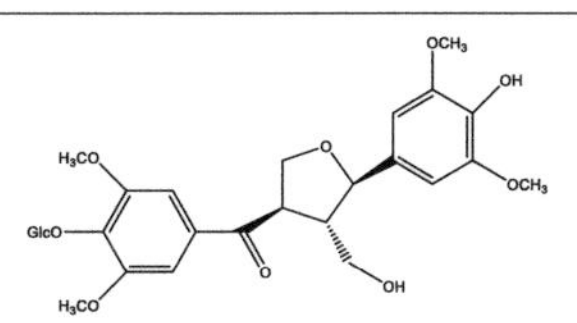 **88.** Officinalioside (OFD)	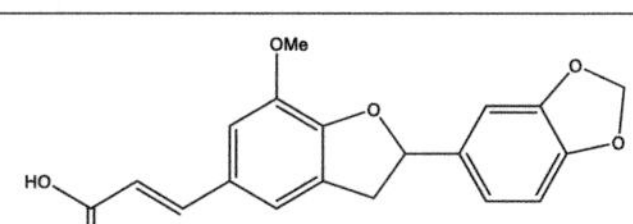 **89.** 5-((E)-2-carboxyvinyl)-7-methoxy-2-(3′,4′-methylenedioxyphenyl) Benzofuran	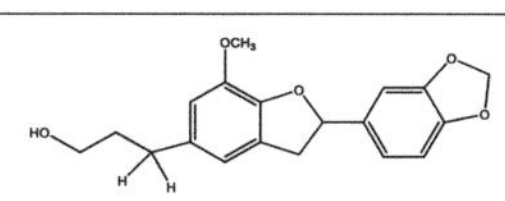 **90.** Egonol

Wikstroemia scytophylla		
Styrax argentifolius	*Schisandra sphenanthera*	
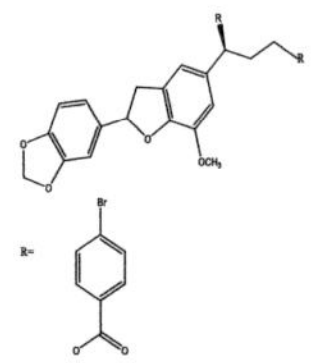 **91.** (−)-Machicendiol	 **92.** Schisphenlignan M	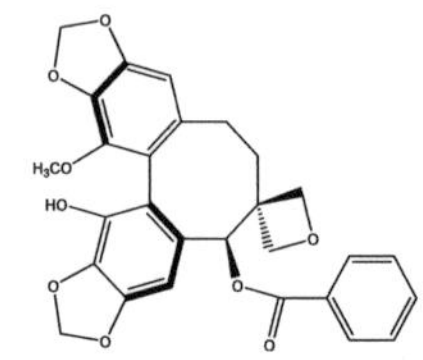 **93.** Schisphenlignan N
Schisandra sphenanthera		
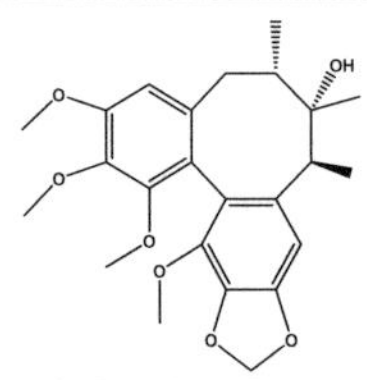 **94.** Gomisin G	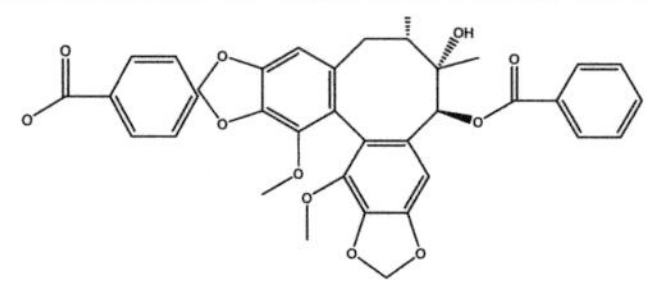 **95.** Schisantherin D	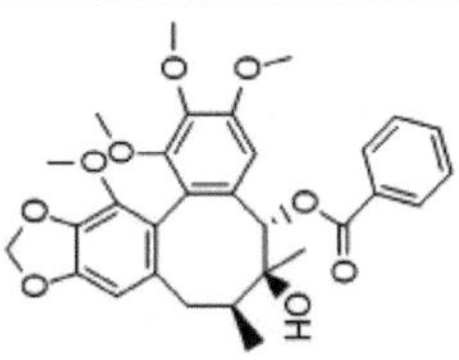 **96.** Schisantherin A
Shisandra sphenanthera		
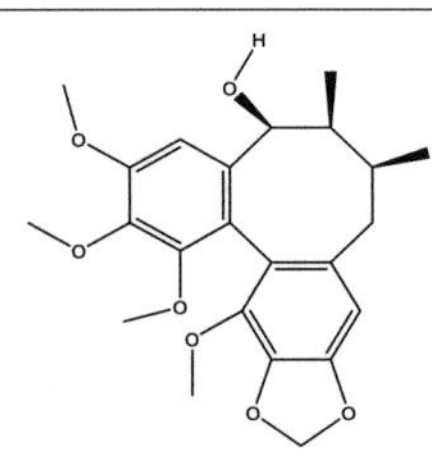 **97.** Epigomisin O	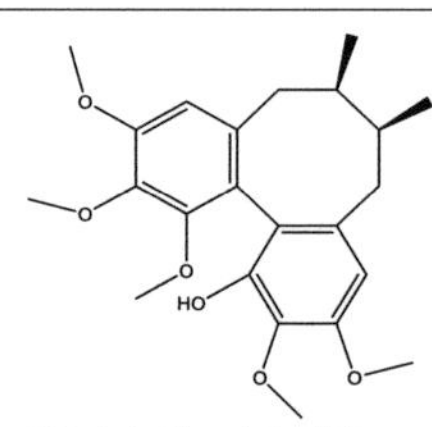 **98.** (+)-Gomisin K3	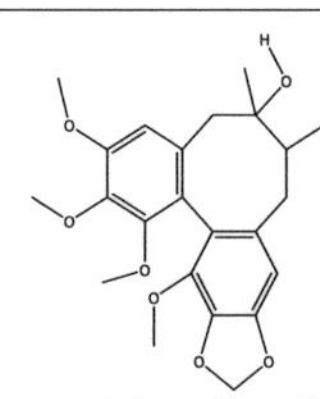 **99.** Schisanhenol B
Schisandra sphenanthera	*Sigesbeckia glabrescens*	
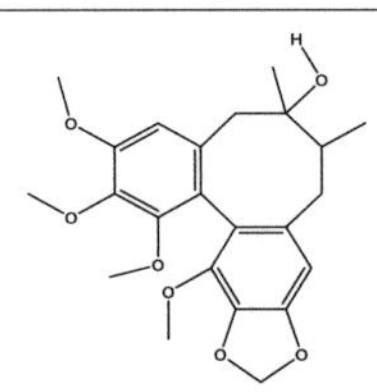 **100.** Gomisin A	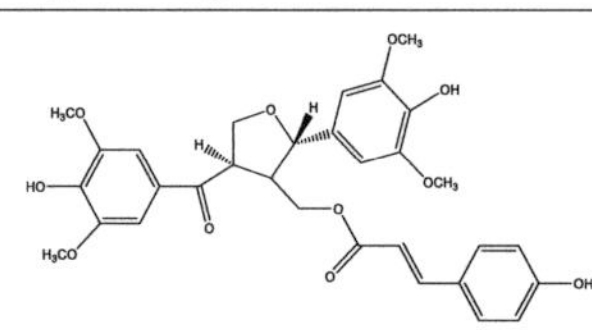 **101.** Glalignin B	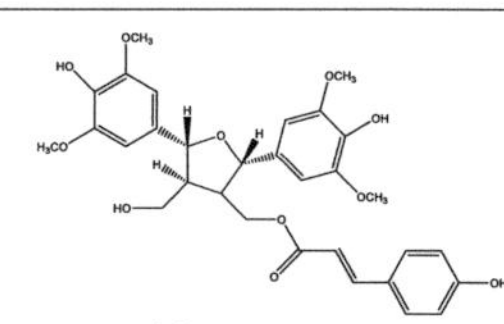 **102.** Glalignin C
Sigesbeckia glabrescens		
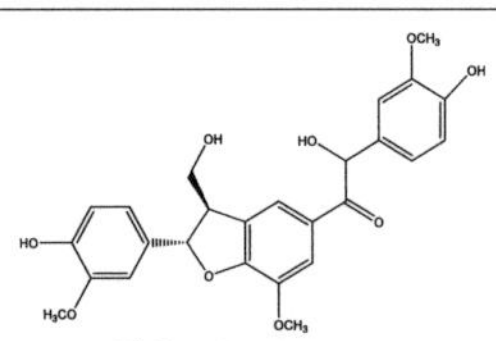 **103.** Glalignin E	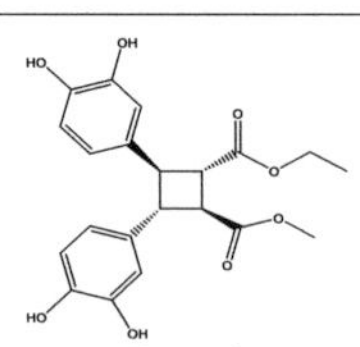 **104.** Glaneolignin A	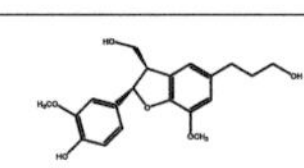 **105.** Dihydrodehydrodiconiferyl alcohol

Wikstroemia scytophylla		
Sigesbeckia glabrescens	*Jurinea macrocephala*	*Piper cubeba*
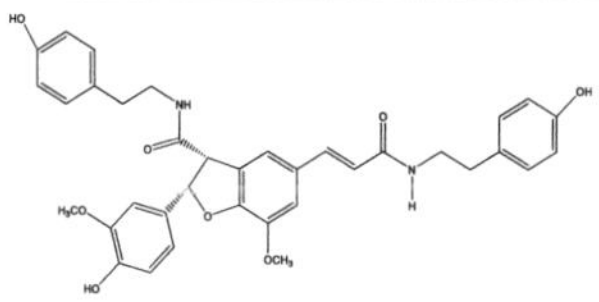	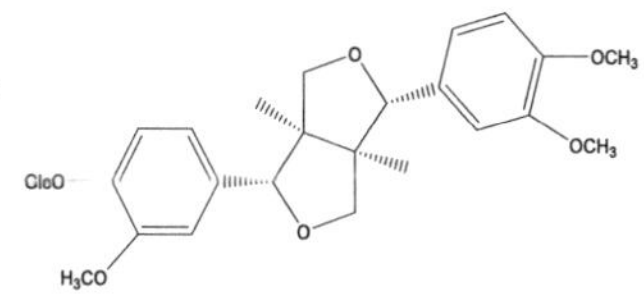	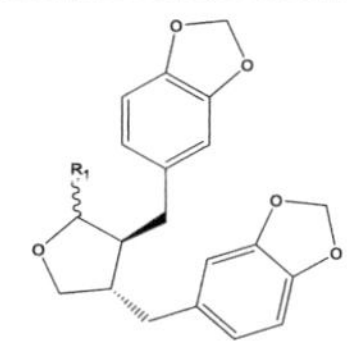 **108.** R_1 = MeO **109.** R_1 = OH **110.** R_1 = EtO
106. Tribulusamide A	**107.** Pinoresinol monomethyl ether-b-D-glucoside (PMG)	**108.** Methylcubebin (MB) **109.** Cubebin (CB) **110.** Ethylcubebin (EB)
P. cubeba	*Solanum violaceum*	
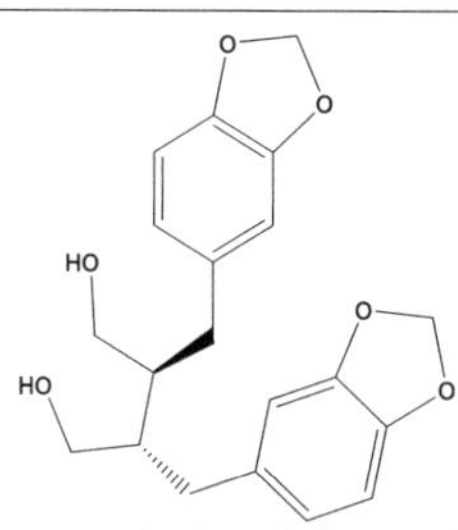 **111.** Dyhydrocubebin (DB)	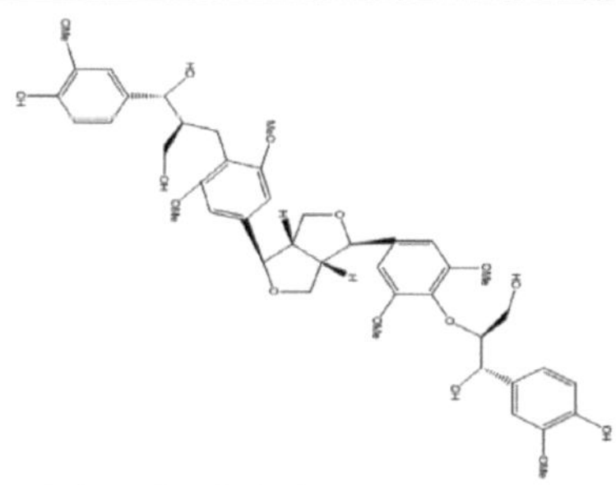 **112.** (1S,2S)-1-(4-hydroxy-3-methoxyphenyl)-2-[2-methoxy-4-[(2S,3R, 4R)-tetrahydro-4-[(4-hydroxy-3-methoxyphenyl)methyl]-3-(hydroxymethyl)- 2-furanyl] phenoxy]-1,3-propanediol (MFP)	

Table 2.
Lignans and neolignans structures.

among others, but the comparability of these studies depends on the type of assay with which the findings are reported. Therefore, the assay selection is of great importance in understanding the toxicity profile of lignans, as an approximation of their cytotoxic potential if used in humans.

The aim of this research was to present an overview of the anticancer activity of lignans *in vitro* and *in vivo* studies (**Table 1**), with the type of assay described in the international literature in the last 5 years, as well as their structures (**Table 2**).

2. Discussion

Lignans act as antioxidants and play an important role in protection against herbivores, pathogenic fungi, and bacteria [59]. These lignans have positive effects on different diseases, such as cancer and type 2 diabetes.

The lignans present in the feed diet might be metabolized by the gut microbiota through deglycosylations, p-dehydroxylations, and m-demethylations, but there is no enantiomeric inversion, producing phytoestrogens (molecules with an estrogen-like effect), but there is not enantiomeric inversion; these metabolites are called "mammalian lignans or enterolignans" [60], for example, aglycones of enterolactone and enterodiol, formed from matairesinol and secoisolariciresinol, respectively. Both of these aglycones have antitumor effects against breast, colon, and lung cancer [61].

In this review, we found 112 lignans and norlignans with cytotoxic activity, isolated from plants of 34 families, such as Magnolicea, Lauraceae, and Sauracea, among others. We found that 13 of these lignans have a high activity on several human cancer cell lines.

Only cytotoxicity activity was determined in 92 of these lignans and this effect was evaluated by MTT assay. The antitumor effect of sesamine and honokiol was determined on tumors induced with lymphoma cells and squamous cells carcinoma respectively.

In the treatment of cancer, there are used compounds that produce cell death in two ways: apoptosis and direct toxicity, then the new therapies are focused on substances to induce apoptotic cancer cell death [62]. In this review, we found 16 lignans that promote cell death by apoptosis.

The apoptotic cell death could occur by the disruption of the mitochondrial membrane, which is a crucial signaling pathway in the induction of apoptosis diminishing the levels of ATP, inhibiting ERK and p38 MAPK signaling. Bcl-2 (antiapoptotic protein) protein family control apoptosis by regulating mitochondrial membrane permeability while Bax is an inducer of apoptosis. Caspase-9 is activated, promoting the cleavage of caspase-3 and PARP, which contributes to apoptosis and ultimately cell death. Lignans 23 y 35 induced apoptosis by this route [29, 20].

MMP-9 is an overexpressed proteolytic enzyme in cancer cells that acts as a precursor to the action of other endopeptidases. This enzyme is a new target for cancer therapy owing to its pivotal role in metastatic tumors. Compounds 41, 42, and 43 inhibit the overexpression of MMP-9 [32].

In vitro test flow cytometry is used for the investigation and diagnosis of diseases such as cancer. In the different studies reported in this review, this technique was used to find out: the percentage of viable cancer cells, the characteristics of the cells such as size and shape, tumor markers, cell cycle analysis, and type of cell death [63]. In **Table 1**, it is shown that compounds 35, 47, 51, 61, and 112 induced apoptotic death of cancer cells by this technique.

Tubulin and its assembly product, microtubules, are among the most successful targets in cancer chemotherapy. It is currently known that podophyllotoxin and its commercial derivatives Etoposide and Teniposide exert their mechanism of action in cancer cells by altering Topoisomerase II and tubulin [64]. Williams et al. (2017) found that Bifidenone lignan also acts at the microtubule level of NCI-H460 cells, causing the inhibition of tubulin polymerization and therefore the arrest of the G2 / M phase of the cell cycle [32].

Arctigenin (ATN) is a dibenzylbutirolactone lignan isolated from the fruit of Arctium lappa and exhibited a cytotoxic effect on different breast cancer cell lines (MDA-MB-231, MDA-MB-435S, MDA-MB-453, and MDA-MB-468). In ER-positive MCF-7 cells, ATN inhibited downstream effector molecules of the target of rapamycin (TOR), decreasing the expression of estrogen receptor-α (Erα) and inducing autophagy.

Another way for cell death: Autophagy is a self-degradative process, which involves the enzymatic breakdown of different cytoplasmatic components. This process promotes the elimination of damaged or harmful components [65].

In vitro, this lignan inhibited the migration and invasion of MDA-MB-231 by downregulation of MMP-2, MMP-9, and heparinase expression [66].

(−)-Trachelogenin (TA) belongs to the dibenzylbutyrolactone lignan class and has been isolated from different plants, such as Trachelospermi caulis, *T. asiaticum, T.*

Jasminoides, and Combretum fruticosum. This lignan has different pharmacological activities, such as anti-inflammatory [67], antidepressant, and anticancer effects [68]. TA did not induce apoptosis but induced autophagic death, mediated by increased LC3; its possible mechanism of induced autophagic cell death involves cytoplasmic vacuolization and formation of autophagosomes mediated by increasing LC3 activation, promoting changes in the expression of Beclin-1 levels [24].

4-O-methylhonokiol (MH) is a neolignan, a type of phenolic compound. It is found in the bark of *Magnolia grandiflora*, *Magnolia virginiana* flowers, and *Magnolia officinalis*. MH induced cytotoxicity on human oral carcinoma cells (OSCC PE/CA-PJ41). Its anticancer activity is due to its capacity to induce ROS-mediated alteration of MMP, mitochondrial apoptosis, and cell cycle arrest [25], and to inhibit neuroinflammation, amyloidogenesis, and memory impairment [69]. MH protected against diabetic cardiomyopathy in type 2 diabetic mice [70]. It also inhibited NkKB activity on human colon cancer cells and cell cycle arrest, and induced apoptosis [71]. Additionally, MH induced apoptosis on oral squamous cancer cells (OSCC) via Sp1 [72].

Deoxypodophyllotoxin (DPT) was isolated from plants of the genus Podophyllum and has also been obtained from other species, such as *Athriscus sylvestris*, *Juniperus oblonga,* and *Cupressus macrocarpa*. DPT presented high toxicity and some side effects, so its use is limited [73]. In vitro, DPT reduced the cell proliferation of NB cells, MDA-MB-231, and A549 lines, induced apoptosis and cell cycle arrest, reduced the expression of pCNA, and increased intracellular free calcium levels that promoted NB cell death.

Matairesinol (MT) was isolated from *Juniperus oblonga* and exhibited anti-inflammatory [74] and cytotoxic activity against neuroblastoma cell lines, with and without tetracycline-inducible MYCN over-expression, and induced apoptosis and cell cycle arrest [39]. MT ameliorated experimental autoimmune uveitis [75] and showed angiogenic activity in vivo and in vitro. This compound also inhibited the proliferation of human umbilical vein endothelial cells (HUVECs) [76].

Other lignans with significant anticancer activity are: methoxypinoresinol, which is a furanoid lignan isolated from the leaves of *Calotropis gigantea*; honokiol was isolated from *Magnolia officinalis*; trachelogenin isolated from *Combretum fruticosum*; bifidenone, which is isolated from *Beilschmiedia* sp.; hedyotol-B, which was isolated from the stems of *Herpetospermum pedunculosum*; bejolghotin G, H, and I, which were isolated from the leaves and twigs of *Cinnamomum bejolghota*. These compounds have been isolated recently, and they are the subject of few pharmacological studies.

The most studied cancer cell lines were lung, hepatocellular carcinoma, colon, and breast. The cell lines diversity was colon cancer, breast cancer, human melanoma, and pancreatic cancer. These cell lines had the highest number of reports.

The lignans and neolignans with middle activity in lung cancer cells were: 12–20, 63–68, 112, colon cancer cells: 12–20, 63–68, 80–85,112, hepatocellular carcinoma cells: 12–20, 69, 70, 80–85, 112, and breast cancer cells: 11, 51, 63–68, 107,112.

In this review, we found that the less studied cancer cells were ovarian, gastric, endocervical adenocarcinoma cells, cholangiocarcinoma, laryngeal, leukemia, neuroblastoma, pancreatic cancer, prostate cancer, renal cancer, and osteosarcoma.

This review shows that various lignans and neolignans could be promising candidates for the treatment of different types of cancer.

Conflict of interest

The authors declare that they have no competing interests.

Author details

Ana Laura Esquivel-Campos[1], Salud Pérez-Gutiérrez[1], Leonor Sánchez-Pérez[2], Nimsi Campos-Xolalpa[1] and Julia Pérez-Ramos[1]*

1 Biological Systems Department, Universidad Autonoma Metropolitana-Xochimilco, México City, México

2 Health Care Department, Universidad Autonoma Metropolitana-Xochimilco, México City, México

*Address all correspondence to: jperez@correo.xoc.uam.mx

IntechOpen

© 2022 The Author(s). Licensee IntechOpen. This chapter is distributed under the terms of the Creative Commons Attribution License (http://creativecommons.org/licenses/by/3.0), which permits unrestricted use, distribution, and reproduction in any medium, provided the original work is properly cited.

References

[1] Ferlay J, Colombet M, Soerjomataram I, Mathers C, Parkin DM, Piñeros M, et al. Estimating the global cancer incidence and mortality in 2018: GLOBOCAN. International Journal of Cancer. 2018;**144**:1941-1953. DOI: 10.1002/ijc.31937

[2] World Health Organization. Assessing national capacity for the prevention and control of noncommunicable diseases: Report of the 2019 global survey. 2020; Available from: https://apps.who.int/iris/handle/10665/331452 [Accessed: November 16, 2021]

[3] Ji N, Jiang L, Deng P, Xu H, Chen F, Liu J, et al. Synergistic effect of honokiol and 5-fluorouracil on apoptosis of oral squamous cell carcinoma cells. Journal of Oral. 2017;**46**:201-207. DOI: 10.1111/jop.12481

[4] Talapatra SK, Talapatra B. Shikimic acid pathway. In: Chemistry of Plant Natural Products. 1st ed. Berlin, Heidelberg: Springer; 2014. pp. 625-674. DOI: 10.1007/978-3-642-45410-3_13

[5] Suzuki S, Umezawa T. Biosynthesis of lignans and norlignans. Journal of Wood Science. 2007;**53**(4):273-284. DOI: 10.1007/s10086-007-0892-x

[6] Moss GP. Nomenclature of lignans and neolignans (IUPAC Recommendations 2000). Pure and Applied Chemistry. 2000;**72**:1493-1523. DOI: 10.1351/pac200072081493

[7] Ríos JL, Giner RM, Prieto JM. New findings on the bioactivity of lignans. Studies in Natural Products Chemistry. 2002;**26**:183-292. DOI: 10.1016/S1572-5995(02)80008-4

[8] Dar AA, Arumugam N. Lignans of sesame: purification methods, biological activities and biosynthesis–A review. Bioorganic Chemistry. 2013;**50**:1-10. DOI: 10.1016/j.bioorg.2013.06.009

[9] Jiang ZH, Liu YP, Huang ZH, Wang TT, Feng XY, Yue H, et al. Cytotoxic dihydrobenzofuran neolignans from *Mappianthus iodoies*. Bioorganic Chemistry. 2017;**75**:260-264. DOI: 10.1016/j.bioorg.2017.10.003

[10] Sung MK, Lautens M, Thompson LU. Mammalian lignans inhibit the growth of estrogen-independent human colon tumor cells. Anticancer Research. 1998;**18**:1405-1408 PMID: 9673348

[11] Lei JP, Yuan JJ, Pi SH, Wang R, Tan R, Ma CY, et al. Flavones and lignans from the stems of wikstroemia scytophylla Diels. Pharmacognosy Magazine. 2017;**13**:488. DOI: 10.4103/pm.pm_275_16

[12] Li DQ, Wang D, Zhou L, Li LZ, Liu QB, Wu YY, et al. Antioxidant and cytotoxic lignans from the roots of *Bupleurum chinense*. Journal of Asian Natural Products Research. 2017;**19**: 519-527. DOI: 10.1080/10286020.2016.1234456

[13] Bodede O, Shaik S, Singh M, Moodley R. Phytochemical analysis with antioxidant and cytotoxicity studies of the bioactive principles from Zanthoxylum capense (Small knobwood). Anti-Cancer Agents in Medicinal Chemistry (Formerly Current Medicinal Chemistry-Anti-Cancer Agents). 2017;**17**:627-634. DOI: 10.2174/1871520616666160627091939

[14] Meng Z, Liu H, Zhang J, Zheng Z, Wang Z, Zhang L, et al. Sesamin promotes apoptosis and pyroptosis via autophagy to enhance antitumour effects

on murine T-cell lymphoma. Journal of Pharmacological Sciences. 2021;**147**: 260-270. DOI: 10.1016/j.jphs.2021.08.001

[15] Dang PH, Awale S, Nhan NT. Phytochemical and cytotoxic studies on the leaves of *Calotropis gigantea*. Bioorganic & Medicinal Chemistry Letters. 2016;**27**:2902-2906. DOI: 10.1016/j.bmcl.2017.04.087

[16] Han JH, Jeong HJ, Lee HN, Kwon YJ, Shin HM, Choi Y, et al. Erythro-austrobailignan-6 down-regulates HER2/EGFR/integrinβ3 expression via p38 activation in breast. cancer. Phytomedicine. 2017;**24**:24-30. DOI: 10.1016/j.phymed.2016.11.009

[17] Zhang RF, Feng X, Su GZ, Yin X, Yang XY, Zhao YF, et al. Noralashinol B, a norlignan with cytotoxicity from stem barks of *Syringa pinnatifolia*. Journal of Asian Natural Products Research. 2017; **19**:416-422. DOI: 10.1080/10286020.2017.1307188

[18] Feng T, Cao W, Shen W, Zhang L, Gu X, Guo Y, et al. Arctigenin inhibits STAT3 and exhibits anticancer potential in human triple-negative breast cancer therapy. Oncotarget. 2017;**8**:329. DOI: 10.18632/oncotarget.13393

[19] Maxwell T, Lee KS, Kim S, Nam KS. Arctigenin inhibits the activation of the mTOR pathway, resulting in autophagic cell death and decreased ER expression in ER-positive human breast cancer cells. International Journal of Oncology. 2018; **52**:1339-1349. DOI: 10.3892/ijo.2018.4271

[20] Baba Y, Shigemi Z, Hara N, Moriguchi M, Ikeda M, Watanabe T, et al. Arctigenin induces the apoptosis of primary effusion lymphoma cells under conditions of glucose deprivation. International Journal of Oncology. 2018; **52**:505-517. DOI: 10.3892/ijo.2017.4215

[21] Huang KJ, Kuo CH, Chen SH, Lin CY, Lee YR. Honokiol inhibits in vitro and in vivo growth of oral squamous cell carcinoma through induction of apoptosis, cell cycle arrest and autophagy. Journal of Cellular and Molecular Medicine. 2018;**22**:1894-1908. DOI: 10.1111/jcmm.13474

[22] Jiraviriyakul A, Songjang W, Kaewthet P, Tanawatkitichai P, Bayan P, Pongcharoen S. Honokiol-enhanced cytotoxic T lymphocyte activity against cholangiocarcinoma cells mediated by dendritic cells pulsed with damage-associated molecular patterns. World Journal of Gastroenterology. 2019;**25**: 3941. DOI: 10.3748/wjg.v25.i29.3941

[23] Yang M, Zhang Y, Chen L, Chen Y. A new (propylphenyl) bibenzyl derivative from *Dendrobium williamsonii*. Natural Product Research. 2018;**32**:1699-1705. DOI: 10.1080/14786419.2017.1396599

[24] Moura AF, Lima KSB, Sousa TS, Marinho-Filho JDB, Pessoa C, Silveira ER, et al. In vitro antitumor effect of a lignan isolated from *Combretum fruticosum*, trachelogenin, in HCT-116 human colon cancer cells. Toxicology In Vitro. 2018;**47**:129-136. DOI: 10.1016/j.tiv.2017.11.014

[25] Xiao S, Chen F, Gao C. Antitumor activity of 4-O-Methylhonokiol in human oral cancer cells is mediated via ROS generation, disruption of mitochondrial potential, cell cycle arrest and modulation of Bcl-2/Bax proteins. Journal of BUON. 2017;**22**:1577-1581 PMID: 29332355

[26] Williams RB, Martin SM, Lawrence JA, Norman VL, O'Neil-Johnson M, Eldridge GR, et al. Isolation and identification of the novel tubulin polymerization inhibitor bifidenone. Journal of Natural Products. 2017;**80**:

616-624. DOI: 10.1021/acs.jnatprod.6b00893

[27] Dibwe DF, Sun S, Ueda JY, Balachandran C, Matsumoto K, Awale S. Discovery of potential antiausterity agents from the Japanese cypress *Chamaecyparis obtusa*. Bioorganic & Medicinal Chemistry Letters. 2017;**27**: 4898-4903. DOI: 10.1016/j.bmcl.2017.09.034

[28] Kim SH, Son KM, Kim KY, Yu SN, Park SG, Kim YW, et al. Deoxypodophyllotoxin induces cytoprotective autophagy against apoptosis via inhibition of PI3K/AKT/mTOR pathway in osteosarcoma U2OS cells. Pharmacological Reports. 2017;**69**: 878-884. DOI: 10.1016/j.pharep.2017.04.007

[29] Ma ZJ, Lu L, Yang JJ, Wang XX, Su G, Wang ZL, et al. Lariciresinol induces apoptosis in HepG2 cells via mitochondrial-mediated apoptosis pathway. European Journal of Pharmacology. 2018;**821**:1-10. DOI: 10.1016/j.ejphar.2017.12.027

[30] Gigliarelli G, Zadra C, Cossignani L, Robles Zepeda RE, Rascón-Valenzuela LA, Velázquez-Contreras CA, et al. Two new lignans from the resin of *Bursera microphylla* A. gray and their cytotoxic activity. Natural Product Research. 2018; **32**:2646-2651. DOI: 10.1080/14786419.2017.1375922

[31] Liu Y, Yang Y, Tasneem S, Hussain N, Daniyal M, Yuan H, et al. Lignans from Tujia ethnomedicine Heilaohu: Chemical characterization and evaluation of their cytotoxicity and antioxidant activities. Molecules. 2018;**23**: 2147. DOI: 10.3390/molecules23092147

[32] Zhu Y, Huang RZ, Wang CG, Ouyang XL, Jing XT, Liang D, et al. New inhibitors of matrix metalloproteinases 9 (MMP-9): Lignans from Selaginella moellendorffii. Fitoterapia. 2018;**130**: 281-289. DOI: 10.1016/j.fitote.2018.09.008

[33] Lee TK, Lee D, Yu JS, Jo MS, Baek SC, Shin MS, et al. Biological evaluation of a new lignan from the roots of rice (Oryza sativa). Chemistry & Biodiversity. 2018;**15**(11):e1800333. DOI: 10.1002/cbdv.201800333

[34] Jeong M, Kim HM, Lee JS, Choi JH, Jang DS. (−)-Asarinin from the roots of asarum sieboldii induces apoptotic cell death via caspase activation in human ovarian cancer cells. Molecules. 2018;**23**: 1849. DOI: 10.3390/molecules23081849

[35] Lou LL, Yaoaz GD, Wang J, Zhao WY, Wang XB, Huang XX, et al. Enantiomeric neolignans from *Picrasma quassioides* exhibit distinctive cytotoxicity on hepatic carcinoma cells through ROS generation and apoptosis induction. Bioorganic & Medicinal Chemistry Letters. 2018;**28**:1263-1268. DOI: 10.1016/j.bmcl.2018.03.043

[36] de Sousa FS, Nunes EA, Gomes KS, Cerchiaro G, Lago JHG. Genotoxic and cytotoxic effects of neolignans isolated from *Nectandra leucantha* (Lauraceae). Toxicology In Vitro. 2019;**55**:116-123. DOI: 10.1016/j.tiv.2018.12.011

[37] Flegkas A, Milosević Ifantis T, Barda C, Samara P, Tsitsilonis O, Skaltsa H. Antiproliferative activity of (−)-rabdosiin isolated from *Ocimum sanctum* L. Medicine. 2019;**6**:37. DOI: 10.3390/medicines6010037

[38] Wang GK, Jin WF, Zhang N, Wang G, Cheng YY, Morris-Natschke SL, et al. Kalshiolin A, new lignan from *Kalimeris shimadai*. Journal of Asian Natural Products Research. 2020;**22**: 489-495. DOI: 10.1080/10286020.2019.1592164

[39] Qiao Y, Sunada NK, Hatada AE, Lange I, Khutsishvili M, Alizade V, et al. Potential anti-neuroblastoma agents from Juniperus oblonga. Biochemical and Biophysical Research Communications. 2019;**516**:733-738. DOI: 10.1016/j.bbrc.2019.06.123

[40] Han J, Chen X, Liu W, Cui H, Yuan T. Triterpenoid saponin and lignan glycosides from the traditional medicine Elaeagnus angustifolia flowers and their cytotoxic activities. Molecules. 2020;**25**: 462. DOI: 10.3390/molecules25030462

[41] Ma Y, Wang H, Wang R, Meng F, Dong Z, Wang G, et al. Cytotoxic lignans from the stems of Herpetospermum pedunculosum. Phytochemistry. 2019; **164**:102-110. DOI: 10.1016/j. phytochem.2019.05.004

[42] Yang Y, Liu Y, Daniyal M, Yu H, Xie Q, Li B, et al. New lignans from roots of Kadsura coccinea. Fitoterapia. 2019; **139**:104368. DOI: 10.1016/j. fitote.2019.104368

[43] Tran HNK, Cao TQ, Kim JA, Woo MH, Min BS. Anti-inflammatory and cytotoxic activities of constituents isolated from the fruits of *Ziziphus jujuba* var. *inermis* Rehder. Fitoterapia. 2019; **137**:104261. DOI: 10.1016/j. fitote.2019.104261

[44] Guo R, Lv TM, Shang XY, Yao GD, Lin B, Wang XB, et al. Racemic neolignans from *Crataegus pinnatifida*: Chiral resolution, configurational assignment, and cytotoxic activities against human hepatoma cells. Fitoterapia. 2019;**137**:104287. DOI: 10.1016/j.fitote.2019.104287

[45] Rao L, You YX, Su Y, Fan Y, Liu Y, He Q, et al. Lignans and neolignans with antioxidant and human cancer cell proliferation inhibitory activities from *Cinnamomum bejolghota* confirm its functional food property. Journal of Agricultural and Food Chemistry. 2020; **68**:8825-8835. DOI: 10.1021/acs. jafc.0c02885

[46] Al-Sayed E, Ke TY, Hwang TL, Chen SR, Korinek M, Chen SL, et al. Cytotoxic and anti-inflammatory effects of lignans and diterpenes from Cupressus macrocarpa. Bioorganic & Medicinal Chemistry Letters. 2020;**30**: 127127. DOI: 10.1016/j.bmcl.2020.127127

[47] Zhang L, Wang XL, Wang B, Zhang LT, Gao HM, Shen T, et al. Lignans from Euphorbia hirta L. Natural Product Research. 2020:**36**:1478-6427. DOI: 10.1080/14786419.2020.1761358

[48] Nguyen LH, Vu VN, Thi DP, Tran VH, Litaudon M, Roussi F, et al. Cytotoxic lignans from fruits of Cleistanthus tonkinensis. Fitoterapia. 2020;**140**:104432. DOI: 10.1016/j. fitote.2019.104432

[49] Shang XY, Guo R, Yu XQ, Lin B, Huang XX, Yao GD, et al. Enantiomeric 8-O-4′-type neolignans from Crataegus pinnatifida exhibit cytotoxic effect via apoptosis and autophagy in Hep3B cells. Bioorganic Chemistry. 2020;**104**:104267. DOI: 10.1016/j.bioorg.2020.104267

[50] Pootaeng-On Y, Charoensuksai P, Wongprayoon P, Jiajaroen S, Chainok K, Rayanil K. Miliusins; cytotoxic neolignans from the leaves of *Miliusa sessilis*. Phytochemistry. 2020;**176**:112417. DOI: 10.1016/j.phytochem.2020.112417

[51] Su XM, Liang Q, Zhang XM, Yao ZY, Xu WH. Four new chemical constituents from *Piper pleiocarpum*. Fitoterapia. 2020;**143**:104544. DOI: 10.1016/j. fitote.2020.104544

[52] Wang X, Zhao Y, Dong X, Wu X, Yu HY, Zhang LH, et al. Amides and lignans from *Solanum lyratum*.

Phytochemistry Letters. 2021;**45**:25-29. DOI: 10.1016/j.phytol.2021.07.002

[53] Tra NT, Van Tuyen N, Van Cuong P, Ha NTT, Anh LTT, Son NT. Chemical constituents from the leaves of Styrax argentifolius HL Li and their biological activities. Phytochemistry Letters. 2021; **41**:70-73. DOI: 10.1016/j.phytol.2020.11.003

[54] Huang S, Liu Y, Li Y, Fan H, Huang W, Deng C, et al. Dibenzocyclooctadiene lignans from the root bark of *Schisandra sphenanthera*. Phytochemistry Letters. 2021;**45**: 137-141. DOI: 10.1016/j.phytol.2021.08.015

[55] Gao XX, Gao YN, Wang DD, Hu GS, Yan T, Jia JM, et al. Six novel lignanoids with complex structures from *Sigesbeckia glabrescens* Makino with their cytotoxic activities. Fitoterapia. 2021;**148**:104799. DOI: 10.1016/j.fitote.2020.104799

[56] Atabaki V, Pourahmad J, Hosseinabadi T. Phytochemical compounds from *Jurinea macrocephala* subsp. *elbursensis* and their cytotoxicity evaluation. South African Journal of Botany. 2021;**137**:399-405. DOI: 10.1016/j.sajb.2020.11.011

[57] Gusson-Zanetoni JP, Monteiro da Silva JSG, Picão TB, Cardin LT, Prates J, Sousa SO, et al. Effect of *Piper cubeba* total extract and isolated lignans on head and neck cancer cell lines and normal fibroblasts. Journal of Pharmacological Sciences. 2021;**148**:93-102. DOI: 10.1016/j.jphs.2021.09.002

[58] Kaunda JS, Qin XJ, Zhu HT, Wang D, Yang CR, Zhang YJ. Previously undescribed pyridyl-steroidal glycoalkaloids and 23S, 26R-hydroxylated spirostanoid saponin from the fruits of Solanum violaceum ortega and their bioactivities. Phytochemistry. 2021;**184**:112656. DOI: 10.1016/j.phytochem.2021.112656

[59] Harmatha J, Dinan L. Biological activities of lignans and stilbenoids associated with plant-insectchemical interactions. Phytochemistry Reviews. 2003;**2**:321-330. DOI: 10.1023/B:PHYT.0000045494.98645.a3

[60] Tsao R. Chemistry and biochemistry of dietary polyphenols. Nutrients. 2010; **2**:1231-1246. DOI: 10.3390/nu2121231

[61] Kirsch V, Bakuradze T, Richling E. Toxicological testing of syringaresinol and enterolignans. Current Research in Toxicology. 2020;**1**:104-110. DOI: 10.1016/j.crtox.2020.09.002

[62] Gerl R, Vaux DL. Apoptosis in the development and treatment of cancer. Carcinogenesis. 2005;**2005**(26):263-270. DOI: 10.1093/carcin/bgh283

[63] Henry CM, Hollville E, Martin SJ. Measuring apoptosis by microscopy and flow cytometry. Methods. 2013;**61**:90-97. DOI: 10.1016/j.ymeth.2013.01.008

[64] Guerram M, Jiang ZZ, Zhang LY. Podophyllotoxin, a medicinal agent of plant origin: Past, present and future. Chinese Journal of Natural Medicines. 2012;**10**:161-169. DOI: 10.3724/SP.J.1009.2012.00161

[65] Glick D, Barth S, Macleod KF. Autophagy: Cellular and molecular mechanisms. The Journal of Pathology. 2010;**221**(1):3-12. DOI: 10.1002/path.2697

[66] Lou C, Zhu Z, Zhao Y, Zhu R, Zhao H. Arctigenin, a lignan from Arctium lappa L., inhibits metastasis of human breast cancer cells through the downregulation of MMP-2/−9 and heparanase in MDA-MB-231 cells. Oncology Reports. 2017;**37**:179-184. DOI: 10.3892/or.2016.5269

[67] Shin HS, Bae MJ, Jung SY, See HJ, Kim YT, Do JR, et al. Enhancing effect of trachelogenin from *Trachelospermi caulis* extract on intestinal barrier function. Biological and Pharmaceutical Bulletin. 2015;**38**:1707-1713. DOI: 10.1248/bpb.b15-00332

[68] Kuehnl S, Schroecksnadel S, Temml V, Gostner JM, Schennach H, Schuster D, et al. Lignans from Carthamus tinctorius suppress tryptophan breakdown via indoleamine 2, 3-dioxygenase. Phytomedicine. 2013; **20**:1190-1195. DOI: 10.1016/j.phymed.2013.06.006

[69] Lee YJ, Choi DY, Choi IS, Kim KH, Kim YH, Kim HM, et al. Inhibitory effect of 4-O-methylhonokiol on lipopolysaccharide-induced neuroinflammation, amyloidogenesis and memory impairment via inhibition of nuclear factor-kappaB in vitro and in vivo models. Journal of Neuroinflammation. 2012;**9**:1-19. DOI: 10.1186/1742-2094-9-35

[70] Zheng Z, Ma T, Guo H, Kim KS, Kim KT, Bi L, et al. 4-O-methylhonokiol protects against diabetic cardiomyopathy in type 2 diabetic mice by activation of AMPK-mediated cardiac lipid metabolism improvement. Journal of Cellular and Molecular Medicine. 2019;**23**:5771-5781. DOI: 10.1111/jcmm.14493

[71] Oh JH, Ban JO, Cho MC, Jo M, Jung JK, Ahn B, et al. 4-O-methylhonokiol inhibits colon tumor growth via p21-mediated suppression of NF-κB activity. The Journal of Nutritional Biochemistry. 2012;**23**: 706-715. DOI: 10.1016/j.jnutbio.2011.03.013

[72] Cho JH, Lee RH, Jeon YJ, Shin JC, Park SM, Choi NJ, et al. Role of transcription factor Sp1 in the 4-O-methylhonokiol-mediated apoptotic effect on oral squamous cancer cells and xenograft. The International Journal of Biochemistry & Cell Biology. 2015;**64**: 287-297. DOI: 10.1016/j.biocel.2015.05.007

[73] Hu S, Zhou Q, Wu WR, Duan YX, Gao ZY, Li YW, et al. Anticancer effect of deoxypodophyllotoxin induces apoptosis of human prostate cancer cells. Oncology Letters. 2016;**12**:2918-2923. DOI: 10.3892/ol.2016.4943

[74] Kuehnl S, Schroecksnadel S, Temml V, Gostner JM, Schennach H, Schuster D, et al. Lignans from Carthamus tinctorius suppress tryptophan breakdown via indoleamine 2, 3-dioxygenase. Phytomedicine. 2013; **20**:190-1195. DOI: 10.1016/j.phymed.2013.06.006

[75] Li X, Gao Q, Yang L, Han M, Zhou C, Mu H. Matairesinol ameliorates experimental autoimmune uveitis by suppression of IRBP-specific Th17 cells. Journal of Neuroimmunology. 2020;**345**: 577286. DOI: 10.1016/j.jneuroim.2020.577286

[76] Lee B, Kim KH, Jung HJ, Kwon HJ. Matairesinol inhibits angiogenesis via suppression of mitochondrial reactive oxygen species. Biochemical and Biophysical Research Communications. 2012;**421**:76-80. DOI: 10.1016/j.bbrc.2012.03.114

Chapter 11

Plant Secondary Metabolites: Therapeutic Potential and Pharmacological Properties

Muhammad Zeeshan Bhatti, Hammad Ismail and Waqas Khan Kayani

Abstract

Plants are an essential source for discovering novel medical compounds for drug development, and secondary metabolites are sources of medicines from plants. Secondary metabolites include alkaloids, flavonoids, terpenoids, tannins, coumarins, quinones, carotenoids, and steroids. Each year, several new secondary metabolites are extracted from plants, providing a source of possibilities to investigate against malignant illnesses, despite certain natural chemicals having distinct anticancer activities according to their physicochemical features. Secondary metabolites found in plants are frequently great leads for therapeutic development. However, changes in the molecular structure of these compounds are improving their anticancer activity and selectivity and their absorption, distribution, metabolism, and excretion capacities while minimizing their toxicity and side effects. In this section, we will discuss the most significant breakthroughs in the field of plant secondary metabolites, some of which are currently in clinical use and others that are in clinical trials as anticancer drugs. This study gives an up-to-date and thorough summary of secondary plant metabolites and their antioxidant, antibacterial, and anticancer effects. Furthermore, antioxidant and antibacterial, and anticancer effects of secondary metabolites are addressed. As a result, this article will serve as a thorough, quick reference for people interested in secondary metabolite antioxidants, anticancer, and antibacterial properties.

Keywords: plant secondary metabolites, pharmacological, anticancer, antioxidant, antimicrobial

1. Introduction

Plants are essential in pharmacological research and drug development, not only when bioactive substances are used as therapeutic agents directly, but also as starting materials for drug production or as models for pharmacologically active molecules.

 IntechOpen

Secondary metabolites differ depending on the plant species. Secondary metabolites are molecules produced by plants that remain unknown in their roles in growth, photosynthesis, reproduction, and other primary processes. Secondary compounds are widely employed in plants, primarily in Asia [1]. Secondary metabolites boost human immunity because pharmaceuticals are mainly based on plant components. Secondary compounds in plants can serve as medicinal for humans [2]. Several criteria have been considered to classify secondary metabolites, including chemical structure, composition, solubility, and biosynthetic pathway [3].

1.1 Phenolics

Plants' most critical secondary metabolites and bioactive chemicals are flavonoids and phenolic acids [4]. They're also a natural antioxidant capable of scavenging free superoxide radicals, slowing the aging process, and lowering cancer risk. Flavonoids have been shown to reduce blood glucose levels in people. Phenolic acids Flavonoids have been found in several investigations [5]. Phenolic acid is a well-known class of secondary metabolites with a wide range of pharmacological effects. Phenolics are reported for various biological functions. Some of the effects of phenolics include enhancing bile secretion, lowering blood cholesterol and lipid levels, and antibacterial activity against bacteria such as *staphylococcus aureus* [6]. Antiulcer, anti-inflammatory, antioxidant, cytotoxic and antitumor, antispasmodic, and antidepressant properties are all found in phenolics and flavonoids [1, 4, 7]. Multiple glycoprotein VI signaling pathway components prevented collagen-stimulated platelet activation by dietary polyphenolic substances, particularly quercetin [8].

1.2 Phenolic acids

The phrase "phenolic acids" refers to phenolic compounds that have only one carboxylic acid group [9]. They are found in a different plant-based diet, with the most significant amounts in seeds, fruit skins, and vegetable leaves [10]. Plant phenolic acids are an essential part of the human diet because of their high antioxidant capacity and other health advantages. According to epidemiological studies, a diet with high antioxidant vegetables and fruits lowers the incidence of several oxidative disorders like cancer, diabetes, and cardiovascular disease. They also induced protective enzymes that positively affect signaling pathways, indicating indirect antioxidant activity [11]. Phenolic acids influence the action of glucose and insulin receptors. They increase the GLUT2 glucose transporter levels in insulin-producing pancreatic cells and stimulate GLUT4 transportation *via* the PI3K/Akt and AMP-derived kinase pathways. Ferulic and chlorogenic acids, for example, demonstrated the precise transporter activation mechanism and acted as anti-diabetic drugs [9, 12–14]. Among all the phenolic chemicals found in feces water, phenolic acids are the most prevalent [15]. They have antibacterial properties and can also be used as food preservatives. Phenolic acids and their derivatives play an essential role in cancer prevention and treatment [4, 9, 15]. Plant phenolics may be able to help in this area. Natural products or derivatives accounted for more than half of anticancer prescription medications approved globally between the 1940s and 2006, and numerous clinical trials are still ongoing [16]. They halt the creation of DNA adducts, thwart the synthesis of genotoxic compounds, and inhibit the mutagen's activity [9, 17]. Most phenolics work at different locations to treat or inhibit various cancers [18].

1.3 Flavonoids

Flavonoids are a type of polyphenolic chemical that occurs naturally. It's one of the most prevalent combinations found in vegetables, fruits, and beverages made from plants. Flavonoids are dietary supplements that promote health and prevent disease. It is now measured as an essential part of a wide range of nutraceutical, pharmacological, medical, and other products [19]. Aside from their antioxidant properties, flavonoids have a wide range of biological activities that contribute to human health [20]. Anti-inflammatory, antiulcer, antiviral, anticancer, antidiabetic, and cytotoxic actions are only a few examples. Flavonoids have shown various dietary benefits on antioxidant activity in multiple studies. Flavonoids also protect cell membranes from lipid peroxidation-induced damage. As a result, flavonoids play an essential role as antioxidants in oxidative stress-related illnesses [21]. Inflammatory disorders such as leukemia, asthma, sepsis, atherosclerosis, sclerosis, allergic rhinitis, psoriasis, rheumatoid arthritis, ileitis/colitis, and others have been linked to flavonoids. To eradicate foreign pathogens and restore wounded tissues, recruitment of inflammatory cells and release of RNS, ROS, and proinflammatory cytokines. Inflammation is usually quick and self-limiting, but abnormal resolution and protracted inflammation can lead to various chronic diseases [22].

Flavonoids similarly inhibit phosphodiesterases involved in cell activation. According to a different study, flavonoid-rich extracts from plants have antibacterial properties [22]. According to numerous studies, natural flavonoids have been exceptional antiviral action since the 1940s. They aid in the blockage of several enzymes involved in the virus's life cycle. According to many studies, flavonoids such as hesperetin, quercetin, and naringin have anti-dengue action [23]. Flavonoids have a prominent effect on the immunological implications that occur through the genesis and progression of cancer. They can affect various biological signals in cancer, including vascularization, apoptosis, cell proliferation, and cell differentiation. Flavonoids mainly increase carcinogenicity's start and promotion stages and influence expansion and hormonal activity [19, 20].

1.4 Terpenes

Terpenes are a diverse group of secondary metabolites in plants, with over 40,000 distinct compounds [24]. Terpenes are categorized based on how many isoprene units they contain. Terpenes are combinations of volatile molecules with characteristic odors found in the flowers and fruits of many plants, including mint, lemon, ginger, eucalyptus, and great basil [25]. They have a variety of biological roles and are involved in plant's metabolism. Terpenes are photosynthetic pigments, electron carriers, plant growth regulators, are part of cell membranes, and participate in protein glycosylation in the central metabolism [24, 26]. They combine as defense chemicals, poisonous substances, and food deterrents in the secondary metabolism of insects [1].

1.5 Saponins

Saponins, glycosides extensively distributed in plants, are a varied group of molecules that includes a triterpenoid or steroidal aglycone with one or more sugar chains [27]. Because their immune-enhancing qualities have been utilized as adjuvants in vaccine formulations since the 1950s [28]. Ginseng dammarane sapogenins' chemopreventive and chemotherapeutic properties have encouraged the creation of anticancer medicines at various stages of development [29]. Maturation inhibitors are novel HIV medicines researched using betulinic acid derivatives [30]. Inflammation, infection, alcoholism, pre- and postmenopausal symptoms, cerebrovascular and cardiovascular diseases such as hypertension and coronary heart disease, prophylaxis, and dementia, ultraviolet damage including cataract, gastric ulcer, gastritis, and duodenal ulcer have all been treated with saponin-containing pharmaceutical compositions or plant extracts [27, 30, 31]. Saponins have also been patented for use as adjuvants to improve the absorption of bioactive chemicals and medications [32]. Plants that contain saponins, such as yucca, ginseng, chestnut, licorice, and sarsaparilla, have been utilized in traditional medicine for ages to prevent and treat various disorders by numerous cultures [31].

1.6 Tannins

Tannins are phenolic chemicals that are found practically everywhere in plants. Fruit, the bark of trees, wood, and as well as in numerous wild plants and herbs, and forestry and agriculture [33], contain them. Chestnut tannin, is a renowned member of the commercial hydrolyzable tannins family, has been recommended as an antibacterial or a way to reduce mycotoxins [34]. Other uses for tannins, including ellagitannins and gallotannins, include treating bacterial infections, regulating cytotoxins production, antihistamine, antiasthma, and avoiding rhinitis, as well as blocking HIV propagation in human cells [33, 35]. There have also been reports on the usefulness of several tannin-derived chemicals in treating obesity, arteriosclerosis, and thrombosis, decreasing triglycerides, preventing Staphylococcus aureus and other gram-positive bacteria, and leukemia [33]. Patents have also been published on the non-commercial use of tannins to treat cognitive, neurological, and metabolic diseases, diabetes II and obesity, hypertension, and hypercholesterolemia [35]. *Acacia mearnsii* and *Acacia nilotica* tannins are among the condensed tannins. Both claim that tannins have antipyretic properties, with the first claiming antidiarrheic properties [36, 37]. Above the typical anthelmintic activity of tannins, Quebracho wood commercial tannin from Argentina also has anthelmintic activity. Sumac tannins, a combined condensed and hydrolyzable tannin have been suggested to possess anti-inflammatory, antimicrobial, and immunomodulatory potentials [33, 35]. Extensive applications for the cure of blood pressure, hypertension, and, most notably, hemorrhoidal disorders have been commercially available.

1.7 Lignans

The word "Lignan" refers to a class of dimeric phenylpropanoids containing two C6-C3 phenylpropanoids are linked by a C8 phenylpropanol. Lignans can be found in over 60 different types of vascular foliage. Lignans are a nonflavonoid polyphenol subclass [38]. They have high functional importance, and eating a diet rich in them can lower your risk of cardiovascular disease. Lignans can be found in barley, flaxseed, wheat bran, almonds, legumes, sesame seeds, fruits, and vegetables. A 12-year study published in 1889 found that those with elevated enterolactone levels had a decreased incidence of heart failure compared with low levels [39]. Clinical trials have demonstrated that adding diets with 30–50 grams of flaxseed per day for 4–12 weeks reduced LDL cholesterol by 8%–14% [40]. Another possible study looked at the influence of dietary lignan on breast cancer risk; women who consumed dietary lignan had a 17% minor risk of breast cancer than those in the lowest quartile [41].

According to the study report, women who consume many dietary lignans have a lower risk of endometrial cancer. Enterodiol and enterolactone have been shown to reduce the risk of hormone-related cancers [42]. Lignans are hypotensive, anticarcinogenic, cardiac-protective, lower cholesterol, and lengthen the food's time in the stomach [43]. Because lignans have antioxidant properties, they can reduce oxidative stress and reduce the risk of diabetes-I. In type II diabetes, it can also block the phosphoenolpyruvate carboxykinase, which activates glucogenesis in the liver [44]. For decades, silymarin has been used to cure liver, spleen, and gallbladder illnesses. Hepatoprotective, antioxidant, anti-inflammatory, anticarcinogenic, and antidiabetic activities are found in silymarin [45].

1.8 Hydroxybenzoic acid

In the last ten years, at least three decades, hydroxybenzoic acids have been shown to have biological activity among the diversity of natural phenolic acids. Grapefruit, olive oil, and medlar fruit are all sources of 3-hydroxybenzoic acid [46]. It's a glycosylating enzyme [47]. Carrots, oil palm, grapes, and various other plants have been shown to contain p-hydroxybenzoic acid, including satinwood, peroba, yellow-leaf tree, taheebo, southern catalpa, red sandalwood, chinese chaste tree, betel palm, cuban royal palm, and medlar [46]. Antifungal, antimutagenic, antisickling, estrogenic, and antibacterial properties have been discovered. The freshwater green alga responds to p-Hydroxybenzoic acid by growing faster [48, 49].

Khadem and Marles [46] have summarized the pharmaceutical activities of different hydroxybenzoic acids as mentioned in the following. Pyrocatechuic acid is a radical scavenger, a siderophore, and an antioxidant. Gentisic acid reduces LDL oxidation in humans and is an anti-inflammatory, analgesic, antiarthritic, antirheumatic, and cytostatic drug. Resorcylic acid is a nematicidal substance. For dandruff, ichthyosis, acne, psoriasis, and other skin disorders, salicylic acid has anti-inflammatory, keratolytic, antipyretic, antiseptic, analgesic, and antifungal characteristics. It acts as a hormonal modulator of plant tolerance to disease assaults and environmental stress. 6-Methylsalicylic acid is a toxin found in plants. It works as an antimicrobial and antifeeding agent. Thyroid peroxidase is inhibited by -resorcylic acid. Orsellinic acid has antibacterial properties. Antifungal, anti-inflammatory, antihepatotoxic, antioxidant, cytotoxic, free radical scavenger, apoptotic, chemopreventive, neuroprotective, platelet aggregation inhibitor, and LDL oxidation inhibitor are some of the bioactivities of protocatechuic acid. In addition to its antisickling and anthelmintic properties, vanillic acid has been shown to reduce hepatic fibrosis during liver injury. It's also reported to be a 5′-nucleotidase inhibitor in snake venom. Antibacterial and antioxidant properties are found in isovanillic acid. Syringic acid possesses antibacterial and hepatoprotective properties in addition to being an antioxidant. Digallic acid is cytotoxic and anti-apoptotic. It has antigenotoxic and antioxidant properties as well. For lower plants, it has growth inhibitory and dormancy-inducing properties. Lunularic acid also exhibits antifungal, antialgicidal, and antihyaluronidase properties. Hydrangeic acid has anti-diabetic properties, lowering blood sugar, triglyceride, and free fatty acid levels. Anacardic acid is effective against the larvae of the Colorado potato beetle (Leptinotarsa decemlineata).

Anti-Helicobacter pylori action has been discovered in an anacardic acid combination. Ginkgolic acid suppresses protein SUMOylation in addition to its anticancer and antitubercular properties. SUMO proteins (small ubiquitin-related modifier proteins) regulate various cellular activities linked to cancer and neurological illnesses.

Turgorins are thought to be chemicals that regulate thigmotactic and nyctinastic leaf movement. Current research has discovered that plant hormones do not control nyctinastic leaf movement but rather compounds that differ depending on the plant species. Platensimycin is a gram-positive bacterium (MRSA) inhibitor that inhibits cellular lipid production. Cannabidiolic acid inhibits cyclooxygenase-2 selectively and has antiproliferative properties. Cajaninstilbene acid contains anti-triglyceride and anti-glycemic properties. Cajaninstilbene acid, in addition to being an antioxidant, may be helpful for postmenopausal osteoporosis. It also had impermeability, anti-inflammatory, and analgesic properties [46–49].

1.9 Gallic acid

Tallow-tree, the mangosteen related bridelia, garcinia densivenia, sappanwood, cinnabar ebony, elephant-apple, peroba, guava, water-berry, staghorn sumac, tamarisk, grape, witch-hazel, and red toon all contain gallic acid [46]. It's been used as a styptic and astringent. Gallic acid has antineoplastic and bacteriostatic effects and is antimelanogenic and antioxidant [50]. Evening primrose phenolic fractions containing gallic acid demonstrated antitumor efficacy. It is reported for anticancer effects [51]. Gallic acid is also thought to have the anti-angiogenic properties of sweet leaf tea extract. In the mammalian intestine, gallic acid inhibits sucrase and some disaccharidases. As an anti-HSV-2 agent, Gallic acid showed promise [52]. It inhibits cell survival, invasion, proliferation, and angiogenesis of glioma cells, making it a potential treatment for brain tumors. On the other hand, Tannins have cytotoxic effects on cells other than tumor cells. Apoptosis and necrosis were used to kill Gallic acid-mediated cervical cancer cells [53]. Many gallic acid derivatives have antioxidant and antibacterial properties in nature [46].

1.10 Ellagic acid

Ellagic acid is a polyphenol extractive (tannin) present in various dicotyledons. Ellagic acid is mainly found as ester-linked with sugars in the composition of tannins, which are secondary metabolites in higher plants [54]. The authors note the principal active component for ellagic acid's considerable antioxidant, anti-inflammatory, and gastroprotective activities [55]. Furthermore, ellagic acid's involvement in the GABAergic system, inhibition of acetylcholinesterase, aldose reductase, suppression of proinflammatory markers, protein tyrosine phosphatases, and interaction with the serotonergic and adrenergic systems offer a solid basis for potential advances in the treatment of a variety of medical complications [55, 56]. Recent research suggests that ellagic acid can operate as an acetylcholinesterase inhibitor, raising acetylcholine levels in the brain. As a result, there is the potential to partially mitigate or repair cognitive dysfunctions in neurodegenerative diseases like Alzheimer's [57]. Lastly, one of the ellagic acid's most well-known effects, melanogenesis suppression, has been linked to the antioxidant properties of the compound [58]. Ellagic acid and its derivatives can be used in the supplement and functional food industries because of its anti-inflammatory properties in different cell systems. The development of medications necessitates additional investigation since delivery mechanisms will largely determine ellagic acid bioavailability [59].

1.11 Stilbenes

Stilbenes are phenylpropanoids with a 1,2-diphenylethylene backbone belonging to a small phenylpropanoid category. Transresveratrol is the fundamental unit of most plant stilbenes [60]. Stilbenes are natural antifungal, antiviral, antibacterial, antifungal, and antiviral; they have been demonstrated to have anti-inflammatory characteristics, estrogen receptor agonist properties, and impacts on cell proliferation, cell signaling pathways, and apoptosis [61, 62]. The majority of natural stilbenes are in the trans form. Resveratrol is the only stilbene that has been thoroughly researched and found to have potent anticancer, anti-inflammatory, and antioxidant properties. Pterostilbene has been demonstrated to have anti-diabetic characteristics [63]. Antitubulin properties have been reported for combretastatin [64]. Rhapontigenin has strong inhibitory potential on histamine release, responsible for various allergic reactions. In vitro, resveratrol and rhaponticin can prevent platelet aggregation [65].

1.12 Hydroxycinnamic Acids

The most extensive family of hydroxycinnamic acids comprises phenylalanine and tyrosine and has three-carbon side chains, e.g., p-coumaric, ferulic, caffeic, and sinapic acids. Hydroxycinnamic acids can also be found as amides and esters. Although these forms have been described for industrial and biological potential, there is no evidence to support their use as cosmeceutical components [66]. They have various physiological effects, including anti-inflammatory, antioxidant, antibacterial, anti-melanogenic, and anti-collagenase activity, which drive a surge in using hydroxycinnamic acids in skincare formulations. Antioxidant, antibacterial, anticancer, anti-inflammatory, antiplatelet aggregation, and other intriguing health effects have been discovered on coumaric acid and its derivatives [24]. Caffeic acid is produced via coumaric acid's hydroxylation and possesses anticancer, anti-inflammatory, antibacterial, and antidiabetic effects [67]. Ferulic acid has shown antioxidant, anticancer, UV-absorbing, and anti-inflammatory effects, and it is now being used in cosmetic emulsions for topical application [9]. Antioxidant, anticancer, anti-inflammatory, and antibacterial activities of rosmarinic acid have been discovered [68]. Numerous studies have shown anti-inflammatory, antidiabetic, antiviral, antioxidant, and anti-tyrosinase properties of chlorogenic acid [69]. Fruits and vegetables also contain sinapic acid [70].

1.13 Curcuminoids

Curcuminoids are phenolic chemicals used for spice, color, culinary additives, and medicinal agents. Curcuminoids have exhibited various pharmaceutical effects in preclinical cell culture and animal investigations, including antioxidant, neuroprotective, anticancer, anti-inflammatory, anti-acidogenic, radioprotective, and arthritis [71]. Curcuminoids have also been shown to have a potential therapeutic effect in various chronic disorders, including colon, lung, breast cancer, and inflammatory bowel disease [72]. Ex vivo AChE assay revealed dose-dependent inhibition of curcuminoids and their components in the frontal brain and hippocampus. In scopolamine-induced amnesia, their effect on memory was prominent and was comparable in memory-enhancing impact [73].

Curcuminoids have shown significant antioxidant activity in several in vitro and in vivo studies. They can help individuals with b-thalassemia/Hb E disease reduce oxidative damage. Curcuminoids are antioxidative polyphenols with radiomodulatory characteristics, which allow them to protect non-cancerous cells while radiosensitizing tumor cells [74]. Human cancer cell lines were used to test the antiproliferative effects of curcuminoids and two turmerones substances derived from the rhizome of C. longa. Curcuminoids and turmerone both reduced cancer cell proliferation in a dose-dependent manner. Curcuminoids, turmerone, and Arturmerone's immunomodulatory effects highlighted the potential for curcuminoids and turmerones to be used as chemopreventive agents [75]. Turmeric's curcuminoids and other vital components inhibited the virulence features of Streptococcus mutants' biofilms, for example, bacterial adhesion, acidogenicity, and aciduricity, without killing the target bacteria. These substances can be used to prevent the production of dental biofilms and, as a result, dental caries. Aqeel et al. [76] evaluated the antiacanthamoebic potential of resveratrol and curcuminoids utilizing adhesion and cytotoxicity experiments using primary human brain microvascular endothelial cells, which contribute to the blood-brain barrier. Amoeba binding was reduced by 57% and 73%, respectively, when organisms were pre-exposed to 100 mg resveratrol and DMC, whereas cytotoxicity of host cells was decreased by 86%. According to the findings, resveratrol and DMC have potent anti-acanthamoeba properties [71].

2. Antioxidant activity of secondary metabolites

Secondary metabolites are organic compounds biosynthesized within an organism and not considered necessary for their growth, development, and reproduction. They are not involved in metabolic reactions and are considered neutral, especially in primary metabolic responses. However, they are generally regarded as the compounds of defense of an organism against environmental stresses and predators, signaling molecules, and involved in various molecular interactions like symbiosis, competition, and metal ions transport [77, 78]. They are engaged in improving health as many secondary metabolites act as antibiotics, anabolics, immunomodulators, and growth promoters. Some act as nutraceuticals, fighting against diseases (directly) and aiding the body to fight (indirectly). Some are pesticides, insecticides, and pheromones and displayed established health-promoting effects and significant roles as disease eradicators [79]. More than two million secondary metabolites are known to date, and they are generally classified into alkaloids, flavonoids, polyphenols, phytosterols, and terpenoids. However, McMurry [80] classified them into five main classes: terpenoids and steroids, fatty acid-derived substances and polyketides, alkaloids, nonribosomal polypeptides, and enzyme cofactors. Secondary metabolites are reported mainly from plants (80%). However, many bacterial, fungal, and aquatic organisms like corals, tunicates, snails, and sponges are also reported to contain these compounds [81].

The majority of the secondary metabolites are plant-based (especially tannins, terpenoids, alkaloids, and flavonoids) and represent many vital functions in medicines, culinary, cosmetics, tannery industry, *etc*. However, besides that, the data is scarce regarding the antioxidant activities of the pure secondary metabolites; some of the compounds that are proven antioxidants in nature are mentioned in **Table 1**.

As a result of metabolism, many free radicals are also generated within the living organisms' bodies and are regarded as reactive oxygen species (ROS). These ROS cause oxidative damage to the bodies of living organisms, and the antioxidant species mitigate them by reducing oxidative damage. Hence, they are considered as the first line of defense. Peroxidases and metal chelating proteins help reduce oxidative stress damage together with free radical scavengers like vitamins C and E [83, 84]. There are a few examples of synthetic antioxidants which are used in industry. However, they are not believed to be safe, so the requirement for the antioxidants from natural sources increases, e.g., plants [85].

Secondary metabolite	Category
Chrysin	Flavones
Apigenin	
Naringin and Naringenin	Flavonones
Taxifolin	
Eriodictyol	
Hesperidin	
Isosakuranetin	
Quercetin	Flavonols
Kaempferol	
Rutin	
Astilbin	Flavononols
Engeletin	
Genistin	
Taxifolin	
Daidzin and Daidzein	Isoflavones
Genistein	
(+)-Catechin, (+)-Gallocatechin, (−)-Epicatechin and (−)-Epigallocatechin,	Flavanols
(−)-Epicatechin gallate and (−)-Epigallocatechin gallate	
Cyanidin	
Epigenidin	Anthocyanidins
Delphinium	
Pelargonidin	

Source: Adapted and modified from Naczk and Shahidi [82]

Table 1.
Known natural secondary metabolites with proven antioxidant activities.

Naturally biosynthesized secondary metabolites with enormous antioxidant activity of phenolic nature include flavonoids, terpenes, phenolic acids, lignans, stilbenes, tocopherols, tannins, *etc*. these compounds are biosynthesized in plants having a strong antioxidant potential through which living organisms are somehow protected from various diseases [86]. Among secondary metabolites, alkaloids in an organism protect it from biotic stresses, and phenolics play a protective role against oxidative stresses being strong antioxidants. Plants can protect from UV radiation because they contain phenylpropanoids [87]. Polyphenolic compounds are considered a prime group responsible for antioxidant activities [88, 89].

The food consumed containing phenolic compounds displays an antioxidant role due to these antioxidant compounds (**Figure 1**) [90]. Terpenoids are a broad category of secondary metabolites regarded as strong antioxidants and used mostly in perfumery [91]. Stilbenes are phytoalexins biosynthesized in plants to overcome stresses are reported for antioxidant properties and resveratrol; for example, they are an active constituent of many medications. Isoflavones are polyphenolic biomolecules,

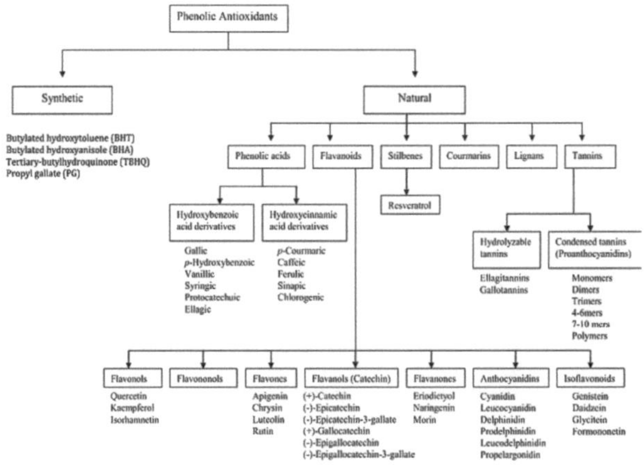

Figure 1.
Phenolic antioxidants. Adapted from Shahidi and Ambigaipalan [85].

biosynthesized in the Fabaceae family, especially in soybean in the form of glycosides, and exhibit antioxidant activities. Tannins are complex derivatives of phenolic acids, are found in many plant species, and are enormously effective antioxidants with promising cytotoxic and antiparasitic properties [92, 93].

There are few antioxidants synthesized and allowed to be used in the food industry, including butylated hydroxytoluene (BHT), butylated hydroxyanisole (BHA), octyl gallate (OG), propyl gallate (PG), dodecyl gallate (DG) and tertiary-butylhydroquinone (TBHQ) [94] to delay lipid oxidation and as processing agent of oils and fats [95].

3. Antimicrobial activity of secondary metabolites

Secondary metabolites include alkaloids, flavonoids, terpenoids, and other phenolic compounds; these molecules are linked to plant defense processes and protect against many diseases. Secondary metabolites are involved in antibacterial and antifungal activities [96].

3.1 Antibacterial properties of secondary metabolites

Bacterial infections are considered a significant public health problem worldwide. Bacterial infection can also occur due to multi-drug resistance, which leads to mortality and morbidity [97]. For that reason, antibiotic resistance has become a global concern. The increase in the multi-drug resistance of bacteria threatens the therapeutic efficacy of several drugs. Using different solvent systems, numerous researchers have studied plants' antibacterial activities of leaves, flowers, stems, roots, and fruits [98]. Therefore, new antibacterial drugs are needed to treat various diseases with low toxicity and less price. For that purpose, secondary metabolites from plants are currently considered to develop new drugs because they are rich in natural compounds.

Gallic acid and its derivatives are potential antibacterial agents that reduce bacterial diseases. Gallic acid and methyl gallate have shown significant antibacterial activity against *Salmonella* [99]. Phenolic compounds such as stilbenes, tannins, and isoflavones inhibited the growth of *Bacillus* and *E. coli* bacteria [100, 101]. Anacardic acid analogs extracted from the *A. Ovest* with various side chains exhibited antibacterial activity against *S. aureus* and *S. pyogenes*. On the other hand, alkaloids are used as scaffolding substructures in other antibacterial drugs, such as linezolid and trimethoprim. The alkaloid cocsoline from *Epinetrumvillosum* has broad antibacterial activity by inhibiting *Shigella*, *Campylobacter jejuni*, and *C. coli* stains [102, 103]. Squalamine, a polyamine alkaloid extracted from the tissue of the squalus shark acanthosis, revealed broad-spectrum steroidal antibiotic with potent bactericidal properties against both gram-positive and gram-negative bacterial stains [104]. Other alkaloids such as solanine, solasodine, and B-solamarine were isolated from *Solanum dulcamara* L. have shown antibacterial activity against *S. aureus* [97]. Bis-indole alkaloids from marine invertebrates have demonstrated antibacterial activity against *S. aureus* and *methicillin-resistant S. aureus* (MRSA) [105]. Berberine and hydrastine alkaloids were isolated from Goldenseal have been showing substantial antibacterial activity, particularly against *S. pyogenes* and *S. aureus* [106]. Cocsoline alkaloid isolated from Epinetrumvillosum (Exell) possesses antibacterial activity against *Shigella* strains, *Campylobacter jejuni*, and *C. coli* [107]. Tetrahydroanthraquinones are also exhibiting antibacterial activity. *Pseudomonas aeruginosa* and other gram-positive bacteria were suppressed by Altersolanols A–C and E. The antibacterial activity of tetrahydroanthraquinones is due to the presence of the hydroxy group at the C-5 position [108]. Coniothranthraquinone 1 has demonstrated antibacterial activity against *S. aureus*, while trichodermaquinone had antibacterial activity against MRSA [109, 110]. Deoxybostrycin and bostrycin have significant antibacterial properties against *S. aureus*, *E. coli*, *Pseudomonas aeruginosa*, *Sarcina ventriculi*, and *Bacillus subtilis* [111].

3.2 Antifungal properties

Resistance to antifungal drugs has been spread in recent years. Resistance to antifungal drugs has led to increased morbidity and mortality. Since the molecular mechanisms in humans and fungi are so similar, there is always the possibility that the fungal cytotoxic agent is toxic to host cells. As a result, patients with compromised immune systems, such as transplants, cancer patients, and diabetics, who do not respond effectively to current antifungal treatments, need new antifungal therapies. Antifungal drugs currently used to treat fungal infections have significant side effects such as itching, diarrhea, vomiting, *etc*. In addition, it is less effective because of the development of drug resistance by the many fungi [112, 113]. The alkaloids protoberberine jatrorrhizine, isolated from *Mahonia aquifolium,* were the most potent inhibitory antifungal activity [114]. (+)-Cocsoline is a bisbenzylisoquinoline alkaloid isolated from epinetrumvillosum whose antifungal action has been demonstrated [115]. The alkaloids N-ethylhydrasteinehydroxylactam and 1-methoxyberberine chloride isolated from *Corydalis longipes* have been shown to have significant inhibitory action [116]. *Glaucium oxylobum* produced four alkaloids: dicentrine, glaucine, protopine, and alpha-allocryptopin exhibited antifungal activity against *Microsporumgypseum*, *Microsporumcanis*, *T. mentagrophytes*, and *Epidermophytonfloccosum* [117]. Canthin-6-one and 5-methoxy-canthin-6-one from *Zanthoxylumchiloperone* var. *angustifolium* are antifungal against *Candida albicans*, *Aspergillus fumigatus*, and *T. mentagrophytes* [118]. Frangulanine, a cyclic peptide alkaloid, and waltherione A, quinolinone alkaloids derived from *Melochiaodorata* have been shown to antifungal activity against a wide range of pathogenic fungi [119]. Additionally, anodic alkali aninolinate has been shown to have antifungal action [120]. Two antimyctic fructoxin alkaloids have been identified from the root of Dictamnusdasycarpus. 3-Methoxisampangin from cleistopholispatens significantly inhibits *C. albicans*, *A. fumigatus*, and *C. neoformans* [121]. A new alkaloid, 2-(3,4-dimethyl-2,5-dihydro-1H-pyrrol-2-yl)-1-methylethyl pentanoate, was isolated from the plant *Datura metel* has shown *in-vitro* and *in-vivo* action against *Aspergillus* and *Candida species* [122]. Fungi toxic action was demonstrated for alkaloids isolated from *Ruta graveolens* L., Tomadini Glycoalkaloids isolated from tomatoes, cannabinoid alkaloid, isoquinoline, methaqualone, flavonol, and gallic acid [123, 124].

4. Anticancer potential of secondary metabolites

Cancer is the cause of death worldwide; experts are developing new therapies less likely to cause side effects. Cancer is one of the most severe health concerns, despite substantial advances in cancer therapy [125]. Several new secondary metabolites from plants are discovered each year, opening new avenues for research in the fight against cancer. Plant secondary metabolites have substantially contributed to this topic, which has been at the heart of herbal medicines. Plant's secondary metabolites have been shown to have anticancer effects, such as the ability to reduce cancer cell growth and development, kill cancer cells, and fight against multi-drug resistance in certain malignancies [126]. Plant secondary metabolites are thought to be helpful in drug development. The secondary plant metabolites are presently used in clinical and undergoing clinical trials as anticancer therapies [127, 128].

For thousands of years, humans have used herbs to treat certain diseases. Researchers are particularly interested in generating anticancer drugs from the plant's secondary metabolites. Plant secondary metabolites such as flavonoids, polyphenols, anthraquinones, triterpenoids, alkaloids, terpenoids, quinones, and others play an essential role in cancer prevention [129]. Flavonoids (6,7,30-trimethoxy-3,5,40-trihydroxy-flavone and 5,40-dihydroxy-3,6,30-trimethoxy-flavone 7-O- -d-glucoside) isolated from *Chrysosplenium nudicaule Spearmary* was reported as cytotoxic and antitumor activities in cancer cell growth of human leukemia and gastric cancer cell lines [130, 131]. The agathisflavone induces apoptosis and antiproliferative effect on the development of leukemia cells. Citrus flavonoids have a profound inhibitory effect on the development of leukemia cells. Other research suggests that quercetin may act as an antiproliferative agent by inhibiting cell proliferation, growth, and cell cycle termination [132, 133]. Studies of Kaempferol and quercetin have shown antiproliferative action by inhibiting the development of the human colon (HT-29, COLO 201, and LS-174T), breast (MCF-7 ADRr), and ovarian (OVCA 433) cancer cell lines [134–136]. In addition, quercetin inhibited the G1 phase of the cell cycle in human leukemic T-cells and human gastric cancer cells [137, 138]. In a human oral squamous carcinoma cell line (SCC-25), quercetin had a biphasic effect on cell growth and proliferation [139]. On the other hand, *in-vivo* research on quercetin has yielded consistent findings, indicating a promising chemopreventive drug against skin cancer [140]. In contrast, kaempferol treatment of the human lung cancer cell line A549 resulted in a dosage and time-dependent decrease in cell survival and DNA synthesis. While Kaempferol dramatically decreased the number of breast cancer cells (MCF-7) viable estrogen receptor-positive [141, 142].

Phenolic compounds are one of the most diverse and widespread groups of plant metabolites, and they have a wide range of biological roles in regulating carcinogenesis [143]. Polyphenols have several advantages as anticancer drugs, including high accessibility, minimal toxicity, and broad biological effects. The main advantage of polyphenols as anticancer drugs is cytotoxic effects on malignant cells growth [144, 145]. Many polyphenols have an anticancer effect in various cancer models, regardless of their different modes of action [146, 147]. Polyphenols of strawberries, including anthocyanins, Kaempferol, quercetin, coumaric acid esters, and ellagic acid esters, have been shown to inhibit the development of human oral and breast colon and prostate cancer cell lines [148]. The primary polyphenol of green tea, epigallocatechin-3-gallate (EGCG), is anticancer in various cancer types [149]. Researchers suggested that EGCG regulation may stimulate the production of reactive oxygen species and inhibit angiogenesis in cancer cells by regulating different pathways, such as AMP-activated protein kinase, epidermal growth factor receptor, insulin-like growth factor receptor, extracellular signal-regulated kinase, cyclin D1, Akt, STAT3, Wnt, and mTOR signaling in cancer cells [150–152]. A key ingredient of *Plumbago zeylanica* naphthoquinone has been shown *in-vitro* and *in-vivo* anticancer effective against various malignancies, including breast, pancreatic, lung, prostate, melanoma, and leukemia [153]. Cardanol, anacardic acid, and methyl cardol have been shown to decrease the cell growth of Hela cells and pituitary adenoma cells [154, 155]. In addition, anacardic acid-induced polymerase breakage, cell arrest, and regulation of apoptosis and anti-apoptotic proteins [156]. Furthermore, *in-vivo* investigations have confirmed plant-derived phenolic compounds' anticancer activity [157]. Colon, lung, breast, liver, prostate, stomach, esophagus, small intestine, pancreas mammary gland, and skin cancers are using xenograft animal models [158]. In another study of cyanidin-3-glucoside (C3G), the major anthocyanin in blackberry was investigated for the inhibition of 7,12-dimethylbenz[a]anthracene (DMBA)-12-O-tetradecanolyphorbol-13-acetate (TPA)-induced skin papillomas in an animal model [159]. Similarly, natural anthraquinones, such as rhein and emodin, have antitumor properties [160]. Tetrahydroanthraquinones, a kind of anthraquinone, inhibit cell proliferation, invasion, metastasis, and angiogenesis by apoptosis and cell cycle arrest. Altersolanol A (tetrahydroanthraquinone) has anticancer properties against bladder, colon, and stomach cancer. Moreover, Altersolanol A anticancer efficacy is linked to its pro-apoptotic and antiinvasive properties. A study reported that Altersolanol A has anticancer potential by reducing angiogenesis *in-vitro* and *in-vivo* [161, 162]. In addition, Altersolanol F reduced the viability of colorectal and cervical cancer cells, while Altersolanol N has cytotoxic effect against murine cancer cell line (L5178Y) [163, 164]. Likewise, several investigations have demonstrated catechins as antiproliferative properties in breast, colon, melanoma, and prostate cancer cells [165–167].

Isoquinoline alkaloid is a major alkaloid class with an anticancer effect in different cancer cells. Isoquinoline alkaloids are naturally isolated from the roots, and the bark of *Coptis chinensis* are important sources of [168]. Studies found that protoberberines (isoquinoline alkaloids) have significant anticancer potential in the treatment of gastric cancer [169]. Similarly, berberine alkaloid has been reported to have anticancer effects by suppressing the ERK/JNK/p38 MAPK/mTOR/p70 ribosomal S6 protein kinase and PI3K/Akt signaling pathways in cancer studies [170]. Tetrandrine (TET), a natural bis-benzylisoquinoline alkaloid, has shown anticancer activity against cancer cell lines. Tetrandrine-mediated cytotoxicity of chemotherapeutic drugs used to treat gastric cancer, including paclitaxel, 5-FU, oxaliplatin, and docetaxel [171, 172]. Piperlongumine, an amide alkaloid, has been shown anticancer by the intracellular ROS, p38/JNK signaling pathway [173, 174]. Hersutin alkaloid has been shown to induce apoptosis in HER2-positive and p53-mutated breast cancer cells [175].

Oxymatrine, a natural alkaloid isolated from the roots of *Sophora chrysophylla*, exhibits anticancer activity in human cervical cancer cells [176].

Terpenes are a broad category of secondary metabolites that include low polarity fragrant scaffolds and isoprene derivatives with various pharmacological activities, including anticancer activity. Triterpenoids have previously been shown to have anticancer properties in both *in-vitro* and *in-vivo* by nuclear factor-κβ (NF-κβ) and STAT3 signaling pathways [177]. The anticancer and narcotic activities of costunolide, a sesquiterpene lactone isolated from *Saussurea lappa,* have been demonstrated in gastrointestinal diseases [178]. Thymoquinone has been shown to slow the progression of diseases such as leukemia, breast adenocarcinoma, colorectal, pancreatic, prostate, and hepatic cancer [179]. The anticancer efficacy of thymoquinone against gastric cancer cells. Several other studies have shown that the combination of thymoquinone with 5-fluorouracil and cisplatin significantly improves the chemotherapeutic-induced anticancer effects in gastric cancer. Furthermore, thymoquinone has been shown to inhibit the Janus kinase (JAK)/STAT3 signaling pathway [180].

5. Conclusions

This study shows that plant cells produce a variety of compounds, mainly secondary metabolites, for defense mechanisms against bacteria, fungi, antioxidants, and cancer. Secondary metabolites with antibacterial, antifungal, antioxidant, and anticancer effects are sources of natural bioactive molecules, which control disease-causing pathogens in plants and humans. In addition, the different plant families have shown a unique combination of secondary metabolites; therefore, exhibiting different antibacterial, antifungal, antioxidant, and anticancer activities. The emerging research on identifying secondary metabolites is ongoing, and further research is encouraged to advance our knowledge about these compounds. Secondary metabolites can help treat infectious diseases that have increased resistance to current antibiotics. They can offer alternative medical therapy to individuals, particularly in developing nations where people may not access health care.

Acknowledgements

The authors are thankful to the Department of Biological Sciences, National University of Medical Sciences, Rawalpindi, Pakistan, for supporting this study. We also apologize to the authors of many exciting studies omitted due to limited information.

Conflict of interest

The authors declare that they have no conflict of interest.

DOI: http://dx.doi.org/10.5772/intechopen.103698

Author details

Muhammad Zeeshan Bhatti[1]*, Hammad Ismail[2] and Waqas Khan Kayani[3]

1 Department of Biological Sciences, National University of Medical Sciences, Rawalpindi, Pakistan

2 Department of Biochemistry and Biotechnology, University of Gujrat, Gujrat, Pakistan

3 Faculty of Basic and Applied Sciences, Department of Biotechnology, University of Kotli, Azad Jammu and Kashmir, Pakistan

*Address all correspondence to: zeeshan.bhatti@numspak.edu.pk

IntechOpen

© 2022 The Author(s). Licensee IntechOpen. This chapter is distributed under the terms of the Creative Commons Attribution License (http://creativecommons.org/licenses/by/3.0), which permits unrestricted use, distribution, and reproduction in any medium, provided the original work is properly cited.

References

[1] Guerriero G, Berni R, Muñoz-Sanchez JA, Apone F, Abdel-Salam EM, Qahtan AA, et al. Production of plant secondary metabolites: Examples, tips and suggestions for biotechnologists. Genes. 2018;**9**(6):309

[2] Singh R. Medicinal plants: A review. Journal of Plant Sciences. 2015;**3**(1):50-55

[3] Tiwari R, Rana C. Plant secondary metabolites: A review. International Journal of Engineering Research and General Science. 2015;**3**(5):661-670

[4] Velu G, Palanichamy V, Rajan AP. Phytochemical and pharmacological importance of plant secondary metabolites in modern medicine. In: Roopan SM, Madhumitha G, editors. Bioorganic Phase in Natural Food: An Overview. 1 ed. Springer International Publishing AG, part of Springer Nature Springer, Cham; 2018. p. 135-156

[5] Ghasemzadeh A, Ghasemzadeh N. Flavonoids and phenolic acids: Role and biochemical activity in plants and human. Journal of Medicinal Plant Research. 2011;**5**(31):6697-6703

[6] Saxena M, Saxena J, Nema R, Singh D, Gupta A. Phytochemistry of medicinal plants. Journal of pharmacognosy and phytochemistry. 2013;**1**(6):168-82

[7] Lin M, Han P, Li Y, Wang W, Lai D, Zhou L. Quinoa secondary metabolites and their biological activities or functions. Molecules. 2019;**24**(13):2512

[8] Yao Y, Chen Y, Adili R, McKeown T, Chen P, Zhu G, et al. Plant-based food cyanidin-3-glucoside modulates human platelet glycoprotein VI signaling and inhibits platelet activation and thrombus formation. The Journal of nutrition. 2017;**147**(10):1917-1925

[9] Kumar N, Goel N. Phenolic acids: Natural versatile molecules with promising therapeutic applications. Biotechnology Reports. 2019;**24**:e00370

[10] Pattnaik M, Pandey P, Martin GJ, Mishra HN, Ashokkumar M. Innovative technologies for extraction and microencapsulation of bioactives from plant-based food waste and their applications in functional food development. Food. 2021;**10**(2):279

[11] Stevenson D, Hurst R. Polyphenolic phytochemicals–just antioxidants or much more? Cellular and Molecular Life Sciences. 2007;**64**(22):2900-2916

[12] Ong KW, Hsu A, Tan BKH. Anti-diabetic and anti-lipidemic effects of chlorogenic acid are mediated by ampk activation. Biochemical Pharmacology. 2013;**85**(9):1341-1351

[13] Świderska E, Strycharz J, Wróblewski A, Szemraj J, Drzewoski J, Śliwińska A. Role of PI3K/AKT pathway in insulin-mediated glucose uptake. Blood Glucose Levels. 2018;**1**:1-18

[14] Vinayagam R, Jayachandran M, Xu B. Antidiabetic effects of simple phenolic acids: A comprehensive review. Phytotherapy Research. 2016;**30**(2):184-199

[15] Cueva C, Moreno-Arribas MV, Martín-Álvarez PJ, Bills G, Vicente MF, Basilio A, et al. Antimicrobial activity of phenolic acids against commensal, probiotic and pathogenic bacteria. Research in Microbiology. 2010;**161**(5): 372-382

[16] Efferth T, Li PC, Konkimalla VSB, Kaina B. From traditional Chinese medicine to rational cancer therapy. Trends in Molecular Medicine. 2007;**13**(8):353-361

[17] Abdal Dayem A, Choi HY, Yang G-M, Kim K, Saha SK, Cho S-G. The anti-cancer effect of polyphenols against breast cancer and cancer stem cells: Molecular mechanisms. Nutrients. 2016;**8**(9):581

[18] Choi J, Jiang X, Jeong JB, Lee S-H. Anticancer activity of protocatechualdehyde in human breast cancer cells. Journal of Medicinal Food. 2014;**17**(8):842-848

[19] Karak P. Biological activities of flavonoids: An overview. International Journal of Pharmaceutical Sciences and Research. 2019;**10**(4):1567-1574

[20] Tiwari SC, Husain N. Biological activities and role of flavonoids in human health–A. Indian Journal of Science Research. 2017;**12**(2):193-196

[21] Bernatoniene J, Kopustinskiene DM. The role of catechins in cellular responses to oxidative stress. Molecules. 2018;**23**(4):965

[22] Pan M-H, Lai C-S, Ho C-T. Anti-inflammatory activity of natural dietary flavonoids. Food & Function. 2010;**1**(1):15-31

[23] Zandi K, Teoh B-T, Sam S-S, Wong P-F, Mustafa MR, AbuBakar S. Antiviral activity of four types of bioflavonoid against dengue virus type-2. Virology Journal. 2011;**8**(1):1-11

[24] González Mera IF, González Falconí D, Morera CV. Secondary metabolites in plants: Main classes, phytochemical analysis and pharmacological activities. Bionatura. 2019;**4**(4):1000-1009

[25] Olivoto T, Nardino M, Carvalho IR, Follmann DN, Szareski V, Jardel I, et al. Plant secondary metabolites and its dynamical systems of induction in response to environmental factors: A review. African Journal of Agricultural Research. 2017;**12**(2):71-84

[26] Loreto F, Dicke M, Schnitzler JP, Turlings TC. Plant volatiles and the environment. Plant, Cell & Environment. 2014;**37**(8):1905-1908

[27] Güçlü-Üstündağ Ö, Mazza G. Saponins: Properties, applications and processing. Critical Reviews in Food Science and Nutrition. 2007;**47**(3):231-258

[28] Cibulski SP, Rivera-Patron M, Mourglia-Ettlin G, Casaravilla C, Yendo ACA, Fett-Neto AG, et al. Quillaja brasiliensis saponin-based nanoparticulate adjuvants are capable of triggering early immune responses. Scientific Reports. 2018;**8**(1):1-15

[29] Wang C-Z, Cai Y, Anderson S, Yuan C-S. Ginseng metabolites on cancer chemoprevention: An angiogenesis link? Diseases. 2015;**3**(3):193-204

[30] Saklani A, Kutty SK. Plant-derived compounds in clinical trials. Drug Discovery Today. 2008;**13**(3-4):161-171

[31] Roopashree K, Naik D. Saponins: Properties, applications and as insecticides: A review. Bioscience Trends. 2019;**8**:1-14

[32] Liao Y, Li Z, Zhou Q, Sheng M, Qu Q, Shi Y, et al. Saponin surfactants used in drug delivery systems: A new application for natural medicine components. International Journal of Pharmaceutics. 2021;**603**:120709

[33] Pizzi A. Tannins medical/pharmacological and related applications:

A critical review. Sustainable Chemistry and Pharmacy. 2021;**22**:100481

[34] Miele S, Tegli S, Izquierdo CG, Cerboneschi M, Bargiacchi E. Hydrolysable tannins in agriculture. Tannins-Structural Properties, Biological Properties and Current Knowledge. 2019

[35] Fraga-Corral M, Otero P, Echave J, Garcia-Oliveira P, Carpena M, Jarboui A, et al. By-products of agri-food industry as tannin-rich sources: A review of Tannins' biological activities and their potential for valorization. Food. 2021;**10**(1):137

[36] Subhan N, Burrows GE, Kerr PG, Obied HK. Phytochemistry, ethnomedicine, and pharmacology of Acacia. Studies in Natural Products Chemistry. 2018;**57**:247-326

[37] Cock IE. Medicinal and aromatic plants–Australia. Ethnopharmacology, Encyclopedia of Life Support Systems (EOLSS). 2011

[38] Nadeem M, Taj Khan I, Khan F, Ajmal SM. Lignans and flavonolignans. In: Silva AS, Nabavi SF, Saeedi M, Nabavi SM, editors. Recent advances in natural products analysis. Elsevier Inc.: Elsevier; 2020. p. 98-11

[39] Park D, Huang T, Frishman WH. Phytoestrogens as cardioprotective agents. Cardiology in Review. 2005;**13**(1):13-17

[40] Patade A, Devareddy L, Lucas EA, Korlagunta K, Daggy BP, Arjmandi BH. Flaxseed reduces total and LDL cholesterol concentrations in Native American postmenopausal women. Journal of Women's Health. 2008;**17**(3):355-366

[41] Touillaud MS, Thiébaut AC, Fournier A, Niravong M, Boutron-Ruault M-C, Clavel-Chapelon F. Dietary lignan intake and postmenopausal breast cancer risk by estrogen and progesterone receptor status. Journal of the National Cancer Institute. 2007;**99**(6):475-486

[42] Ganorkar PM, Jain RK. Flaxseed-a nutritional punch. International Food Research Journal. 2013;**20**(2):519-525

[43] Imran M, Ahmad N, Anjum FM, Khan MK, Mushtaq Z, Nadeem M, et al. Potential protective properties of flax lignan secoisolariciresinol diglucoside. Nutrition Journal. 2015;**14**(1):1-7

[44] Wall S, Chow J, Huang J, Jiang M, Yang A, Jones K. Phosphoenolpyruvate carboxykinase: Possible therapeutic targets for insulin resistant Type-II Diabetes. The FASEB Journal. 2015;**29**:LB67

[45] Vargas-Mendoza N, Madrigal-Santillán E, Morales-González Á, Esquivel-Soto J, Esquivel-Chirino C, y González-Rubio MG-L, Gayosso-de-Lucio JA, Morales-González JA. Hepatoprotective effect of silymarin. World Journal of Hepatology. 2014;**6**(3):144

[46] Khadem S, Marles RJ. Monocyclic phenolic acids; hydroxy-and polyhydroxybenzoic acids: Occurrence and recent bioactivity studies. Molecules. 2010;**15**(11):7985-8005

[47] Sircar D, Mukherjee C, Beuerle T, Beerhues L, Mitra A. Characterization of p-hydroxybenzaldehyde dehydrogenase, the final enzyme of p-hydroxybenzoic acid biosynthesis in hairy roots of Daucus carota. Acta Physiologiae Plantarum. 2011;**33**(5):2019-2024

[48] Dias MI, Sousa MJ, Alves RC, Ferreira IC. Exploring plant tissue culture to improve the production of phenolic compounds: A review. Industrial Crops and Products. 2016;**82**:9-22

[49] Manuja R, Sachdeva S, Jain A, Chaudhary J. A comprehensive review on biological activities of p-hydroxy benzoic acid and its derivatives. International Journal of Pharmaceutical Science Research. 2013;**22**(2):109-115

[50] Badhani B, Sharma N, Kakkar R. Gallic acid: A versatile antioxidant with promising therapeutic and industrial applications. RSC Advances. 2015;**5**(35):27540-27557

[51] Verma S, Singh A, Mishra A. Gallic acid: Molecular rival of cancer. Environmental Toxicology and Pharmacology. 2013;**35**(3):473-485

[52] Kahkeshani N, Farzaei F, Fotouhi M, Alavi SS, Bahramsoltani R, Naseri R, et al. Pharmacological effects of gallic acid in health and diseases: A mechanistic review. Iranian Journal of Basic Medical Sciences. 2019;**22**(3):225

[53] Lu Y, Jiang F, Jiang H, Wu K, Zheng X, Cai Y, et al. Gallic acid suppresses cell viability, proliferation, invasion and angiogenesis in human glioma cells. European Journal of Pharmacology. 2010;**641**(2-3):102-107

[54] Evtyugin DD, Magina S, Evtuguin DV. Recent advances in the production and applications of ellagic acid and its derivatives. A review. Molecules. 2020;**25**(12):2745

[55] Sagun K, Roy VK, Kumar RS, Ibrahim KS, Parimelazhagan T, Kumar NS, et al. Antioxidant potential, anti-inflammatory activity and gastroprotective mechanisms of Mallotus roxburghianus (Muell.) against ethanol-induced gastric ulcers in Wistar albino rats. Journal of Functional Foods. 2017;**36**:448-458

[56] Akileshwari C, Raghu G, Muthenna P, Mueller NH, Suryanaryana P, Petrash JM, et al. Bioflavonoid ellagic acid inhibits aldose reductase: Implications for prevention of diabetic complications. Journal of Functional Foods. 2014;**6**:374-383

[57] Jha AB, Panchal SS, Shah A. Ellagic acid: Insights into its neuroprotective and cognitive enhancement effects in sporadic Alzheimer's disease. Pharmacology, Biochemistry, and Behavior. 2018;**175**:33-46

[58] Ortiz-Ruiz CV, Berna J, Tudela J, Varon R, Garcia-Canovas F. Action of ellagic acid on the melanin biosynthesis pathway. Journal of Dermatological Science. 2016;**82**(2):115-122

[59] Ríos J-L, Giner RM, Marín M, Recio MC. A pharmacological update of ellagic acid. Planta Medica. 2018;**84**(15):1068-1093

[60] Tsopmo A, Awah FM, Kuete V. Lignans and Stilbenes from African Medicinal Plants. In: Kuete V, editor. Medicinal Plant Research in Africa. Oxford: Elsevier; 2013. p. 435-478

[61] Kasiotis KM, Pratsinis H, Kletsas D, Haroutounian SA. Resveratrol and related stilbenes: Their anti-aging and anti-angiogenic properties. Food and Chemical Toxicology. 2013;**61**:112-120

[62] Sirerol JA, Rodríguez ML, Mena S, Asensi MA, Estrela JM, Ortega AL. Role of natural stilbenes in the prevention of cancer. Oxidative Medicine and Cellular Longevity. 2016;**2016**(Article ID 3128951):1-15

[63] Tastekin B, Pelit A, Polat S, Tuli A, Sencar L, Alparslan MM, et al. Therapeutic potential of pterostilbene and resveratrol on biomechanic, biochemical, and histological parameters in streptozotocin-induced diabetic rats. Evidence-based Complementary and Alternative Medicine. 2018;**2018**

[64] Nainwal LM, Alam MM, Shaquiquzzaman M, Marella A, Kamal A. Combretastatin-based compounds with therapeutic characteristics: A patent review. Expert Opinion on Therapeutic Patents. 2019;**29**(9):703-731

[65] Dvorakova M, Landa P. Anti-inflammatory activity of natural stilbenoids: A review. Pharmacological Research. 2017;**124**:126-145

[66] Taofiq O, González-Paramás AM, Barreiro MF, Ferreira IC. Hydroxycinnamic acids and their derivatives: Cosmeceutical significance, challenges and future perspectives, a review. Molecules. 2017;**22**(2):281

[67] Espíndola KMM, Ferreira RG, Narvaez LEM, Silva Rosario ACR, da Silva AHM, Silva AGB, et al. Chemical and pharmacological aspects of caffeic acid and its activity in hepatocarcinoma. Frontiers in Oncology. 2019;**9**:541

[68] Huerta-Madroñal M, Caro-León J, Espinosa-Cano E, Aguilar MR, Vázquez-Lasa B. Chitosan–Rosmarinic acid conjugates with antioxidant, anti-inflammatory and photoprotective properties. Carbohydrate Polymers. 2021;**273**:118619

[69] Bahadori MB, Dinparast L, Zengin G, Sarikurkcu C, Bahadori S, Asghari B, et al. Functional components, anti-diabetic, anti-Alzheimer's disease, and antioxidant activities of Salvia syriaca L. International Journal of Food Properties. 2017;**20**(8):1761-1772

[70] Placines C, Castañeda-Loaiza V, João Rodrigues M, Pereira C, Stefanucci A, Mollica A, et al. Phenolic profile, toxicity, enzyme inhibition, in silico studies, and antioxidant properties of Cakile maritima scop.(Brassicaceae) from southern Portugal. Plants. 2020;**9**(2):142

[71] Amalraj A, Pius A, Gopi S, Gopi S. Biological activities of curcuminoids, other biomolecules from turmeric and their derivatives–A review. Journal of Traditional and Complementary Medicine. 2017;**7**(2):205-233

[72] Kunnumakkara AB, Bordoloi D, Padmavathi G, Monisha J, Roy NK, Prasad S, et al. Curcumin, the golden nutraceutical: Multitargeting for multiple chronic diseases. British Journal of Pharmacology. 2017;**174**(11):1325-1348

[73] Ahmed T, Enam S, Gilani A. Curcuminoids enhance memory in an amyloid-infused rat model of Alzheimer's disease. Neuroscience. 2010;**169**(3):1296-1306

[74] Naik SR, Thakare VN, Patil SR. Protective effect of curcumin on experimentally induced inflammation, hepatotoxicity and cardiotoxicity in rats: Evidence of its antioxidant property. Experimental and Toxicologic Pathology. 2011;**63**(5):419-431

[75] Yue GG, Chan BC, Hon P-M, Lee MY, Fung K-P, Leung P-C, et al. Evaluation of in vitro anti-proliferative and immunomodulatory activities of compounds isolated from Curcuma longa. Food and Chemical Toxicology. 2010;**48**(8-9):2011-2020

[76] Vieira ILBF, de Souza DCP, da Silva CL, Chen LC, Guillo LA. In vitro mutagenicity and blood compatibility of paclitaxel and curcumin in poly (DL-lactide-co-glicolide) films. Toxicology in Vitro. 2013;**27**(1):198-203

[77] Kinghorn AD. The discovery of drugs from higher plants. Discovery of Novel Natural Products with Therapeutic Potential. 1994:81-108

[78] Demain AL, Fang A. The natural functions of secondary metabolites.

History of modern biotechnology I. 2000:1-39

[79] Thirumurugan D, Cholarajan A, Raja S, Vijayakumar R. An introductory chapter: Secondary metabolites. Second Metab—sources Application. 2018:1-21

[80] McMurry JE. Organic chemistry with biological applications. 3rd edition. Stamford, USA: Cengage Learning, Inc; 2014

[81] Berdy J. Bioactive microbial metabolites. The Journal of antibiotics. 2005;**58**(1):1-26

[82] Naczk M, Shahidi F. Extraction and analysis of phenolics in food. Journal of Chromatography A. 2004;**1054**(1):95-111

[83] Halliwell B, Gutteridge JM. Free radicals in biology and medicine: 540 pp. 1989. Price: $98 US hardcover; $39.95 US softbound; $50£ Outside US. Pergamon, 1991

[84] Niki E. Free Radicals in Chemistry and Biochemistry. In: Hiramatsu M, Yoshikawa T, Inoue M, editors. Food and Free Radicals. Boston, MA: Springer US; 1997. p. 1-10

[85] Shahidi F, Ambigaipalan P. Phenolics and polyphenolics in foods, beverages and spices: Antioxidant activity and health effects–A review. Journal of Functional Foods. 2015;**18**:820-897

[86] Hou W-C, Lin R-D, Cheng K-T, Hung Y-T, Cho C-H, Chen C-H, et al. Free radical-scavenging activity of Taiwanese native plants. Phytomedicine. 2003;**10**(2-3):170-175

[87] Dietrich H. Bioactive compounds in fruit and juice. Fruit Processing. 2004;**1**:50-55

[88] Liazid A, Palma M, Brigui J, Barroso CG. Investigation on phenolic compounds stability during microwave-assisted extraction. Journal of Chromatography. A. 2007;**1140**(1-2):29-34

[89] Magalhães AS, Silva BM, Pereira JA, Andrade PB, Valentão P, Carvalho M. Protective effect of quince (Cydonia oblonga Miller) fruit against oxidative hemolysis of human erythrocytes. Food and Chemical Toxicology. 2009;**47**(6):1372-1377

[90] Swallah MS, Sun H, Affoh R, Fu H, Yu H. Antioxidant Potential Overviews of Secondary Metabolites (Polyphenols) in Fruits. International Journal of Food Science. 2020;2020(Article ID 9081686):1-8

[91] Ohloff, G. Scent and Fragrances: The Fascination of Odors and Their Chemical Perspectives; Spinger-Verlag: Berlin, 1994; pp 127-197

[92] van Wyk B-E, Wink M. Phytomedicines, herbal drugs, and poisons: Chicago: The University of Chicago Press; Royal Botanic Gardens, Kew: Kew Publishing; 2015

[93] Wink M, Van Wyk B-E. Mind-altering and Poisonous Plants of the World. Timber Press Portland, 2008

[94] Makahleh A, Saad B, Bari MF. Synthetic phenolics as antioxidants for food preservation. In: Shahidi F, editor. Handbook of Antioxidants for Food Preservation. Woodhead Publishing: Elsevier 2015. p. 51-78

[95] Saad B, Sing YY, Nawi MA, Hashim N, Ali ASM, Saleh MI, et al. Determination of synthetic phenolic antioxidants in food items using reversed-phase HPLC. Food Chemistry. 2007;**105**(1):389-394

[96] Tepe B, Daferera D, Sokmen A, Sokmen M, Polissiou M. Antimicrobial and antioxidant activities of the essential

oil and various extracts of Salvia tomentosa Miller (Lamiaceae). Food Chemistry. 2005;**90**(3):333-340. DOI: 10.1016/j.foodchem.2003.09.013

[97] Kumar P, Sharma B, Bakshi N. Biological activity of alkaloids fromSolanum dulcamaraL. Natural Product Research. 2009;**23**(8):719

[98] Jain C, Khatana S, Vijayvergia R. Bioactivity of secondary metabolites of various plants: a review. International Journal of Pharmaceutical Sciences and Research. 2019;**10**(2):494-504

[99] Kang MS, Oh JS, Kang IC, Hong SJ, Choi CH. Inhibitory effect of methyl gallate and gallic acid on oral bacteria. Journal of Microbiology. 2008;**46**(6):744-750. DOI: 10.1007/s12275-008-0235-7

[100] Alvesalo J, Vuorela H, Tammela P, Leinonen M, Saikku P, Vuorela P. Inhibitory effect of dietary phenolic compounds on Chlamydia pneumoniae in cell cultures. Biochemical Pharmacology. 2006;**71**(6):735-741. DOI: 10.1016/j.bcp.2005.12.006

[101] Morinaga N, Iwamaru Y, Yahiro K, Tagashira M, Moss J, Noda M. Differential activities of plant polyphenols on the binding and internalisation of cholera toxin in vero cells. The Journal of Biological Chemistry. 2005;**280**(24):23303-23309. DOI: 10.1074/jbc.M502093200

[102] Hraiech S, Brégeon F, Brunel JM, Rolain JM, Lepidi H, Andrieu V, et al. Antibacterial efficacy of inhaled squalamine in a rat model of chronic *Pseudomonas aeruginosa* pneumonia. The Journal of antimicrobial chemotherapy. 2012;**67**(10): 2452-2458. DOI: 10.1093/jac/dks230

[103] Parhi A, Kelley C, Kaul M, Pilch DS, LaVoie EJ. Antibacterial activity of substituted 5-methylbenzo[c] phenanthridinium derivatives. Bioorganic & Medicinal Chemistry Letters. 2012;**22**(23):7080-7083. DOI: 10.1016/j.bmcl.2012.09.097

[104] Moore KS, Wehrli S, Roder H, Rogers M, Forrest JN Jr, McCrimmon D, et al. Squalamine: An aminosterol antibiotic from the shark. Proceedings of the National Academy of Sciences of the United States of America. 1993;**90**(4):1354-1358. DOI: 10.1073/pnas.90.4.1354

[105] Zoraghi R, Worrall L, See RH, Strangman W, Popplewell WL, Gong H, et al. Methicillin-resistant Staphylococcus aureus (MRSA) pyruvate kinase as a target for bis-indole alkaloids with antibacterial activities. The Journal of Biological Chemistry. 2011;**286**(52):44716-44725. DOI: 10.1074/jbc.M111.289033

[106] Villinski J, Dumas E, Chai H-B, Pezzuto J, Angerhofer C, Gafner S. Antibacterial activity and alkaloid content of *Berberis thunbergii*, *Berberis vulgaris* and *Hydrastis canadensis*. Pharmaceutical Biology. 2008;**41**(8):551-557. DOI: 10.1080/13880200390500768

[107] Otshudi AL, Apers S, Pieters L, Claeys M, Pannecouque C, De Clercq E, et al. Biologically active bisbenzylisoquinoline alkaloids from the root bark of Epinetrum villosum. Journal of Ethnopharmacology. 2005;**102**(1):89-94. DOI: 10.1016/j.jep.2005.05.021

[108] Yagi A, Okamura N, Haraguchi H, Abot T, Hashimoto K. Antimicrobial tetrahydroanthraquinones from a strain of Alternaria solani. Phytochemistry. 1993;**33**(1):87-91. DOI: 10.1016/0031-9422(93)85401-c

[109] Ng TB, Cheung RC, Wong JH, Bekhit AA, Bekhit AD. Antibacterial products of marine organisms. Applied Microbiology and Biotechnology.

2015;**99**(10):4145-4173. DOI: 10.1007/s00253-015-6553-x

[110] Khamthong N, Rukachaisirikul V, Tadpetch K, Kaewpet M, Phongpaichit S, Preedanon S, et al. Tetrahydroanthraquinone and xanthone derivatives from the marine-derived fungus Trichoderma aureoviride PSU-F95. Archives of Pharmacal Research. 2012;**35**(3):461-468. DOI: 10.1007/s12272-012-0309-2

[111] Feng S, Wang W. Bioactivities and structure-activity relationships of natural tetrahydroanthraquinone compounds: A review. Frontiers in Pharmacology. 2020;**11**:799. DOI: 10.3389/fphar.2020.00799

[112] Grayer RJ, Kokubun T. Plant–fungal interactions: The search for phytoalexins and other antifungal compounds from higher plants. Phytochemistry. 2001;**56**(3):253-263. DOI: 10.1016/s0031-9422(00)00450-7

[113] Ribera AE, Zuñiga G. Induced plant secondary metabolites for phytopatogenic fungi control: a review. Journal of soil science and plant nutrition. 2012;**12**(4):893-911

[114] Jung HJ, Sung WS, Yeo SH, Kim HS, Lee IS, Woo ER, et al. Antifungal effect of amentoflavone derived from Selaginella tamariscina. Archives of Pharmacal Research. 2006;**29**(9):746-751. DOI: 10.1007/bf02974074

[115] Morteza-Semnani K, Amin G, Shidfar MR, Hadizadeh H, Shafiee A. Antifungal activity of the methanolic extract and alkaloids of Glaucium oxylobum. Fitoterapia. 2003;**74**(5):493-496. DOI: 10.1016/s0367-326x(03)00113-8

[116] Thouvenel C, Gantier JC, Duret P, Fourneau C, Hocquemiller R, Ferreira ME, et al. Antifungal compounds from Zanthoxylum chiloperone var. angustifolium. Phytotherapy Research. 2003;**17**(6):678-680. DOI: 10.1002/ptr.1137

[117] Balls AK, Hale WS, Harris TH. A crystalline protein obtained from a lipoprotein of wheat flour. Cereal Chemistry. 1942;**19**:279-288

[118] Liu SC, Oguntimein B, Hufford CD, Clark AM. 3-Methoxysampangine, a novel antifungal copyrine alkaloid from Cleistopholis patens. Antimicrobial Agents and Chemotherapy. 1990;**34**(4):529-533. DOI: 10.1128/aac.34.4.529

[119] Emile A, Waikedre J, Herrenknecht C, Fourneau C, Gantier JC, Hnawia E, et al. Bioassay-guided isolation of antifungal alkaloids from Melochia odorata. Phytotherapy Research. 2007;**21**(4):398-400. DOI: 10.1002/ptr.2078

[120] Cantrell CL, Schrader KK, Mamonov LK, Sitpaeva GT, Kustova TS, Dunbar C, et al. Isolation and identification of antifungal and antialgal alkaloids from *Haplophyllum sieversii*. Journal of Agricultural and Food Chemistry. 2005;**53**(20):7741-7748. DOI: 10.1021/jf051478v

[121] Salehi B, Sharopov F, Boyunegmez Tumer T, Ozleyen A, Rodriguez-Perez C, Ezzat SM, et al. Symphytum Species: A Comprehensive Review on Chemical Composition, Food Applications and Phytopharmacology. Molecules. 2019;**24**(12):1-33

[122] Bahçeevli AK, Kurucu S, Kolak U, Topçu G, Adou E, Kingston DG. Alkaloids and aromatics of Cyathobasis fruticulosa (Bunge) Aellen. Journal of Natural Products. 2005;**68**(6):956-958. DOI: 10.1021/np0580006

[123] Cretton S, Dorsaz S, Azzollini A, Favre-Godal Q, Marcourt L, Ebrahimi SN, et al. Antifungal Quinoline Alkaloids from Waltheria indica. Journal of Natural Products. 2016;**79**(2):300-307. DOI: 10.1021/acs.jnatprod.5b00896

[124] Singh NV, Azmi S, Maurya S, Singh UP, Jha RN, Pandey VB. Two plant alkaloids isolated from Corydalis longipes as potential antifungal agents. Folia Microbiologica. 2003;**48**(5):605-609. DOI: 10.1007/bf02993466

[125] Ceramella J, Loizzo MR, Iacopetta D, Bonesi M, Sicari V, Pellicano TM, et al. Anchusa azurea Mill. (Boraginaceae) aerial parts methanol extract interfering with cytoskeleton organisation induces programmed cancer cells death. Food & Function. 2019;**10**(7):4280-4290. DOI: 10.1039/c9fo00582j

[126] Solarova Z, Liskova A, Samec M, Kubatka P, Busselberg D, Solar P. Anticancer potential of Lichens' secondary metabolites. Biomolecules. 2020;**10**(1): 1-31. DOI: 10.3390/biom10010087

[127] Yao H, Liu J, Xu S, Zhu Z, Xu J. The structural modification of natural products for novel drug discovery. Expert Opinion on Drug Discovery. 2017;**12**(2):121-140. DOI: 10.1080/17460441.2016.1272757

[128] Guo Z. The modification of natural products for medical use. Acta Pharmaceutica Sinica B. 2017;**7**(2):119-136. DOI: 10.1016/j.apsb.2016.06.003

[129] Yurdacan B, Egeli U, Eskiler GG, Eryilmaz IE, Cecener G, Tunca B. The role of usnic acid-induced apoptosis and autophagy in hepatocellular carcinoma. Human & Experimental Toxicology. 2019;**38**(2):201-215. DOI: 10.1177/0960327118792052

[130] Wang Y, Yang Y, Yang X, Gao L, Yan B. K562 apoptosis induced by flavone from Tibetan medicine Chrysosplenium nudicaule Bunge and its molecular mechanism. The Practical Journal of Cancer. 2005;**20**(4):374-376

[131] Luo Y, Yu H, Yang Y, Tian W, Dong K, Shan J, et al. A flavonoid compound from Chrysosplenium nudicaule inhibits growth and induces apoptosis of the human stomach cancer cell line SGC-7901. Pharmaceutical Biology. 2016;**54**(7):1133-1139. DOI: 10.3109/13880209.2015.1055634

[132] Konan NA, Lincopan N, Diaz IE, de Fatima JJ, Tiba MM, Amarante Mendes JG, et al. Cytotoxicity of cashew flavonoids towards malignant cell lines. Experimental and Toxicologic Pathology. 2012;**64**(5):435-440. DOI: 10.1016/j.etp.2010.10.010

[133] Manthey JA, Grohmann K, Guthrie N. Biological properties of citrus flavonoids pertaining to cancer and inflammation. Current Medicinal Chemistry. 2001;**8**(2):135-153. DOI: 10.2174/0929867013373723

[134] Ranelletti FO, Ricci R, Larocca LM, Maggiano N, Capelli A, Scambia G, et al. Growth-inhibitory effect of quercetin and presence of type-II estrogen-binding sites in human colon-cancer cell lines and primary colorectal tumors. International Journal of Cancer. 1992;**50**(3):486-492. DOI: 10.1002/ijc.2910500326

[135] Scambia G, Ranelletti FO, Benedetti Panici P, Piantelli M, Bonanno G, De Vincenzo R, et al. Quercetin inhibits the growth of a multidrug-resistant estrogen-receptor-negative MCF-7 human breast-cancer cell line expressing type II estrogen-binding sites. Cancer Chemotherapy and Pharmacology. 1991;**28**(4):255-258. DOI: 10.1007/bf00685531

[136] Scambia G, Ranelletti FO, Panici PB, Piantelli M, Bonanno G, De Vincenzo R, et al. Inhibitory effect of quercetin on OVCA 433 cells and presence of type II oestrogen binding sites in primary ovarian tumours and cultured cells. British Journal of Cancer. 1990;**62**(6):942-946. DOI: 10.1038/bjc.1990.414

[137] Matsukawa Y, Yoshida M, Sakai T, Marui N, Matsumoto K, Fujioka A, et al. The effect of quercetin and other flavonoids on cell cycle progression and growth of human gastric cancer cells. Planta Medica. 2007;**56**(06):677-678. DOI: 10.1055/s-2006-961347

[138] Nair HK, Rao KV, Aalinkeel R, Mahajan S, Chawda R, Schwartz SA. Inhibition of prostate cancer cell colony formation by the flavonoid quercetin correlates with modulation of specific regulatory genes. Clinical and Diagnostic Laboratory Immunology. 2004;**11**(1):63-69. DOI: 10.1128/cdli.11.1.63-69.2004

[139] Kaneuchi M, Sasaki M, Tanaka Y, Sakuragi N, Fujimoto S, Dahiya R. Quercetin regulates growth of Ishikawa cells through the suppression of EGF and cyclin D1. International Journal of Oncology. 2003;**22**(1):159-164

[140] Dunnick J, Hailey JR. Toxicity and carcinogenicity studies of quercetin, a natural component of foods. Fundamental and Applied Toxicology. 1992;**19**(3):423-431. DOI: 10.1016/0272-0590(92)90181-g

[141] Nguyen TT, Tran E, Ong CK, Lee SK, Do PT, Huynh TT, et al. Kaempferol-induced growth inhibition and apoptosis in A549 lung cancer cells is mediated by activation of MEK-MAPK. Journal of Cellular Physiology. 2003;**197**(1):110-121. DOI: 10.1002/jcp.10340

[142] Hung H. Inhibition of estrogen receptor alpha expression and function in MCF-7 cells by Kaempferol. Journal of Cellular Physiology. 2004;**198**(2):197-208

[143] Fresco P, Borges F, Diniz C, Marques MP. New insights on the anticancer properties of dietary polyphenols. Medicinal Research Reviews. 2006;**26**(6):747-766. DOI: 10.1002/med.20060

[144] Mojzer EB, Hrncic MK, Skerget M, Knez Z, Bren U. Polyphenols: Extraction methods, antioxidative action, bioavailability and anticarcinogenic effects. Molecules. 2016;**21**(7):1-38. DOI: 10.3390/molecules2107090

[145] Fantini M, Benvenuto M, Masuelli L, Frajese GV, Tresoldi I, Modesti A, et al. In vitro and in vivo antitumoral effects of combinations of polyphenols, or polyphenols and anticancer drugs: Perspectives on cancer treatment. International Journal of Molecular Sciences. 2015;**16**(5):9236-9282. DOI: 10.3390/ijms16059236

[146] Manach C, Scalbert A, Morand C, Remesy C, Jimenez L. Polyphenols: Food sources and bioavailability. The American Journal of Clinical Nutrition. 2004;**79**(5):727-747. DOI: 10.1093/ajcn/79.5.727

[147] Yang CS, Landau JM, Huang MT, Newmark HL. Inhibition of carcinogenesis by dietary polyphenolic compounds. Annual Review of Nutrition. 2001;**21**(1):381-406. DOI: 10.1146/annurev.nutr.21.1.381

[148] Seeram NP, Adams LS, Zhang Y, Lee R, Sand D, Scheuller HS, et al. Blackberry, black raspberry, blueberry, cranberry, red raspberry, and strawberry extracts inhibit growth and stimulate apoptosis of human cancer cells

in vitro. Journal of Agricultural and Food Chemistry. 2006;**54**(25):9329-9339. DOI: 10.1021/jf061750g

[149] Weisburg JH, Weissman DB, Sedaghat T, Babich H. In vitro cytotoxicity of epigallocatechin gallate and tea extracts to cancerous and normal cells from the human oral cavity. Basic & Clinical Pharmacology & Toxicology. 2004;**95**(4):191-200. DOI: 10.1111/j.1742-7843.2004.pto_950407.x

[150] Wu D, Liu Z, Li J, Zhang Q, Zhong P, Teng T, et al. Epigallocatechin-3-gallate inhibits the growth and increases the apoptosis of human thyroid carcinoma cells through suppression of EGFR/RAS/RAF/MEK/ERK signaling pathway. Cancer Cell International. 2019;**19**(1):43. DOI: 10.1186/s12935-019-0762-9

[151] Wang YQ, Lu JL, Liang YR, Li QS. Suppressive effects of EGCG on cervical cancer. Molecules. 2018;**23**(9). 1-17. DOI: 10.3390/molecules23092334

[152] Molaei F, Forghanifard MM, Fahim Y, Abbaszadegan MR. Molecular signaling in tumorigenesis of gastric cancer. Iranian Biomedical Journal. 2018;**22**(4).217-230.DOI: 10.22034/ibj.22.4.217

[153] Liu Y, Cai Y, He C, Chen M, Li H. Anticancer properties and pharmaceutical applications of plumbagin: A review. The American Journal of Chinese Medicine. 2017;**45**(3):423-441. DOI: 10.1142/S0192415X17500264

[154] Rathinasamy K. Antibacterial and anticancer activity of the purified cashew nut shell liquid: Implications in cancer chemotherapy and wound healing. Natural Product Research. 2018;**32**(23):2856-2860. DOI: 10.1080/14786419.2017.1380022

[155] Sukumari-Ramesh S, Singh N, Jensen MA, Dhandapani KM, Vender JR. Anacardic acid induces caspase-independent apoptosis and radiosensitises pituitary adenoma cells. Journal of Neurosurgery. 2011;**114**(6):1681-1690. DOI: 10.3171/2010.12.JNS10588

[156] Salehi B, Gultekin-Ozguven M, Kirkin C, Ozcelik B, Morais-Braga MFB, Carneiro JNP, et al. Antioxidant, antimicrobial, and anticancer effects of anacardium plants: An ethnopharmacological perspective. Frontiers in Endocrinology (Lausanne). 2020;**11**:295

[157] Thomasset S, Teller N, Cai H, Marko D, Berry DP, Steward WP, et al. Do anthocyanins and anthocyanidins, cancer chemopreventive pigments in the diet, merit development as potential drugs? Cancer Chemotherapy and Pharmacology. 2009;**64**(1):201-211. DOI: 10.1007/s00280-009-0976-y

[158] Lala G, Malik M, Zhao C, He J, Kwon Y, Giusti MM, et al. Anthocyanin-rich extracts inhibit multiple biomarkers of colon cancer in rats. Nutrition and Cancer. 2006;**54**(1):84-93. DOI: 10.1207/s15327914nc5401_10

[159] Ding M, Feng R, Wang SY, Bowman L, Lu Y, Qian Y, et al. Cyanidin-3-glucoside, a natural product derived from blackberry, exhibits chemopreventive and chemotherapeutic activity. The Journal of Biological Chemistry. 2006;**281**(25):17359-17368. DOI: 10.1074/jbc.M600861200

[160] Han NN, Li X, Tao L, Zhou Q. Doxorubicin and rhein loaded nanomicelles attenuates multi-drug resistance in human ovarian cancer. Biochemical and Biophysical Research Communications. 2018;**498**(1):178-185. DOI: 10.1016/j.bbrc.2018.01.042

[161] Teiten MH, Mack F, Debbab A, Aly AH, Dicato M, Proksch P, et al. Anticancer effect of altersolanol A, a metabolite produced by the endophytic fungus Stemphylium globuliferum, mediated by its pro-apoptotic and anti-invasive potential via the inhibition of NF-kappaB activity. Bioorganic & Medicinal Chemistry. 2013;**21**(13):3850-3858. DOI: 10.1016/j.bmc.2013.04.024

[162] Pompeng P, Sommit D, Sriubolmas N, Ngamrojanavanich N, Matsubara K, Pudhom K. Antiangiogenetic effects of anthranoids from Alternaria sp., an endophytic fungus in a Thai medicinal plant Erythrina variegata. Phytomedicine. 2013;**20**(10):918-922. DOI: 10.1016/j.phymed.2013.03.019

[163] Bin C, Duoji P, Aiguo X, Lan L, Xun Z, Yongchneg L. Tetrahydroanthraquinone derivatives from Alternaria sp. XZSBG -1. Acta Scientlarum Naturalium Universit Sunyatesni. 2016;**55**:91-95

[164] Aly AH, Edrada-Ebel R, Wray V, Muller WE, Kozytska S, Hentschel U, et al. Bioactive metabolites from the endophytic fungus Ampelomyces sp. isolated from the medicinal plant Urospermum picroides. Phytochemistry. 2008;**69**(8):1716-1725. DOI: 10.1016/j.phytochem.2008.02.013

[165] Valcic S, Timmermann BN, Alberts DS, Wachter GA, Krutzsch M, Wymer J, et al. Inhibitory effect of six green tea catechins and caffeine on the growth of four selected human tumor cell lines. Anti-Cancer Drugs. 1996;**7**(4):461-468. DOI: 10.1097/00001813-199606000-00011

[166] Seeram NP, Zhang Y, Nair MG. Inhibition of proliferation of human cancer cells and cyclooxygenase enzymes by anthocyanidins and catechins. Nutrition and Cancer. 2003;**46**(1):101-106. DOI: 10.1207/S15327914NC4601_13

[167] Adhami VM, Ahmad N, Mukhtar H. Molecular targets for green tea in prostate cancer prevention. Journal of Nutrition. 2003;**133**(7 Suppl):2417s-2424s

[168] Sun C, Li J, Wang X, Duan W, Zhang T, Ito Y. Preparative separation of quaternary ammonium alkaloids from Coptis chinensis Franch by pH-zone-refining counter-current chromatography. Journal of Chromatography. A. 2014;**1370**:156-161. DOI: 10.1016/j.chroma.2014.10.043

[169] Kumar A, Chopra K, Mukherjee M, Pottabathini R, Dhull DK. Current knowledge and pharmacological profile of berberine: An update. European Journal of Pharmacology. 2015;**761**:288-297. DOI: 10.1016/j.ejphar.2015.05.068

[170] Zhang Q, Wang X, Cao S, Sun Y, He X, Jiang B, et al. Berberine represses human gastric cancer cell growth in vitro and in vivo by inducing cytostatic autophagy via inhibition of MAPK/mTOR/p70S6K and Akt signaling pathways. Biomedicine & Pharmacotherapy. 2020;**128**:110245. DOI: 10.1016/j.biopha.2020.110245

[171] Bhagya N, Chandrashekar KR. Tetrandrine: A molecule of wide bioactivity. Phytochemistry. 2016;**125**:5-13. DOI: 10.1016/j.phytochem.2016.02.005

[172] Wei J, Liu B, Wang L, Qian X, Ding Y, Yu L. Synergistic interaction between tetrandrine and chemotherapeutic agents and influence of tetrandrine on chemotherapeutic agent-associated genes in human gastric cancer cell lines. Cancer Chemotherapy and Pharmacology. 2007;**60**(5):703-711. DOI: 10.1007/s00280-007-0416-9

[173] Prasad S, Tyagi AK. Historical spice as a future drug: Therapeutic potential of piperlongumine. Current Pharmaceutical Design. 2016;**22**(27):4151-4159. DOI: 10.2174/1381612822666160601103027

[174] Zhang P, Shi L, Zhang T, Hong L, He W, Cao P, et al. Piperlongumine potentiates the anti-tumor efficacy of oxaliplatin through ROS induction in gastric cancer cells. Cellular Oncology (Dordrecht). 2019;**42**(6):847-860. DOI: 10.1007/s13402-019-00471-x

[175] Lou C, Yokoyama S, Saiki I, Hayakawa Y. Selective anticancer activity of hirsutine against HER2-positive breast cancer cells by inducing DNA damage. Oncology Reports. 2015;**33**(4):2072-2076. DOI: 10.3892/or.2015.3796

[176] Li M, Su BS, Chang LH, Gao Q, Chen KL, An P, et al. Oxymatrine induces apoptosis in human cervical cancer cells through guanine nucleotide depletion. Anti-Cancer Drugs. 2014;**25**(2):161-173. DOI: 10.1097/cad.0000000000000012

[177] Yadav VR, Prasad S, Sung B, Kannappan R, Aggarwal BB. Targeting inflammatory pathways by triterpenoids for prevention and treatment of cancer. Toxins (Basel). 2010;**2**(10):2428-2466. DOI: 10.3390/toxins2102428

[178] Lee MG, Lee KT, Chi SG, Park JH. Costunolide induces apoptosis by ROS-mediated mitochondrial permeability transition and cytochrome C release. Biological & Pharmaceutical Bulletin. 2001;**24**(3):303-306. DOI: 10.1248/bpb.24.303

[179] El-Far AH. Thymoquinone anticancer discovery: Possible mechanisms. Current Drug Discovery Technologies. 2015;**12**(2):80-89. DOI: 10.2174/1570163812666150716111821

[180] Arshad S, Naveed M, Ullia M, Javed K, Butt A, Khawar M, et al. Targeting STAT-3 signaling pathway in cancer for development of novel drugs: Advancements and challenges. Genetics and Molecular Biology. 2020;**43**(1):e20180160. DOI: 10.1590/1678-4685-GMB-2018-0160

Chapter 12

Chiral Inversion of Active Compounds in Plant Extract

Ngoc-Van Thi Nguyen

Abstract

Chiral inversion is always mediated by enzymes and varies with solvent, pH and temperature. Considerable attention should be paid to the mechanism of the inversion reaction and its pharmacological and toxicological results. This chapter will discuss the mechanism of chiral inversion of plants in secondary metabolize and its importance in creating pharmacology consequences. Plant stereoisomers of alkaloids and flavonoids exhibit a wide range of pharmacological effects. Recent advances in chiral analysis for the herbal plants in clinical research & forensic toxicology by experiments in which one enantiomer was given to the experiment subjects in a specific situation. Demonstration of metabolic chiral inversion may have consequences for the development of a new pharmaceutical entity. Hence, it helps a better understanding of chiral compounds in plants, facilitating the application for drug development from medicinal herbs and thereby reducing bioanalytical and toxicology workload.

Keywords: chiral inversion of plants, eutomer, distomer, racemization

1. Introduction

Chiral inversion is the process by which enzymes modify the three-dimensional structure of a molecule by converting one enantiomer to its antipode [1]. Racemization occurs when isomerization leads in the creation of a racemic mixture. As a result, chiral inversion influences drug stability throughout drug discovery and development. Biological activity, toxicity, shelf-life and dosage of the compound are affected by the stability of the drug [2]. The process of chiral inversion is affected by a lot of variables, consequently, the strength of chiral inversion under different situations and in various substances can vary significantly. The primary elements that were acknowledged to play a vital part in the process of chiral inversion were reported to be interspecies differences and tissue types. Some recent researches have demonstrated that additional variables, such as administration route or interaction with other xenobiotics, can also impact enantiomeric conversion.

Plants create a vast diversity of physiologically active metabolites, many of which have stereochemical variants on the same molecular scaffold. These alterations in stereochemistry have a significant influence on biological function. Notably, plant stereoisomers of alkaloids and flavonoids exhibit a wide range of pharmacological effects. Alkaloids are cyclic chemical molecules with a negative oxidation state of nitrogen. They are found throughout the flora and play an important function in plant protection, sprouting, and encouraging plant development. Plants containing alkaloids are frequently employed as traditional remedies, and these chemicals typically have specific pharmacological actions. The majority of alkaloids are in a chiral form which is often appeared in products as racemic compounds, while their enantiomers have been proved to have different pharmacological actions [3]. Flavonoids are a vast category of polyphenolic chemicals with a benzo—pyrone structure that is found in all plants. Phenylpropanoid is their produce's pathway. Recent interest in these chemicals has been sparked because of the possible health advantages of these antioxidant polyphenolic compounds [4]. The relevance of racemic flavanones stereospecific pomological disposition has been determinated and described in the last 20 years. The majority of these studies report on the measurement of flavanones in citrus fruit juices and herbs [5].

Can all chiral compounds undergo chiral inversion? Maybe no, many compounds still can be considered stable in metabolize process. Why are some enantiomers of plants inverted by enzymes and others are not attacked? The reason lies in the structure. The intent of this chapter is to provide a comprehensive, rather than an exhaustive, appraisal of chiral bio-inversion. This chapter will discuss enzymatic chiral inversion of plants in secondary metabolize and its importance create pharmacology effect. Therefrom, it helps a better understanding of chiral compounds in plants, facilitating the application for drug development from medicinal herbs.

2. Mechanism of inversion

Under selective conditions, racemization or enantiomerization defined as the chiral conversion of enantiomer into its antipode may present in many plants metabolizing. When the chiral molecule enantiomers in herbals interact with a chiral macromolecule-like enzyme, they generate a pair of diastereoisomeric complexes that vary energetically. It is not surprising, then, that the results of enzyme-mediated reactions performed on a pair of enantiomers may differ in type and/or extent. Indeed, given the structure of the enzyme-substrate complex, it is plausible to believe that enantioselectivity is the rule rather than the exception in metabolism. Likewise, the binding of a prochiral substrate to an enzyme may orient two enantiotopic groups differently about the enzyme catalytic site, causing these two groups to become diastereotopic within the enzyme-substrate complex. It's simple to see how the production of a chiral metabolite from a prochiral substrate may result in stereoselectivity for one isomeric product [6].

At the substrate and product levels, xenobiotic metabolic reactions exhibit two forms of stereoselectivity. As a result, they can be classed according to their stereoselectivity or, if such selectivity is complete, their stereospecificity. Caution while using this latter phrase, because the ability to determination "specificity" is clearly dependent on the analytical approach of the research. The words substrate and product "stereospecificity" were initially introduced to the enzyme-mediated reduction of ketones by Prelog [7], and were later extended to drug metabolic processes by Jenner and Testa [8]. Substrate stereoselectivity is the preferred metabolism of one of two stereoisomers over the other, whereas product stereoselectivity is the preferential production of one stereoisomer over the other stereoisomers that may exist. These two "selectivities" may be so closely related that substrate-product stereoselectivity, i.e., the selective metabolism of one of a pair of enantiomers to form one of several possibly diastereoisomeric products, may also be seen. If the enantiomeric composition of medication or metabolite is detected in the analysis process, data collected from in vivo research on stereochemistry in plants must be regarded with caution (**Table 1**).

2.1 Alkaloids

Plants are thought to generate over 12,000 distinct alkaloids, which may be classified based on their carbon skeleton structures. Many catalytic stages in alkaloid biosynthesis in plants are catalyzed by enzymes from various protein families.

Since the discovery of morphine in 1806, the complex relationships between opium poppy and the human condition have fueled substantial study into the production of morphinan alkaloids [23]. During the 1960s, significant progress toward route elucidation was made, which supported a major theory [24] that morphine was

Plant	Enzyme	Stereoselective	Stereospecificity	Ref.
Alkaloid Compound				
Opium poppy	1,2-dehydroreticuline reductase		(R)-reticuline	[9]
Catharanthus roseus	tetrahydroalstonine synthase		(3S,19S,20S)-Tetrahydroalstonine	[10]
Claviceps purpurea	Dimethylallyl tryptophan synthase	L-tryptophan		[11]
Hyoscyamus niger	Hyoscyamine 6β-hydroxylase	L-hyoscyamine		[12]
Berberis koetineana	Tetrahydrobenzyliso-quinoline *N*-methyltransferase		(*R*)-tetrahydropapaverine	[13]
Flavonoid Compound				
Soybean	Chalcone isomerase		(2S)-flavanone	[14]
Dahlia variabilis	Flavanone 4-reductase	(2S)-flavanone	(2S, 4R)-flavan-4-ol	[15]
Citrus unshiu	Flavonol synthase	(2R,3R)-dihydroflavonol		[16]
Glycyrrhiza echinata	Flavanone 2-hydroxylase	(2S)-flavanone		[17]
Ginko biloba Pseudotsuga menziesii	Dihydroflavonol 4-reductase	(2R,3R)-dihydroflavonol	(2R, 3S, 4S)-flavan-2,3-trans-3,4-cis-diol	[18]
Medicago truncatula	Anthocyanidin reductase		(2R, 3R)-flavan-3-ol	[19]
Medicago sativa	Isoflavone reductase		(2R)-isoflavanone	[20]
Pisum sativum	Hydroxymaackiain-3-Omethyltransferase	(+)-6a-hydroxymaackiain		[21]
Desmodium uncinatum	Leucoanthocyanidin 4-reductase	flavan-2,3-trans-3,4-cisdiol	(2R, 3S)-flavan-3-ol	[22]

Table 1.
Stereoselective and/or specific enzymes of alkaloid and flavonoid compound biosynthesis in plant extract.

generated by 1-benzylisoquinoline alkaloid metabolism [25]. Because only the (R)-conformer could undergo additional phenol coupling to the morphinan scaffold, (S)-reticuline emerged as the primary 1-benzylisoquinoline intermediate, with its stereochemical inversion to (R)-reticuline thought to be a critical gateway reaction [26].

The pathway makes use of opium poppy reticuline epimerase, a multi-domain protein composed of the P450 CYP82Y2 linked to an aldo-keto reductase (AKR). CYP82Y2 (1,2-dehydroreticuline synthase, DRS) catalyzes the conversion of (S)-reticuline to 1,2-dehydroreticuline, which is then converted to (R)-reticuline by the AKR module (1,2-dehydroreticuline reductase, DRR) [26]. A second P450 called CYP719B1 then tranforms (R)-reticuline into salutaridine [27, 28]. This procedure includes (R)-reticuline twisting, reorientating and oxidative C–C bond coupling stimulated by CYP719B1 (**Figure 1**).

Catharanthus roseus, a medicinal plant, creates three of these isomers: ajmalicine (raubasine), tetrahydroalstonine, and 19-epi-ajmalicine (mayumbine) (**Figure 2**) [30]. These heteroyohimbines are produced from deglycosylated strictosidine (strictosidine aglycone), as are the bulk of monoterpene indole alkaloids [31]. A glucose unit removal from strictosidine by strictosidine glucosidase (SGD) leads to the formation of a reactive dialdehyde intermediate that can rearrange to generate a variety of isomers [32]. The stability of these isomers by enzyme-catalyzed reduction is thought to be the first step toward the vast chemical variety found in monoterpene indole alkaloids. The tetrahydroalstonine synthase (THAS) is a zinc-dependent medium-chain dehydrogenase/reductase (MDR) that manufactures the heteroyohimbine tetrahydroalstonine (**Figure 2**) [33]. Although, these studies showed that THAS is an important enzyme for the heteroyohimbine production, the mechanism by which this enzyme controls the stereoselectivity of the reduction remained unexplained. Moreover, the fact that strictosidine aglycone is also a predrug of some alkaloid scaffolds so constitutes a major branch point in the monoterpene indole alkaloid biosynthesis process [29].

2.2 Flavonoids

Most flavonoid biosynthesis enzymes are extremely stereoselective and/or stereospecific; nonetheless, this assertion is based on just one or a few published findings for numerous enzymes. Flavonoids are produced by the phenylpropanoid pathway, which begins with the enzyme L-phenylalanine ammonia-lyase deamination of phenylalanine (PAL). D-phenylalanine is not a substrate for PAL; it is selective for the naturally occurring L-isomer of phenylalanine [34]. The process mediated by chalcone–flavanone isomerase (CHI), which sets the stereochemistry at C-2 of the flavonoid heterocyclic ring, maybe the most stereo-chemically crucial in flavonoid biosynthesis. CHI is a chemically and structurally well-characterized enzyme that creates (2S)-flavanones from chalcones (**Figure 3**) [14, 35].

Unlike other flavonoid enzymes such as PAL or CHI, the 2-oxoglutarate-dependent dioxygenases flavonol synthase (FLS) and anthocyanidin synthase (ANS) have wide substrate and product selectivities in vitro (both take flavanone, dihydroflavonol, and leucoanthocyanidin as substrates). Prescott et al. have reported a detailed structural and in vitro research on recombinant flavonol synthase from Arabidopsis thaliana, with a focus on the stereochemistry of substrate and product, have provided information on how they catalyze reactions with their real substrates in vivo [36]. FLS and ANS prefer substrates with natural C-2 and C-3 stereochemistry [(i.e. (2R,3R)- dihydroquercetin for FLS and (2R,3S, 4R/S)- leucoanthocyanin for ANS], but hydroxylate both (2R)- and

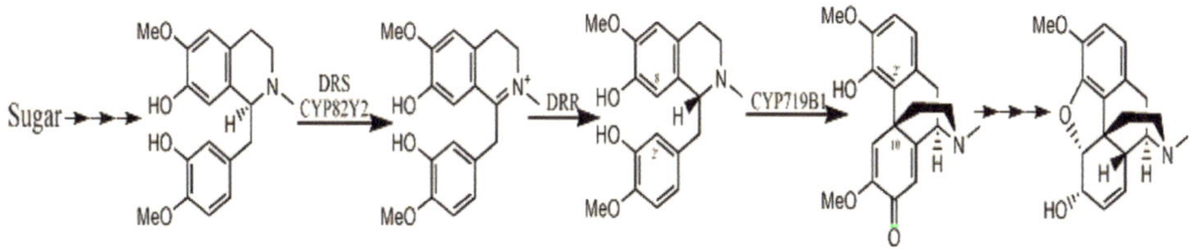

Figure 1.
Proposed chiral inversion of (S)-reticulin to (R)-reticulin catalyzed by 1,2- dehydroreticuline reductase (DRR) and 1,2-dehydroreticuline reductase (DRR) in opium poppy *[9].*

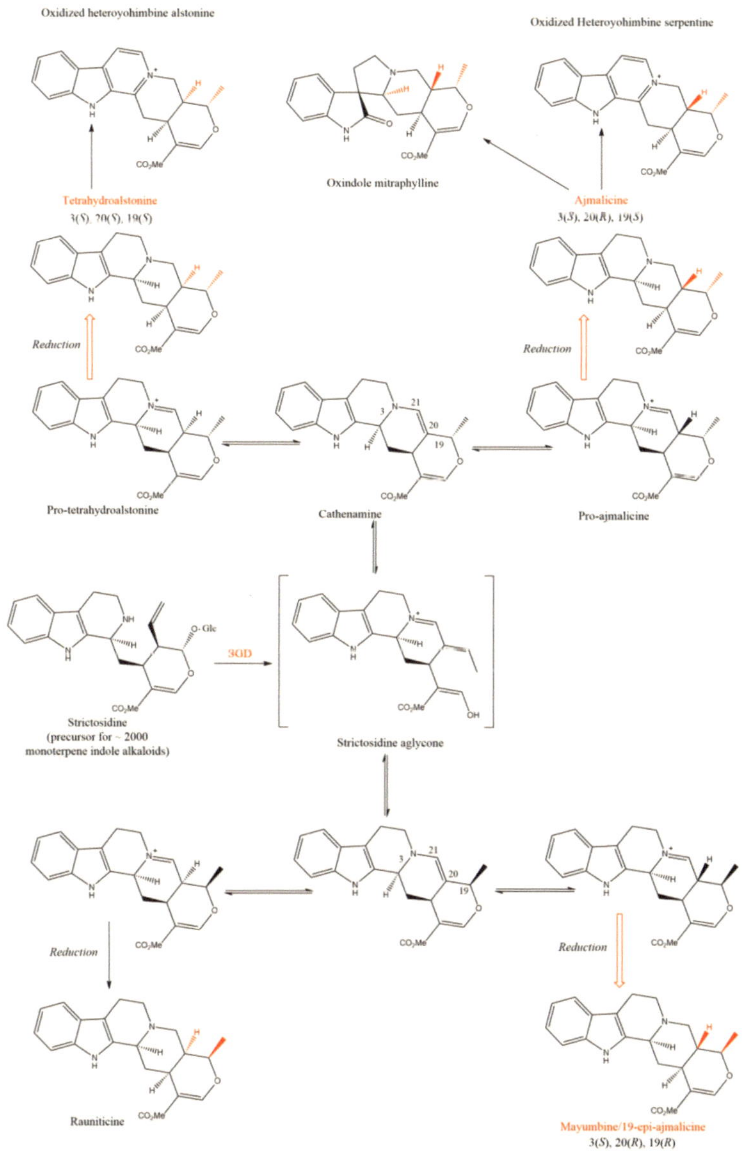

Figure 2.
Heteroyohimbine alkaloid biosynthesis. Red highlighted compounds indicate the three diastereomers identified in Catharanthus roseus. *Alkaloids derived from heteroyohimbines are also illustrated [29].*

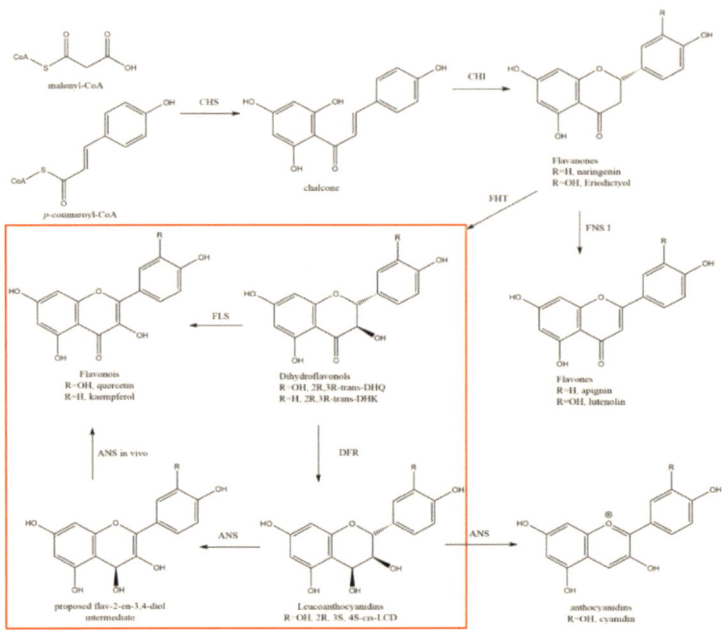

Figure 3.
General outline of the flavonoid pathway (PAL: Phenylalanine ammonia-lyase, CHS: Chalcone synthase, CHI: Chalcone isomerase, FHT: Flavanone 3β-hydroxylase, FNS I: Flavone synthase I, FLS: Flavonols synthase, DFR: Dihydroflavonols reductase, ANS: Anthocyanidin synthase). Chiral inversion in flavonoid metabolizes was highlighted by red frame.

(2S)-naringenin equally well in vitro, indicating that the C-3 hydroxyl group is important in biasing substrate selectivity [37].

The flavan-3-ols (+)-catechin and (−)-epicatechin serve as the foundation for proanthocyanidins (condensed tannins), a family of molecules of great interest due to their effects on animal health [38]. The C-2 and C-3 stereochemistries of (+)-catechin (2,3-trans) are identical to those of flavonoid pathway intermediates, and a pathway leading from (2R, 3S, 4S)-leucoanthocyanidin to (+)-catechin, catalyzed by leucoanthocyanidin reductase (LAR), has been illustrated and affirmed by the cloning of a leucoanthocyanidin reducta [22]. LAR belongs to the Reductase–Epimerase–Dehydrogenase protein family, which also includes isoflavone reductase and similar homologs (**Figure 4**) [39].

The process catalyzed by anthocyanidin reductase (ANS) and anthocyanidin reductase (ANR) leads from leucocyanidin to (−)-epicatechin [40]. By operating on an achiral intermediate, ANR, an enzyme with limited sequence similarity to dihydroflavonol reductase, can introduce the 2,3-cis stereochemistry (anthocyanidin). Mechanisms for this reaction have been hypothesized, and it is plausible that more ANR-like enzymes with the potential to introduce different stereochemistries exist (**Figure 5**) [41].

3. Factors affecting chiral inversion

Chiral inversion is always mediated by enzymes and varies with solvent, pH and temperature. When a molecule has two or more elements of chirality, one of which is configurationally labile, enantiomerization can occur. Many studies have been

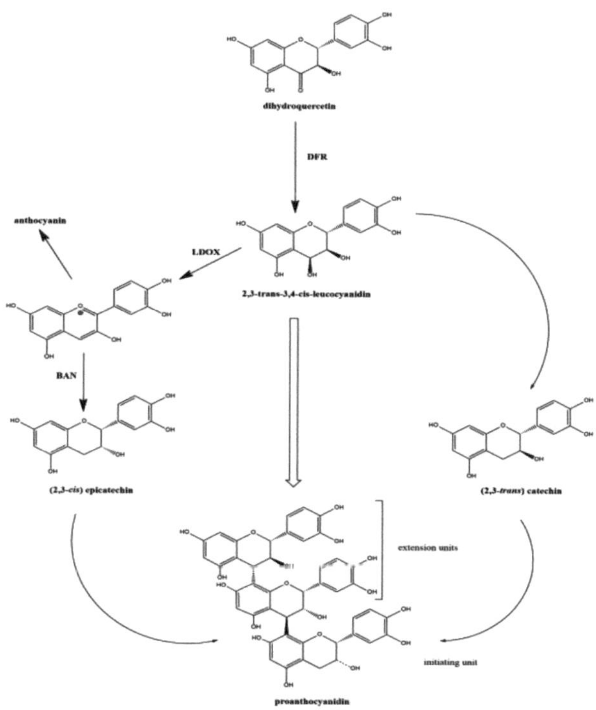

Figure 4.
The pro-anthocyanidin pathway showing the LAR reaction.

reported about the chiral compounds inversion such as: evodiamine in Evodia rutaecarpa [42], ephedrine and atropine (**Figures 6** and 7) [43].

Chiral inversion is always mediated by enzymes. One of the most valuable synthetic features of enzymes is their ability to discriminate between enantiomers of racemic substrates [44]. The ratio of stereoisomers is changed mainly by stereospecificity and stereoselectivity of enzymes transformation. The stereoselectivity and stereospecificity of enzymes change dramatically the ratio of drug enantiomers and metabolites enantiomers in biological systems. The enzyme-mediated chiral inversion can be affected by determining expression, substrate affinity and activity of the enzyme. The difference of species and tissue can be different in the rate of the chiral inversion occursion as well as of the routes and mechanisms of inversion [2].

On another hand, the development of strategies that improves the stereoselectivity of enzyme-catalyzed resolutions has been researching. Modification of the substrate, recycling of the product and changing of the reaction conditions are the three most common ways. From now, even enzymes with modest stereoselectivity can be used successfully [44]. Configurational stability depends mainly on the structure and the conditions, especially with solvent, pH and temperature [2].

According to Ngoc Van Thi Nguyen et al. (2013) research, extraction conditions are also can affect the enantiomerization while this study investigated the optimization of the extraction procedure, more specifically the solvent, pH and temperature [42].

3.1 Solvent

In the metabolic chiral inversion research, avoiding spontaneous or chemical racemization of enantiomers is one of the important things [2]. The organic solvent

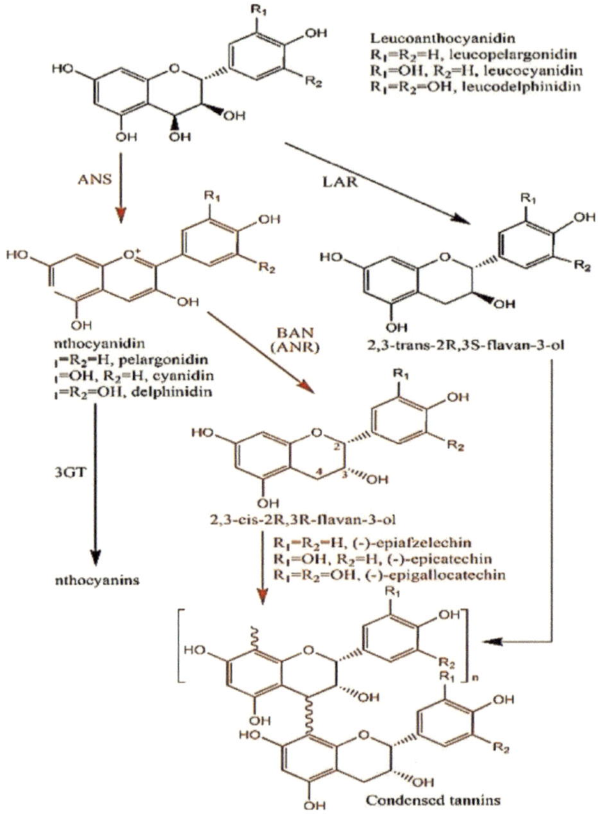

Figure 5.
Pathway for CT biosynthesis placing BAN immediately downstream of ANS.

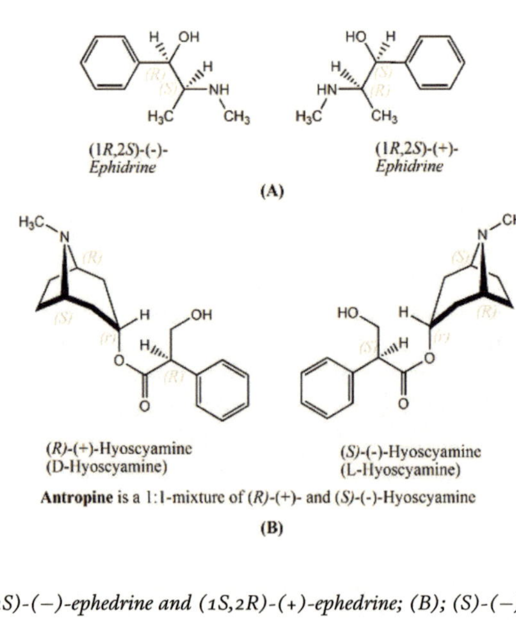

Figure 6.
Structures of (A): (1R,2S)-(−)-ephedrine and (1S,2R)-(+)-ephedrine; (B); (S)-(−)- hyoscyamine and (R)-(+)-hyoscyamine.

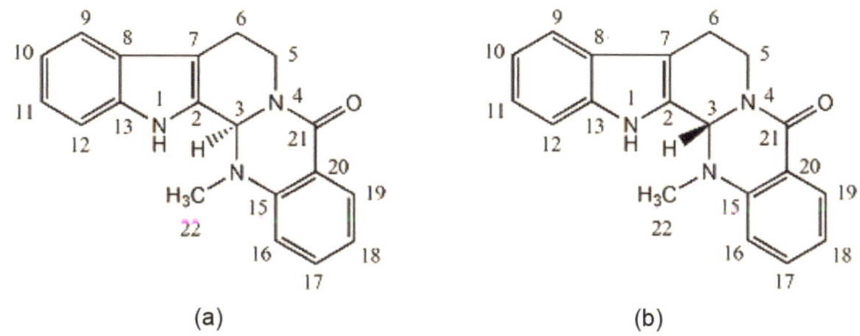

Figure 7.
Chemical structure of (1a) R-(−)-evodiamine, (1b) S-(+) evodiamine.

characterism is one other parameter that can significantly interfere with this chiral inversion [45]. The study of Yang SK [46] has shown that racemization half-lives t1/2 of enantiomeric oxazepam were 4.8 min in methanol, while it was 840 min in diethyl ether, and 5000 min in hexane, 4500 min in acetonitrile, etc.

3.2 pH

Based on the result of the study of Glass Amanda M. et al. (2012), the data collectively prove that pH has a minute effect on the chiral inversion rate (**Figure 8**) [48].

The pH effect on proton extraction to give the enolate-form of CoA-thioester resulting in chiral inversion [47]. Chiral inversion and sufficient emission intensity were observed at basic pH 8 and 8.9, respectively, whereas only little emission was observed under neutral to acidic conditions.

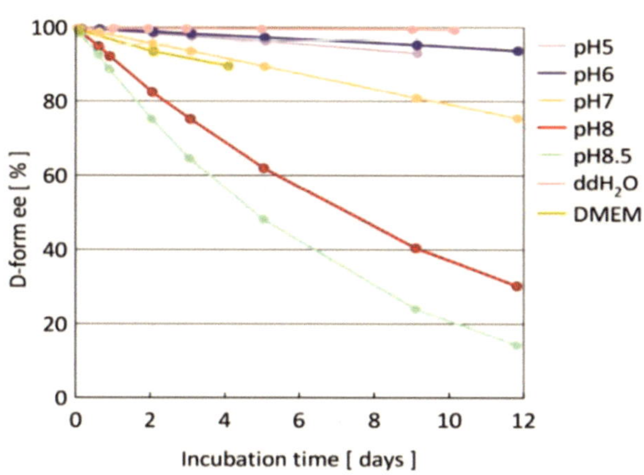

Figure 8.
Time-dependent changes of D-luciferin substrate. Luciferin racemization under various pHs of 150 mM GTA buffer, under ddH$_2$O, and under a medium was monitored for 12 days. The results are highlighted by colors: pH 5 (purple), pH 6 (blue), pH 7 (orange), pH 8 (wine red), pH 8.5 (green), ddH$_2$O (pink) and DMEM (brown). Even under acidic to neutral conditions, obvious racemization that could not be ignored for long-term experiments were observed. The best condition for inhibiting racemization to maintain D-luciferin optical purity was dissolution in ddH$_2$O [47].

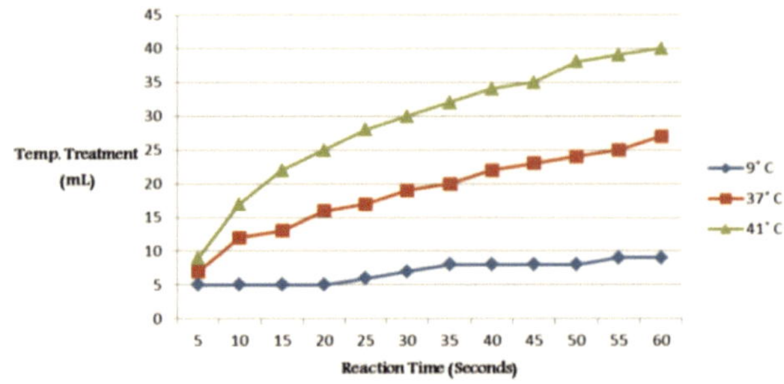

Figure 9.
The effect of temperature on enzyme activity [49].

3.3 Temperature

Enzyme activity is also affected by temperature, which can lead to the chiral inversion efficiency. The research of the effect of temperature on enzyme activity showed that the hydrogen peroxidase activity's best temperature is 41°C. When this condition is decreased to 37° C, the enzyme activity decreased. Continuing to decrease to 9°C can decrease dramatically the activity of the enzyme. The influence on enzyme flexibility is because of the temperature effect on hydrogen bonds and covalent (**Figure 9**) [49].

4. Pharmacological consequences

One of the three majorities of racemic pharmaceuticals are the racemic drugs that only have one eutomer, but the distomer could be transformed into its bioactive antipode by chiral inversion in the body (**Table 2**) [60].

4.1 Alkaloids

Based on many studies about unnatural alkaloid enantiomers, and the results reviewed here the pharmacological effect of natural isomers is enantioselective. However, unnatural enantiomers also have a pharmacological effect of their own which can be discovered in the future. Morphinans of the unnatural (+)-series, in contrast to the (−)-series which are chemically connected with natural morphine, were found to be do not have pharmacological effects as analgesics in vivo, instead, presented useful antitussive properties (**Figure 10**) [62].

(+)- and (−)-spondomine-racemic and dimeric indole alkaloids have been reported in the study of Tian-Yun Jin (2021) [63], especially, (+)-spondomine displayed cytotoxic against the K562 cell line and exhibited Wnt and HIF1. Moreover, all of them were found to be active in promoted angiogenesis and moderate antiinflammation.

Oleracein E (OE) (8,9-dihydroxy-1,5,6,10b-tetrahydro-2H25 pyrrolo[2,1-a] isoquinoline-3-one), an alkaloid possessing tetrahydroisoquinoline and pyrrolidone skeletons. It was reported to have a lot of pharmacological effects such as: antibacterial, anti-inflammatory, anti-aging, anti-hypoxia, anti-oxidant, skeleton-relaxant, hypolipidaemic, analgesic, hypoglycemic, cognition-improvement and neuroprotective functions, especially the optical isomer of (+)-oleracein E (OE) called

Plant	Stereoisomer compound	Pharmacological effect	Test model	Ref.
Alkaloids				
Huperzia serrata	Huperzine A and B	Anticholinesterase activity	Acetylcholinesterase (AChE) inhibitory assay	[50]
Narcissus jonquilla quail	Jonquailine	Anticancer	Human cancer cell line: A549; OE21; Hs683 U373;SKMEL B16F10	[51]
Uncaria rhynchophylla	Speciophylline	Antiplasmodial activity	*Plasmodium falciparum*	[52]
Isatis indigotica	Isatindigotindoline	Inhibitory effects on β-amyloid aggregation	Thioflavin T (ThT)-binding assay	[53]
Flavonoids				
Centaurea maculosa	Trans-flavan-3-ol (+)-catechin	Antibacterial	*Xanthomonas campestris, Pseudomonas fluorescens, Erwinia carotovora*	[54]
Citrus fruit	Narirutin, naringin, hesperidin and neohesperidin	Antioxidant	DPPH assy	[55]
Psidium guajava	Quercetin-3-O-α-L-arabinopyranoside	anti-Streptococcus mutans activity	*S. mutans*	[56]
Rhus retinorrhoea	Persicogenin	Anticancer	MCF-7, HeLa, and HT-29 cells	[57]
Silybum marianum	Silibinin A and Silibinin B	Anticancer	MDA-MB-468 breast cancer cells of the control mice	[58]
Leucosceptrum canum	S-(+)- and R-(−)-leucoflavonines	Aticholinesterase activity	Acetylcholinesterase (AChE) inhibitory assay	[59]

Table 2.
Pharmacologic effect of stereoisomer compound in plant extract.

(−)-trolline has remarkable antibacterial as well as moderate antiviral activity against influenza viruses A and B [64].

4.2 Flavonoids

According to Blair, Lachlan M. (2016) [65], (−)-Foveoglin A (5) exhibited cytotoxicity against a panel of cancer cell lines, while (+)-isofoveoglin (7) and (−)-cyclofoveoglin (8) were weakly cytotoxic, and (+)-foveoglin B (6) was inactive (**Figures 11** and **12**).

Characterize the stereoselective pharmacokinetics of pinocembrin and pinostrobin and their bioactivity in some in vitro investigation with relevant roles in heart disease, colon cancer, and diabetes etiology and pathophysiology [66]. These investigations have revealed that chiral differences in the chemical structure of these compounds result in significant pharmacodynamic differences. Pinocembrin and pinostrobin demonstrated concentration-dependent alpha amylase inhibitory activity. While pinocembrin also has anti-inflammatory antioxidant in the pure S-enantiomer and racemate.

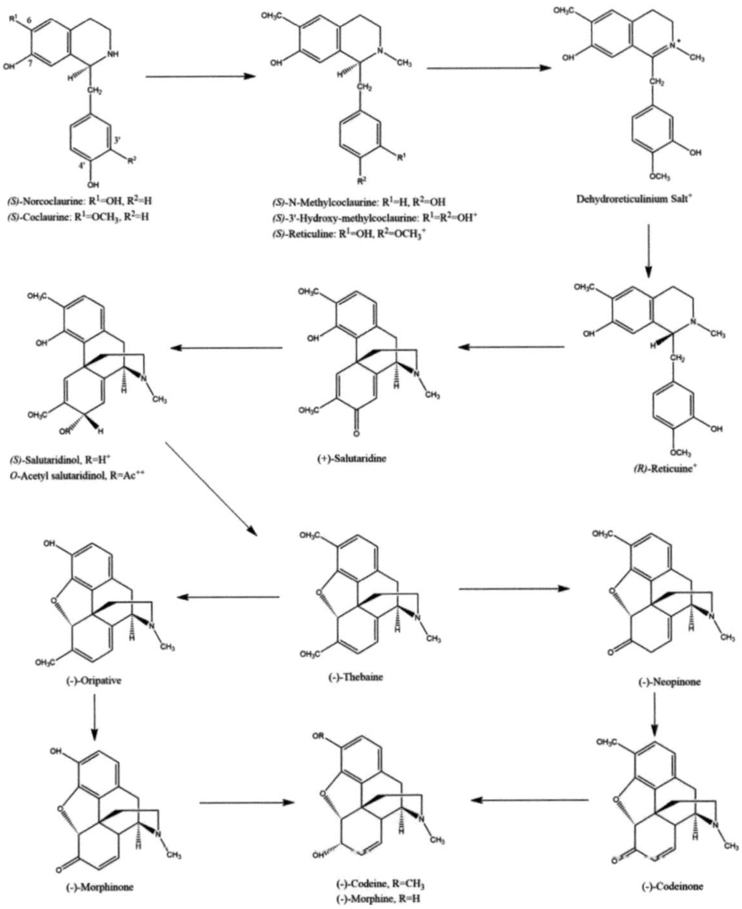

Figure 10.
*Biosynthesis of morphine in plants. * These metabolic conversions are highly stereoselective [61].*

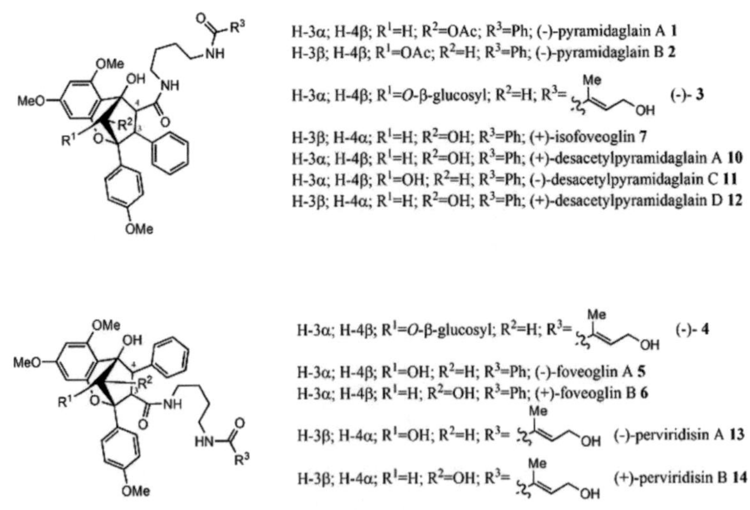

Figure 11.
Aglain and aglaforbesin flavoalkaloids 1–7, 10–12 [65].

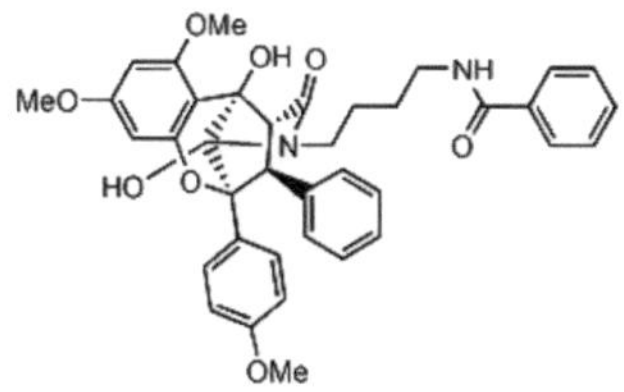

(-)-cyclofoveoglin
8

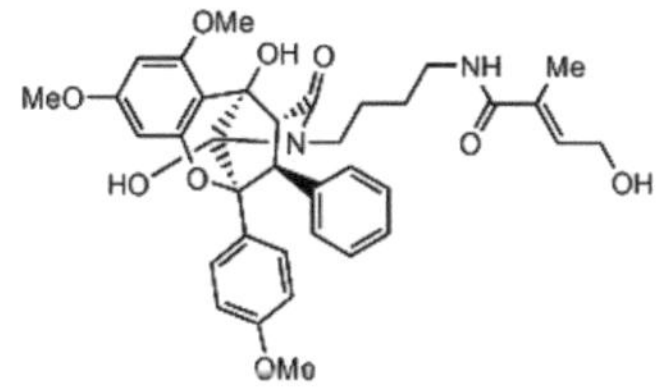

(-)-**9**

Figure 12.
Aglain and aglaforbesin flavoalkaloids 8, 9 [65].

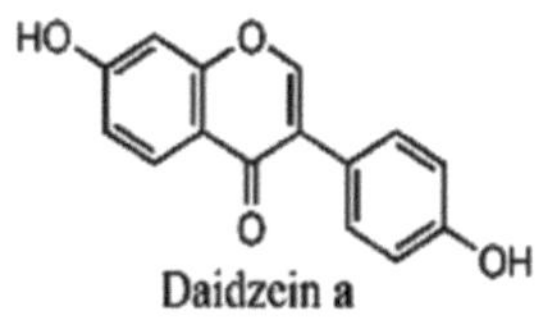

Daidzein a

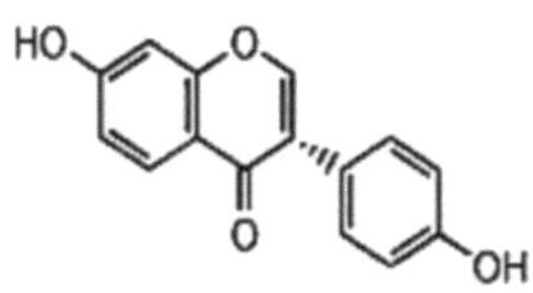

S-(-)equol b

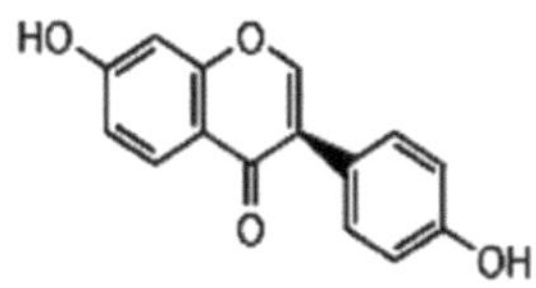

R-(+)equol c

Figure 13.
Chemical structure of daidzein (a) and its metabolites (b and c).

Racemic liquiritigenin is proved to be a dose-dependent inhibition of alpha-amylase enzyme whereas its pure enantiomers did not have this bioactivity. Racemic liquiritigenin showed moderate antiproliferative activity on an HT-29 cancer cell line that was also dose-dependent and had inhibitory effects on the cyclooxygenase-2 enzyme [67].

Racemic liquiritigenin, which was dose-dependent, has been proved its moderate antiproliferative activity on a cancer cell line_ HT-29, and inhibitory effects on the cyclooxygenase-2 enzyme [67]. The nature type of naringenin, hesperetin and hesperidin is S - enantiomer, but both R and S enantiomers can have biological activities such as: antitumor, antioxidant and anti-inflammatory [68]. The two enantiomers of equol: R-(+)-equol and S-(−)-equol have been researched in antitumor activity which shown a significant decrease in the number of palpable tumors presented in animals feeding R-(+)-equol compared to the S-(−)-equol's result (**Figure 13**).

5. Conclusion

Chiral inversion is always mediated by enzymes and varies with solvent, pH and temperature. Considerable attention should be paid to the mechanism of the inversion reaction and its pharmacological and toxicological results. Recent advances in chiral analysis for the herbal plants in clinical research & forensic toxicology by experiments in which one enantiomer was given to the experiment subjects in a specific situation. Demonstration of metabolic chiral inversion demonstration may give an answer for the development of a new pharmaceutical entity. Understanding more about the factors facilitating such interconversions may considerably aid herbal plant development thanks to this feature determination at an early stage and thereby reducing bioanalytical and toxicology workload.

Acknowledgements

The authors would like to express their hearty gratitude to Can Tho University of Medicine and Pharmacy. We also thank all of our colleagues for their excellent assistance.

Conflict of interest

The authors declare no conflict of interest.

Author details

Ngoc-Van Thi Nguyen
Can Tho University of Medicine and Pharmacy, Can Tho City, Vietnam

*Address all correspondence to: nguyenthingocvanct@gmail.com; ntnvan@ctump.edu.vn

IntechOpen

© 2022 The Author(s). Licensee IntechOpen. This chapter is distributed under the terms of the Creative Commons Attribution License (http://creativecommons.org/licenses/by/3.0), which permits unrestricted use, distribution, and reproduction in any medium, provided the original work is properly cited.

References

[1] Khan SJ. Biologically mediated chiral inversion of emerging contaminants. In: Lambropoulou DA, Nollet LML, editors. Transformation Products of Emerging Contaminants in the Environment: Analysis, Processes, Occurrence, Effects and Risks. Chichester, United Kingdom, UK: John Wiley and Sons Ltd; 2014. pp. 261-279

[2] Wsól V, Skálová L, Szotáková B. Chiral inversion of drugs: Coincidence or principle? Current Drug Metabolism. 2004;**5**(6):517-533

[3] Nguyen NVT, Nguyen KNH, Nguyen KT, Kim KH, Aboul-Enein HY. The impact of chirality on the analysis of alkaloids in plant. Pharmacia. 2021; **68**(3):643-656

[4] Kumar S, Pandey AK. Chemistry and biological activities of flavonoids: An overview. ScientificWorldJournal. 2013; **2013**:162750

[5] Yáñez JA, Andrews PK, Davies NM. Methods of analysis and separation of chiral flavonoids. Journal of Chromatography. B, Analytical Technologies in the Biomedical and Life Sciences. 2007;**848**(2):159-181

[6] Caldwell J. Stereochemical determinants of the nature and consequences of drug metabolism. Journal of Chromatography A. 1995;**694**(1):39-48

[7] Prelog V. Specification of the stereospecificity of some oxido-reductases by diamond lattice sections. Pure and Applied Chemistry. 1964;**9**(1): 119-130

[8] Jenner P, Testa B. The influence of stereochemical factors on drug disposition. Drug Metabolism Reviews. 1973;**2**(2):117-184

[9] Girvan HM, Munro AW. Applications of microbial cytochrome P450 enzymes in biotechnology and synthetic biology. Current Opinion in Chemical Biology. 2016;**31**:136-145

[10] Hemscheidt T, Zenk MH. Partial purification and characterization of a NADPH dependent tetrahydroalstonine synthase from Catharanthus roseus cell suspension cultures. Plant Cell Reports. 1985;**4**(4):216-219

[11] Gebler JC, Poulter CD. Purification and characterization of dimethylallyl tryptophan synthase from Claviceps purpurea. Archives of Biochemistry and Biophysics. 1992;**296**(1):308-313

[12] Hashimoto T, Yamada Y. Purification and characterization of hyoscyamine 6 beta-hydroxylase from root cultures of Hyoscyamus niger L. hydroxylase and epoxidase activities in the enzyme preparation. European Journal of Biochemistry. 1987;**164**(2): 277-285

[13] Thomas F, Meinhart HZ. Purification and characterization of three isoforms of S-adenosyl-l-methionine: (R,S)-Tetrahydrobenzylisoquinoline-N-methyltransferase from Berberis koetineana cell cultures. Phytochemistry. 1990;**29**(11):3491-3497

[14] Bednar RA, Hadcock JR. Purification and characterization of chalcone isomerase from soybeans. The Journal of Biological Chemistry. 1988;**263**(20): 9582-9588

[15] Fischer D, Stich K, Britsch L, Grisebach H. Purification and characterization of (+)dihydroflavonol (3-hydroxyflavanone) 4-reductase from

flowers of Dahlia variabilis. Archives of Biochemistry and Biophysics. 1988; **264**(1):40-47

[16] Lukacin R, Wellmann F, Britsch L, Martens S, Matern U. Flavonol synthase from Citrus unshiu is a bifunctional dioxygenase. Phytochemistry. 2003;**62** (3):287-292

[17] Akashi T, Aoki T, Ayabe S. Identification of a cytochrome P450 cDNA encoding (2S)-flavanone 2-hydroxylase of licorice (Glycyrrhiza echinata L.; Fabaceae) which represents licodione synthase and flavone synthase II. FEBS Letters. 1998; **431**(2):287-290

[18] Stafford HA, Lester HH. Flavan-3-ol biosynthesis: The conversion of (+)-Dihydromyricetin to its Flavan-3,4-diol (Leucodelphinidin) and to (+)-Gallocatechin by reductases extracted from tissue cultures of Ginkgo biloba and Pseudotsuga menziesii. Plant Physiology. 1985;**78**(4): 791-794

[19] Xie DY, Sharma SB, Dixon RA. Anthocyanidin reductases from Medicago truncatula and Arabidopsis thaliana. Archives of Biochemistry and Biophysics. 2004;**422**(1):91-102

[20] Paiva NL, Edwards R, Sun YJ, Hrazdina G, Dixon RA. Stress responses in alfalfa (Medicago sativa L.) 11. Molecular cloning and expression of alfalfa isoflavone reductase, a key enzyme of isoflavonoid phytoalexin biosynthesis. Plant Molecular Biology. 1991;**17**(4):653-667

[21] Preisig CL, Matthews DE, Vanetten HD. Purification and characterization of S-Adenosyl-l-methionine:6a-Hydroxymaackiain 3-O-methyltransferase from Pisum sativum. Plant Physiology. 1989;**91**(2):559-566

[22] Tanner GJ, Francki KT, Abrahams S, Watson JM, Larkin PJ, Ashton AR. Proanthocyanidin biosynthesis in plants. Purification of legume leucoanthocyanidin reductase and molecular cloning of its cDNA. The Journal of Biological Chemistry. 2003; **278**(34):31647-31656

[23] Hagel JM, Facchini PJ. Benzylisoquinoline alkaloid metabolism: A century of discovery and a brave new world. Plant & Cell Physiology. 2013;**54**: 647-672

[24] Gulland JMR, R. Constitution of codeine and thebaine. Manchester Literary and Philosophical Society. 1925; **69**:79-68

[25] Kirby GW. Biosynthesis of the morphine alkaloids. Science. 1967;**155**: 170-173

[26] Farrow SC, Hagel JM, Beaudoin GA, Burns DC, Facchini PJ. Stereochemical inversion of (S)-reticuline by a cytochrome P450 fusion in opium poppy. Nature Chemical Biology. 2015; **11**(9):728-732

[27] Galanie S, Thodey K, Trenchard IJ, Filsinger Interrante M, Smolke CD. Complete biosynthesis of opioids in yeast. Science. 2015;**349**(6252): 1095-1100

[28] Gesell A, Rolf M, Ziegler J, Díaz Chávez ML, Huang FC, Kutchan TM. CYP719B1 is salutaridine synthase, the C-C phenol-coupling enzyme of morphine biosynthesis in opium poppy. The Journal of Biological Chemistry. 2009;**284**(36):24432-24442

[29] Stavrinides A, Tatsis E, Caputi L, et al. Structural investigation of heteroyohimbine alkaloid synthesis reveals active site elements that control

stereoselectivity. Nature Communications. 2016;7:1-14

[30] Stoeckigt J, Husson HP, Kan-Fan C, Zenk MH. Cathenamine, a central intermediate in the cell free biosynthesis of ajmalicine and related indole alkaloids. Journal of the Chemical Society, Chemical Communications. 1997;**164–166**

[31] O'Connor SE, Maresh JJ. Chemistry and biology of monoterpene indole alkaloid biosynthesis. Natural Product Reports. 2006;**23**:532-547

[32] Gerasimenko I, Sheludko Y, Ma X, Stöckigt J. Heterologous expression of a Rauvolfia cDNA encoding strictosidine glucosidase, a biosynthetic key to over 2000 monoterpenoid indole alkaloids. European Journal of Biochemistry. 2002; **269**:2204-2213

[33] Stavrinides A, Tatsis EC, Foureau E, Caputi L, Kellner F, Courdavault V, et al. Unlocking the diversity of alkaloids in Catharanthus roseus: Nuclear localization suggests metabolic channeling in secondary metabolism. Chemistry & Biology. 2015;**22**(3): 336-341

[34] Koukol J, Conn EE. The metabolism of aromatic compounds in higher plants. The Journal of Biological Chemistry. 1961;**136**:2692-2698

[35] Jez JM, Bowman ME, Dixon RA, Noel JP. Structure and mechanism of the evolutionarily unique plant enzyme chalcone isomerase. Nature Structural Biology. 2000;**7**(9):786-791

[36] Prescott AG, Stamford NP, Wheeler G, Firmin JL. In vitro properties of a recombinant flavonol synthase from Arabidopsis thaliana. Phytochemistry. 2002;**60**(6):589-593

[37] Turnbull JJ, Nakajima J, Welford RW, Yamazaki M, Saito K, Schofield CJ. Mechanistic studies on three 2-oxoglutarate-dependent oxygenases of flavonoid biosynthesis: Anthocyanidin synthase, flavonol synthase, and flavanone 3beta-hydroxylase. The Journal of Biological Chemistry. 2004; **279**(2):1206-1216

[38] Dixon RA, Xie DY, Sharma SB. Proanthocyanidins–a final frontier in flavonoid research? The New Phytologist. 2005;**165**(1):9-28

[39] Min T, Kasahara H, Bedgar DL, Youn B, Lawrence PK, Gang DR, et al. Crystal structures of pinoresinol-lariciresinol and phenylcoumaran benzylic ether reductases and their relationship to isoflavone reductases. The Journal of Biological Chemistry. 2003;**278**(50):50714-50723

[40] Xie DY, Sharma SB, Paiva NL, Ferreira D, Dixon RA. Role of anthocyanidin reductase, encoded by BANYULS in plant flavonoid biosynthesis. Science. 2003;**299**(5605): 396-399

[41] Marais JPJ, Deavours B, Dixon RA, Ferreira D. The stereochemistry of flavonoids. In: Grotewold E, editor. The Science of Flavonoids. New York, NY: Springer; 2006. DOI: 10.1007/978-0-387-28822-2_1

[42] Nguyen NV, Lee KR, Lee YJ, Choi S, Kang JS, Mar W, et al. Chiral high-performance liquid chromatographic separation of evodiamine enantiomers and rutaecarpine, isolated from Evodiae fructus. Journal of Pharmaceutical and Biomedical Analysis. 2013;**81-82**:151-159

[43] Rajender K, Jürgen M, Ravi B. Enantiomerization study of atropine and its Semipreparative Enantioseparation along with (1RS,2SR)-(±)-ephedrine on

polyacrylamide column using high-performance liquid chromatography. Journal of Liquid Chromatography & Related Technologies. 2015;**38**(1): 111-116

[44] Margolin AL. Enzymes in the synthesis of chiral drugs. Enzyme and Microbial Technology. 1993;**15**(4): 266-280

[45] Pham-Huy C, Villain-Pautet G, Hua H, Chikhi-Chorfi N, Galons H, Thevenin M, et al. Separation of oxazepam, lorazepam, and temazepam enantiomers by HPLC on a derivatized cyclodextrin-bonded phase: Application to the determination of oxazepam in plasma. Journal of Biochemical and Biophysical Methods. 2002;**54**(1–3):287-299

[46] Yang SK, Lu XL. Resolution and stability of oxazepam enantiomers. Chirality. 1992;**4**:443-446

[47] Juri M, Dai-ichiro K, Masatoshi O, Masahiro T, Seiji N, Kazunari A, et al. Biosynthesis-inspired deracemizative production of d -luciferin by combining luciferase and thioesterase. Biochimica et Biophysica Acta (BBA) - General Subjects. 2017;**1861**(8):2112-2118

[48] Glass AM, Krause ME, Laurence JS, Jackson TA. Controlling the chiral inversion reaction of the Metallopeptide Ni-asparagine-cysteine-cysteine with dioxygen. Inorganic Chemistry. 2012; **51**(18):10055-10063

[49] Eed J. Factors affecting enzyme activity. Essai. 2012;**10**(1):19

[50] Jia-Sen L, Yuan-Long Z, Chao-Mei Y, You-Zuo Z, Yan-Yi H, Feng-Wu W, et al. The structures of huperzine a and B, two new alkaloids exhibiting marked anticholinesterase activity. Canadian Journal of Chemistry. 1986;**64**(4): 837-839

[51] Vergura S, Santoro E, Masi M, Evidente A, Scafato P, Superchi S, et al. Absolute configuration assignment to anticancer Amaryllidaceae alkaloid jonquailine. Fitoterapia. 2018;**129**:78-84

[52] Pierrot D, Sinou V, Bun SS, Parzy D, Taudon N, Rodriguez J, et al. Design and synthesis of simplified speciophylline analogues and β-carbolines as active molecules against plasmodium falciparum. Drug Development Research. 2019;**80**(1):133-137

[53] Si-Fan L, Bin L, Yu-Fei X, Le Z, Li-Li L, Xiao-Xiao H, et al. Bioactive spiropyrrolizidine oxindole alkaloid enantiomers from Isatis indigotica fortune. Organic & Biomolecular Chemistry. 2018;**16**(48):9430-9439

[54] Bais HP, Walker TS, Stermitz FR, Hufbauer RA, Vivanco JM. Enantiomeric-dependent phytotoxic and antimicrobial activity of (+/−)-catechin. A rhizosecreted racemic mixture from spotted knapweed. Plant Physiology. 2002;**128**(4):1173-1179

[55] Zheng H, Zhen XT, Chen Y, Zhu SC, Ye LH, Yang SW, et al. In situ antioxidation-assisted matrix solid-phase dispersion microextraction and discrimination of chiral flavonoids from citrus fruit via ion mobility quadrupole time-of-flight high-resolution mass spectrometry. Food Chemistry. 2021; **343**:128422

[56] Prabu GR, Gnanamani A, Sadulla S. Guaijaverin – a plant flavonoid as potential antiplaque agent against Streptococcus mutans. Journal of Applied Microbiology. 2006;**101**(2): 487-495

[57] Saquib Q, Ahmed S, Ahmad MS, Al-Rehaily AJ, Siddiqui MA, Faisal M, et al. Anticancer efficacies of persicogenin and homoeriodictyol isolated from Rhus

retinorrhoea. Process Biochemistry. 2020;**95**:186-196

[58] Jahanafrooz Z, Motamed N, Rinner B, Mokhtarzadeh A, Baradaran B. Silibinin to improve cancer therapeutic, as an apoptotic inducer, autophagy modulator, cell cycle inhibitor, and microRNAs regulator. Life Sciences. 2018;**213**:236-247

[59] Feng L, Zhang Y, Liu YC, Liu Y, Luo SH, Huang CS, et al. Leucoflavonine, a new bioactive racemic flavoalkaloid from the leaves of Leucosceptrum canum. Bioorganic & Medicinal Chemistry. 2019;**27**(2):442-446

[60] Nguyen LA, He H, Pham-Huy C. Chiral drugs: An overview. International Journal of Biomedical Sciences. 2006; **2**(2):85-100

[61] Rainer L, Meinhart HZ. Closure of the oxide bridge in morphine biosynthesis. 1994;**35**(23):3897-3900

[62] Arnold B. Chapter 3: Biological activity of unnatural alkaloid enantiomers. In: Cordell GA, editor. The Alkaloids: Chemistry and Biology. Vol. 50. Academic Press; 1998. pp. 109-139

[63] Jin T, Li P, Wang C, Tang X, Cheng M, Zong Y, et al. Racemic Bisindole alkaloids: Structure, bioactivity, and computational study. Chinese Journal of Chemistry. 2021;**39**(9):2588-2598

[64] Wang PP, Sun HX, Liu CJ, Hu MH, He XQ, Yue S, et al. Racemic oleracein E increases the survival rate and attenuates memory impairment in D-galactose/ $NaNO_2$-induced senescent mice. Phytomedicine. 2016;**23**(5):460-467

[65] Blair LM, Calvert MB, Sperry J. Flavoalkaloids-isolation, biological activity, and Total synthesis. The Alkaloids. Chemistry and Biology. 2017; **77**:85-115

[66] Sayre CL, Alrushaid S, Martinez SE, Anderson HD, Davies NM. Pre-clinical pharmacokinetic and Pharmacodynamic characterization of selected chiral flavonoids: Pinocembrin and Pinostrobin. Journal of Pharmacy & Pharmaceutical Sciences. 2015;**18**(4): 368-395

[67] Alrushaid S, Davies NM, Martinez SE, Sayre CL. Pharmacological characterization of liquiritigenin, a chiral flavonoid in licorice. Research in Pharmaceutical Sciences. 2016;**11**(5): 355-365

[68] Pinto C, Cidade H, Pinto M, Tiritan ME. Chiral flavonoids as antitumor agents. Pharmaceuticals. 2021;**14**(12): 1267

Chapter 13

Secondary Metabolites of Edible Cacti (Cactaceae) from the South American Andes

Frank L. Romero-Orejon, Ana María Muñoz, Luciana de la Fuente-Carmelino, Diana Jimenez-Champi, Eliana Contreras-López, Ivan Best, Luís Aguilar and Fernando Ramos-Escudero

Abstract

The South American Andes hide countless cacti and are part of valuable Andean biodiversity. Within this large family of Cactaceae are edible cacti that are highly valued for their medicinal properties and used as edible fruits. In this review, we will make a description of the overall chemical composition, main phytochemicals found in some edible cacti of the Andean region such as sanky (*Corryocactus brevistylus*), pitahaya (*Hylocereus monacanthus*, *Hylocereus megalanthus*) and tuna or prickly pear (*Opuntia ficus-indica*). In addition, we will include its medicinal and therapeutic properties and its commercial applications and uses as a natural colorant.

Keywords: edible cacti, Andean region, phytochemicals, healthy properties, commercial applications

1. Introduction

Cactaceae are a large family of plants that prosperous in desert and semidesert areas. These plants have been used by ancient civilizations for the treatment and cure of diseases [1]. In addition, they have been used as fodder (dairy cows) [2], medicinal (nutritional qualities and health implications) [3, 4], fruits and vegetables (prickly pear "tuna", and dragon fruit "pitahaya") [5–7] processed products (jams, syrups, concentrated juices, candies, wine, natural colorants, and others) [8, 9]. Cactaceae have been shown to provide significant health benefits because of their dietary fiber, flavonoids, hydroxycinnamic acids, betalains, carotenoids, terpenes, and tannins contents, that show health benefits such as anti-microbial and anti-parasitic, anti-proliferative and cytotoxicity and anti-inflammatory properties and inhibition of enzymes involved in carbohydrate catabolism (α-glucosidase and α-amylase) [10, 11].

Many of the native food fruits that grow in the Andean and Amazonian regions have generated much interest today due to the wide range of nutrients and bioactive compounds they possess. Among edible fruits, we can find Cherimoya (*Annona cherimola* Mill), lucuma (*Pouteria lucuma* (Ruiz & Pav.) Kuntze), goldenberry or cape gooseberry (*Physalis peruviana* L.), saúco (*Sambucus peruviana* H.B.K.), pepino (*Solanum muricatum* Aiton), soursop (*Annona muricata* Linnaeus), asaí or açaí (*Euterpe oleracea* Martius), camu-camu (*Myrciaria dubia* (H.B.K.) McVaugh), Inca peanut or sacha inchi (*Plukenetia volubilis* Linnaeus), sachatomate (*Solanum betaceum* (Cavanilles) Sendtner), guinda (*Prunus serotina*), granadilla (*Passiflora ligularis*), tumbo serrano or banana passion fruit (*Passiflora tripartite* var. mollissima), tuna or prickly pear (*Opuntia ficus-indica*), sanky (*Corryocactus brevistylus*), and pitahaya (*Hylocereus monacanthus, Hylocereus megalanthus*) [12–15].

The objective of this review is to compile the general chemical composition and bioactive phytochemicals of pitahaya, tuna or prickly pear and sanky, as well as the commercial applications. The aim of this paper is also to generate research interest in the valorization of edible cacti (Cactaceae) from the South American Andes and its by-products.

2. Pitahaya

Pitahaya, also known as pitaya or dragon fruit, (*Hylocereus* spp.) is an exotic tropical fruit that belongs to the *Cactaceae* family and *Hylocereeae* tribe [16, 17]. The pitahaya taxonomy is shown in **Table 1**; although it is native to Central and South America, it is currently grown for commercial purposes in Asian countries such as Vietnam, Malaysia, Thailand, and Taiwan [18, 19]. In South America, it is distributed across Venezuela and Bolivia, but the countries having a considerable production are Colombia, Ecuador, and Peru (**Table 2**) [18, 24]. In the Colombian territory, the cultivated hectares are distributed between Boyacá, Quindío, Santander, and Valle del Cauca [18]. In Ecuador, pitahaya is found in the provinces of Pichincha, Morona Santiago, and Loja [22], whereas in Peru, pitahaya is produced in Amazonas, San Martín, Lambayeque, and Junín [16].

The cactus plant of the pitahaya is a climbing, perennial, shrub-like plant that can grow up to 2 m. It is cultivated at an altitude of 500–1900 m above sea level, at an average temperature of 22°C and a relative humidity ranging from 70 to 80% [26].

Scientific name	*Hylocereus* spp.
Kingdom	Plantae
Division	Magnoliophyta
Class	Magnoliopsida
Order	Caryophyllales
Family	Cactaceae
Tribe	Hylocereeae
Genus	Hylocereus
Species	*Hylocereus megalanthus, H. Hylocereus microcladus Backeberg, H. Hylocereus monacanthus*[*]

Adapted from Verona-Ruiz et al. [16].[*]*Some endemic species of South America.*

Table 1.
Hylocereus spp. taxonomy.

Variety	Country	Region	Reference
Selenicereus megalanthus	Colombia	Valle del Cauca	[20]
Hylocereus megalanthus	Colombia	Fusagasugá	[21]
S. megalanthus	Ecuador	Morona Santiago	[22]
S. megalanthus	Peru	San Martín	[23]
S. megalanthus	Colombia	Valle del Cauca	[24]
H. triangularis	Peru	Huancavelica	[25]
S. megalanthus	Ecuador	Pichincha	[26]
S. megalanthus	Peru	Chachapoyas	[27]
S. megalanthus and *Hylocereus polyrhizus*	Colombia	Valle del Cauca	[28]
H. megalanthus	Peru	Cajamarca	[29]
H. megalanthus	Colombia	Ibagué	[30]

Table 2.
Locations of the different varieties of pitahaya are found in the Andean region.

The stems are characterized by their three wavy wings with horny scalloped margins. Each stem segment can grow up to 6 meters long [16]. The cladodes of the pitahaya cactus have 3–5 edges, contributing to its triangular shape. This plant grows wild on trees, rocks, logs, and walls, and is cultivated on trellises to support plant growth [18].

Pitahaya flowers are green on the outside, featuring interior white segments that are approximately 12 inches long and 9 inches wide. The stigma is lobed and green in color. Night-blooming pitahaya flowers open during the early morning and wither at dawn [16]. Pitahaya is characterized by being an ovoid berry 10–12 cm long and 7 cm wide [22]. The peel and pulp of some varieties can change colors, and the peel can vary from yellow to pinkish-red [17]. The fruit can weigh from 200 to 350 grams and contains approximately 650 seeds [18, 24].

Due to the great similarity in their morphological characteristics, different species of this family have been generically called "pitahaya." This made its botanical classification more complex. Within this species, four genera stand out: *Stenecereus* sp., *Selenicerus* sp., *Hylocereus* sp., and *Cereus* sp. [18]. The species that is distinguished by its peels and pulp is *H. undatus*. This pitahaya has white pulp and red-pink peel, and it is the most popular variety worldwide [31]. *H. costaricensis* is a pitahaya with red pulp and peel, and it is also known as *H. polyhizus* [16]. Finally, *H. megalanthus* or *Selenicereus megalanthus* is a pitahaya with yellow peel and white pulp (**Figure 1**), being the most highly produced variety in South America [18].

Figure 1.
Major parts of pitahaya with yellow peel fruits. (A) Fruit, (B) mesocarp, (C) peel, and (D) pitahaya seeds (wet).

Pitahaya is preferred by consumers because of its unique flavor, shape, and pulp color. In addition, it is famous for its low caloric value because it contains small amounts of carbohydrates (9.20 g per 100 g of edible pulp) [16]. In European and Asian markets, pitahaya is categorized as an exotic fruit because of the appearance of its peel and the bittersweet taste of its pulp. In the United States and various European countries, the pitahaya pulp is also in demand as a food ingredient or as a natural food colorant [17]. Conversely, in Malaysia and Indonesia, pitahaya is usually marketed as jams and sweets. In the province of Guangxi (China), red pitahaya is used to make wine [31].

Pitahaya is often described as a sweet fruit; however, this depends on its variety. The yellow pitahaya [*Selenicereus megalanthus* (k. Schum. Ex vaupel) moran] cultivated in Colombia has an acidic and sweet taste owing to the high content of soluble solids, while its organoleptic characteristics are more appealing than other similar species of the genus *Hylocereus* [23].

The betacyanin content in both the pulp and peel usually stands out in this fruit. Similarly, the presence of lycopene, vitamin E, beta carotene, total polyphenols, tannins, and antioxidant compounds has been reported [16, 31]. Essential fatty acids, led by linoleic acid (64.5%), oleic acid (14%), and palmitic acid (14.4%) have been found in the seeds [18, 31]. The beneficial effects of pitahaya range from relief of stomach problems to the amelioration of endocrine disorders and improvements in the digestive system function. The most recognized benefit of pitahaya is the antioxidant capacity attributed to its seeds, the most important antioxidant being linoleic acid because it works as an antioxidant buffer, captures cholesterol, producing a cardiotonic effect [16, 18]. Likewise, linoleic acid has been shown to reduce dyslipidemia and promote wound healing in diabetic rats [18].

2.1 Overall chemical composition

Pitahaya is a sweet-tasting fruit with a pulp of different colors; its weight can reach up to 700 g, with diameters of approximately 15–10 cm. According to Mercado-Silva [32], the pulp color varies between white and red, contains black seeds, and represents 60–80% of the total weight of the fruit depending on the variety. **Table 3** summarizes the nutritional composition of the pulp in three different commercial species. The *H. undatus* variety was added to compare the characteristics of a species outside the Andean region. The moisture (85–89%) and protein content (0.5–0.6 g)

Component	*Hylocereus undatus* (Mexico)	*Selenecereus maegalanthus* (Peru)	*Selenicereus megalanthus* (Colombia)
Water (%)	89	89	85
Proteins (g)	0.5	0.5	0.6
Fats (g)	0.1	0.1	0.4
Carbohydrates (g)	NE	9.1	13
Dietary fibers (g)	0.3	0.3	0.77
Vitamin C (mg)	25.0	8.0	—
Ash (g)	0.5	0.5	0.3

Adapted from Cañar et al. [24], Mercado-Silva [32], and Obregón-La Rosa et al. [23].

Table 3.
Nutritional composition of 100 g of pulp from different species of pitahaya.

do not differ between the varieties. Similarly, the ash content varies from 0.3 g to 0.5 g. However, the Colombian yellow pitahaya had a higher percentage of dietary fiber (0.77 g) when compared with previous reports. This fruit is widely famous for its vitamin C content, which is involved in the formation of collagen, red blood cells, bones, and teeth [20]. However, the table shows that the red variety (*H. undatus*) stands out for its vitamin C content when compared with the yellow variety of the Andean region (*Selenecereus megalanthus*).

Different studies show that the fresh weight of pitahaya increases in direct proportion to the development of the fruit, and the content of soluble solids depends on the ripening stage [22, 32–34]. Following the physicochemical evaluation of pitahaya grown in Morona Santiago (Ecuador), Sotomayor et al. [22] determined that the percentage of peel decreases from 55.9% to 33.4%, while that of the pulp increases from 44.1% to 66.6% between maturity stages 0 and 6. In addition, the flavor of the pitahaya will depend on the maturity during its harvest due to the degradation of polysaccharides, an important factor that determines the concentration of carbohydrates. According to Ochoa Velazco [17], the total soluble solids (TSS) that predominate in pitahaya are glucose and fructose.

The content of TSS in pitahaya is variable (14–16°Brix) and the titratable acidity is usually low (0.2–0.35 m% malic acid/100 g of fresh weight) [32]. In a study conducted by Chauca and Chávez [27], pitahaya from Chachapoyas had a TSS value of 17.4°Brix and an acidity of 0.20% citric acid/100 g fresh weight. Similarly, pitahayas collected from the Cauca valley presented a TSS value of 14.3°Brix and an acidity of 1.35 mg citric acid/100 g fresh weight [24]. Torres Grisales et al. [20] obtained values of 17.7°Brix and titratable acidity of 0.20% citric acid in the *S. megalanthus* variety harvested in the province of Valle del Cauca. Although, the TSS does not determine consumer acceptability, it is an indicator of sweetness, which results from the combination of soluble sugars and organic acids. A TSS combination with high acidity is associated with a better flavor, making the fruit very appealing to consumers [19].

2.2 Bioactive phytochemicals

The fruit of the pitahaya, especially the mesocarp, has a high nutraceutical value and is considered a functional food as it contains healthy bioactive compounds [19]. The pitahaya pulp not only contains sugars and acids but also fiber, vitamin C, pectin, and different pigments [35]. In samples of *S. megalanthus* harvested in Peru, the vitamin C content ranged between 7 and 9 mg/100 g of pulp [23]. Torres Grisales et al. [20] obtained similar results (8 mg ascorbic acid per gram of dry matter) for pitahaya pulp from Valle del Cauca. The pitahaya peel and seeds were also analyzed, showing higher values. The seeds stand out in the results, containing up to 22.0881 mg ascorbic acid per gram of dry matter. Some studies indicate that the content of vitamin C in varieties outside the Andean region is higher (**Table 4**). Thus, the content of vitamin C in the species *H. monacanthus* collected in Fortaleza (Brazil) was found to be 107.02 mg/100 g [36]. Similarly, *H. undatus* from Sao Paulo (Brazil) presented 45.79 mg of vitamin C/100 g [37].

Polyphenols (**Table 4**), carotenoids, tocopherols, and glucosinolates are usually found in fruits and vegetables. A chemoprotective effect has been attributed to these compounds to combat oxidative stress, as well as anti-inflammatory properties benefiting human health [35, 36]. In the case of pitahaya, the total content of phenolic compounds in fruits of *H. megalanthus* from Cajamarca was 16.17 mg of gallic acid equivalent (GAE) per 100 g of pulp [29]. Likewise, these components were evaluated

Species	Location	Bioactive compounds or micronutrients	Concentration	Reference
Selenicereus megalanthus	San Martín, Peru	Vitamin C	8.00 mg/100 g	[23]
Hylocereus megalanthus	Cajamarca, Peru	Total phenolic compounds	16.17 mg GAE/100 g	[29]
H. megalanthus	Fusagasugá, Colombia	Total phenolic compounds	59.10 mg GAE/100 g	[21]
S. megalanthus	Valle del Cauca, Colombia	Total phenolic compounds	1580 mg GAE/100 g	[20]
S. megalanthus	Chachapoyas, Peru	Total phenolic compounds	2.01 mg GAE/g	[27]
H. megalanthus	Ibagué, Colombia	Total phenolic compounds	7.8 mg GAE/100 g	[30]

Table 4.
Bioactive compounds are found in different varieties of pitahaya in the Andean region.

to determine the inhibitory effects on α-amylase and α-glucosidase as a possible complementary therapy for diabetes mellitus. The IC_{50} value in the case of α-amylase was 8692.4 μg/mL and there was no inhibitory effect on α-glucosidase. According to the authors, the antidiabetic effect of the fruits of the *Hylocereus* family is not a result of their ability to inhibit digestive enzymes, but rather a result of their ability to improve insulin resistance and increase gene expression levels of the fibroblast growth factor 21 receptors [29].

In a study by Mejia et al. [21], the phenol content for the *H. megalanthus* variety from Fusagasugá (Colombia) was 59.1 mg gallic acid (GAE)/100 g fresh weight. Compared with the other exotic Colombian fruits assessed, pitahaya showed a moderate phenol content (80 > GAE/100 g). In addition, the antioxidant activity was determined using the 2,2-diphenyl-1-picrylhydrazyl (DPPH) and the 2,2′-azino-bis (3-ethylbenzothiazoline-6-sulphonic acid) (ABTS) methods, as well as the ferric reducing antioxidant power (FRAP) assay relative to fresh weight basis. The results for this fruit were 177.1 μmol of Trolox Equivalent (TE)/100 g, 323.8 TE/100 g, and 811.2 μmol of Fe^{+2}/100 g, respectively. Additionally, the authors evaluated the elimination capacity of reactive nitrogen species. Samples of *H. megalanthus* exhibited a potent peroxyl radical scavenging activity with a value of 2999.77 μmol of TEs/g.

Torres Grisales et al. [20] assessed the content of phenols in all parts of pitahaya and showed that the seeds and the peel contain the highest amount of these compounds. A total of 1580 mg GAE/100 g dry matter were found in the seeds. Additionally, the authors analyzed the antioxidant capacity of phenols extracted from the pulp, peel, and seeds using the ABTS and DPPH assays. High levels of antioxidant activity were shown in all parts of the fruit. Nonetheless, the higher antioxidant activity was found in the seeds, with values of 79.2% and 96% for ABTS and DPPH assays, respectively.

In contrast, the amount of phenolic compounds in the peel and pulp of lyophilized yellow pitahaya (*S. megalanthus*) from Chachapoyas was also determined. In this study, the lowest values in peel and pulp were obtained, with 1.50 and 2.01 mg GAE/g of the sample, respectively. Similarly, the antioxidant activity was lower than expected by the authors, with values of 8.15 and 7.7 μmol TEs/g in pulp and peel, respectively [27].

Component	g/kg of oil
Fatty acid	182 ± 11
Palmitic acid	3 ± 1
Palmitoleic acid	49 ± 3
Stearic acid	239 ± 16
Oleic acid	45 ± 6
Cis-11-vaccenic acid	466 ± 42
Linoleic acid	18 ± 2
Arachidic acid	182 ± 11

Adapted from Villalobos et al. [38].

Table 5.
Fatty acid profile of oil extracted from red pitahaya seeds (g/kg of oil).

The oil content and characteristics of *Hylocereus* seeds have been extensively studied. As mentioned above, pitahaya seeds have a high content of linoleic acid, which represents between % and 2% of the total weight of the fruit [18, 31, 35]. According to Villalobos-Gutiérrez et al. [38], the predominant saturated fatty acids in pitahaya seed oil are palmitic acid, stearic acid, and arachidic acid, which only represent 249 g/kg of total fatty acids. **Table 5** summarizes the amount of total fatty acids found in the red pitahaya variety (*Hylocereus* sp. [Weber] Britton & Rose) from Nicaragua.

Torres Grisales [20] observed that pitahaya seeds have a greater ability to accelerate peristalsis by increasing the amount of feces produced by 55% when compared with biomodels fed a diet based on sunflower seeds. The laxative capacity of the other parts of pitahaya was also evaluated; however, the peel decreased the production of feces. It is important to mention that the consumption of pitahaya pulp and seeds promotes its output, although with a less solid appearance. This could be related to the passage of stool. According to Verona-Ruiz [16], the oligosaccharides present in the pulp serve as a possible source of prebiotics and stimulate the growth and/or activity of specific bacteria in the colon.

2.3 Commercial applications

Numerous studies on different varieties of pitahaya (*H. undatus*, *Hylocereus polyrhizus*, and *H. megalanthus*) have shown the variety of benefits that can confer to human health [19, 39–41]. The peel can be used as raw material for the extraction of pectin, betacyanins, and dietary fiber [42]. In addition, it is used in food packaging and edible coating [43]. The pitahaya pulp has been shown to have nutraceutical properties and can be used to prepare fermented beverages. This would increase the content of phenolic compounds [16, 17, 31]. Finally, pitahaya seeds and their high content of unsaturated fatty acids can be used in the food, cosmetic, or pharmaceutical industry. The oligosaccharides present in the seeds is a potential source of prebiotics with a demonstrated ability to stimulate the growth of lactobacilli and bifidobacteria [16, 35].

Within the region, two studies have been conducted on the production of drinks based on pitahaya pulp. Enriquez Paredes and Ore Areche [25] obtained a functional drink made with malt from *Amaranthus caudatus* L. (kiwicha) and pulp from *Hylocereus triangularis*; both fruits were collected in Huancavelica (Peru). This drink was

Ingredient	Amount
Water	3 L
Pitahaya pulp	1 L
Malted kiwicha flour	100 g
Sugar	200 g
Citric acid	4.50 g
Carboxymethylcellulose	3.70 g

Taken from Enriquez Paredes and Ore Areche [25].

Table 6.
Composition of a drink based on Amaranthus caudatus L. and pulp of Hylocereus triangularis.

	BHC	BH 4%	BH 6%	BH 8%
Phenolic content (mg GAE/100 mL)	33.90 ± 1.27	100.92 ± 2.10	117.59 ± 0.84	112.86 ± 0.11

Taken from Castro Carranza et al. [44].

Table 7.
Phenolic content of the drink based on Hylocereus undatus and vegetable extracts.

successful in a panel of 20 specialists, and its formulation is presented in **Table 6**. The drink reached a pH of 3.7 and 11.50°Brix, with a lack of molds, yeasts, and coliforms. The authors highlighted the protein content of the drink (8.53%), which is highly nutritious and suitable for consumption.

In the study conducted by Castro Carranza et al. [44], a functional drink was developed based on pitahaya (*Hylocereus undatus* (Haw.) Britton & Rose) with extracts of lemon verbena (*Cymbopogon citratus*) and basil (*Ocimum tenuiflorum*). The fruit and leaves for this project were collected in the province of Manabí, Ecuador. The drink was elaborated through the incorporation of the vegetable extracts in different percentages (4, 6 and 8%) to the pitahaya juice. These percentages were reached after mixing the vegetable extracts at a 1:1 volume-volume ratio. The authors determined the total phenolic content of drinks prepared with different percentages of extracts, as shown in **Table 7**.

The results showed that the addition of *C. citratus* and *O. tenuiflorum* extracts can increase the concentration of phenolic content of drinks based on *H. undatus*, thereby enhancing its antioxidant properties. The authors believe that elaborating foods with improved functional properties is possible using the studied fruits to improve the health and quality of life of consumers [44].

3. Tuna or prickly pear

O. ficus-indica (L.) Mill. is a species of the Cactaceae family, genus *Opuntia,* and is commonly known as prickly pear, cactus pear, Barbary fig, Indian fig, nopal, or tuna [45]. The complete taxonomic lineage is shown in **Table 8**. Prickly pear is distributed throughout the American continent, from southern Canada to Patagonia. There are no exact records of its origin, but it is believed to be native to Mexico [35, 46]. Its domestication is centered in arid and semi-arid climates of the regions of Mexico because it has special adaptation mechanisms and a high biomass production capacity. This allows it to grow in harsh conditions, such as high temperatures and nutritionally poor soils. However, prickly pear is also cultivated in tropical and subtropical America and Mediterranean countries [46, 47].

Currently, Mexico is the largest producer of prickly pears in the world, with a production of 356 thousand tons/year [48]. The largest production areas are Zacatecas, Puebla, and Hidalgo, where the harvest takes place from July to September [49]. Another country that stands out for the production of prickly pears is Peru, where up to 100,000 tons of prickly pears are cultivated, with the larger areas of Ayacucho (20%), Huancavelica (15%), Arequipa (15%), Lima (14%), and Apurímac (8%) dedicated to its cultivation [50]. Moving to the Andean region, the next largest producer is Chile, with approximately 800 ha dedicated to the production of red, green, and orange prickly pears. These production areas are concentrated in Metropolitana, Valparaíso, and Coquimbo [51].

Like all cacti, *prickly pear* is a bushy, succulent, branched plant composed of joints or fleshy segments that reach an average height of 3–6 m and have a stem or trunk 60–150 cm wide [52]. The *O. ficus-indica* species is a shrub/arborescent plant that can reach up to 5 meters in height. The root system is fleshy and branched; it develops horizontally and laterally, and can reach 10–15 m from the base of the plant [53]. Its stem is well defined, dark brown, green or gray in color and cylindrical in shape, and is 45 cm long and 20 cm in diameter. Its cladodes are generally elliptical or rhomboid 30–40 cm long and 20–25 cm wide [54].

The fruit is spherical, cylindrical, or elliptical in shape; it can measure 6–10 cm in length and 4–7 cm in diameter. Characterized by being juicy and sweet, its color varies between yellow and red, while the pulp has the color of the peel. The size of the fruit is determined by the number of fertilized and aborted seeds. Its shape is ovoid and fleshy, and the fruit has a leathery pericarp on which tufts of glochids are found [47, 54]. The weight of the fruit varies between 100 and 200 grams, of which 30–40% represents the weight of the peel. During the initial stages of development, the peel is green and can change to greenish-white, yellow, orange, red (**Figure 2**), purple, purplish-yellow, or even violet or dark brown, depending on the growing variety [55]. At present, two

Scientific name	*O. ficus-indica*
Kingdom	Plantae
Division	Magnoliophyta
Class	Magnoliopsida
Order	Caryophyllales
Family	Cactaceae
Subfamily	Opuntioideae
Tribe	Opuntieae
Genus	*Opuntia*
Subgenus	*Opuntia*
Species	*O. ficus-indica* (L.) Mill., 1768

Taken from Guerrero-Beltrán and Ochoa-Velasco [46].

Table 8.
Opuntia ficus-indica taxonomy.

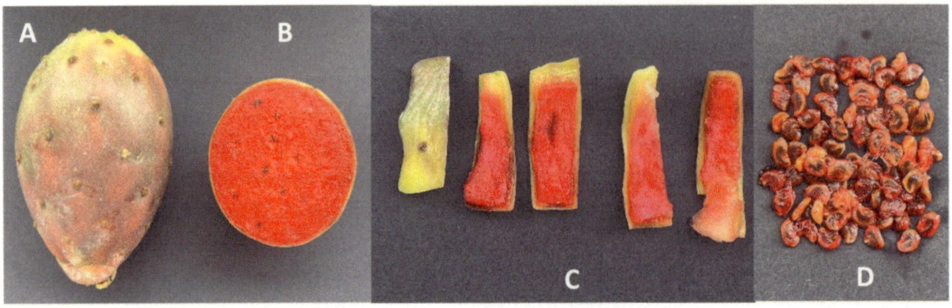

Figure 2.
Major parts of tuna or prickly pears with red peel fruits. (A) Fruit, (B) mesocarp, (C) peel, and (D) tuna seeds (wet).

parts of the plant are used as food, the fruits (tuna) and the cladodes (nopal). Prickly pears can be eaten fresh, after drying in the sun, or in jams. In addition, nopal is consumed mainly in Mexican regions as an ingredient present in salads [56, 57].

This plant is characterized by its richness in polyphenols, vitamins, polyunsaturated fatty acids, and amino acids. The identified compounds and their derivatives have been shown to possess relevant biological activities, including anti-inflammatory, antioxidant, hypoglycemic, antimicrobial, and neuroprotective activities, among others [35]. *Opuntia* flowers come in various colors, but the color progression during bloom ranges from white to yellow and can turn to red, orange, pink, and peach. The flowers are ancillary, also considered by-products because they are generally discarded after the fruit is separated. There is evidence that the floral compounds of *Opuntia* are phenolic pigments and betalain [58].

Conversely, plants of the genus *Opuntia* are used for carminic acid production. This acid is extracted after grinding and desiccating the females of a parasitic worm called "cochineal," which grows on the surface of the succulent branches of these plants [59]. Carmine red dye is used in the food industry, according to European law, as (E-120). Peru produces 90% of the world's cochineal, whereas the Arequipa region accounts for 70% of the national production. Peru produces 1986 tons of dry cochineal per year [56, 59].

3.1 Overall chemical composition

At present, prickly pears can be found in different colors and shapes, and with or without thorns, but there are no obvious differences between traditional and modern cultivars [52]. Their difference lies in the amount of betalains and betaxanthins found, depending on the prickly pear variety [46]. According to Assunção Alves et al. [47], in case of minimal differences between a red, orange, or green prickly pear, these may be due to the cultivation conditions, the state of maturity during harvest, or the analytical methodology used.

According to the information collected by Corzo-Rios et al. [35], prickly pear contains approximately 85% water, 15% sugar, 0.3% ash, and less than 1% of protein. This is similar to that reported by Medina et al. [56] and Gonçalves Albuquerque et al. [53] in red and green prickly pears from Spain and Mexico, respectively. The moisture content varies between 82% and 92%, and the protein and fat content does not exceed 2% and 1% of the total weight, respectively (**Table 9**). Prickly pears do not have a distinctive aroma, but the pulp is very sweet and the sugar components are mainly glucose

Component	Orange prickly pear (Argentina)	Green prickly pear (Mexico)	Red prickly pear (Argentina)
Moisture (%)	81.29	91.10	85.43
Ash (%)	0.53	5.40	2.34
Total soluble solids (%)	14.78	—	10.16
Proteins (%)	1.56	1.70	1.41
Lipids (%)	0.35	0.10	0.22
Fibers (%)	—	1.50	—

Taken from Valero-Galván et al. [60] and Romero et al. [61].

Table 9.
Physicochemical and nutritional composition of Opuntia ficus-indica (L.) Mill. pulp.

and fructose, whose concentration varies from 10 to 17°Brix [52]. In addition, the prickly pear pulp is characterized by a water activity between 97.2% and 99.3%, a pH of 5.2–5.5, and a titratable acidity (% citric acid) that ranges from 0.001 to 0.003 [60].

It is important to highlight the potential of the seeds and peel of the prickly pear. The seeds are distributed within the pulp and constitute between 30% and 40% of the fruit's wet weight. However, neither the seeds nor the peel is used despite being important sources of fatty acids, vitamins, polyphenols, flavonoids, tannins, and fiber [35]. The seeds (6.77 g/100 g dry weight) have been reported to have up to seven times the oil content of the pulp [53]. Besides, Medina [56] informed that the fiber content in the prickly pear's peel varies between 4.86 and 5.65 g/100 g dry weight. These results were obtained from orange and green prickly pears in different regions of Mexico. Likewise, these authors report that the peel and the seeds contain cellulose (71.4% and 83.23% of the total fiber). This is in line with the study conducted by Hernández-Carranza et al. [48] using red prickly pear peel, in which cellulose (34.64% dry weight) predominated in the total dietary fiber.

Different studies compiled by Cota-Sánchez [52] have shown that *Opuntia* fruit is a good source of minerals, specifically calcium, magnesium, potassium, and phosphorus. According to Medina [56], both calcium and potassium are the most predominant, with 26.30 and 158.30 mg/100 mg, respectively. This agrees with the study by Guerrero-Beltrán & Ochoa-Velasco [46], which assessed the chemical composition of prickly pears of different colors (**Table 10**). Potassium was the most abundant mineral, with the exception of the purple prickly pear (19.6 mg/100 g) in each variety, and phosphorus was much higher in green prickly pear when compared with the others (32.5 mg/100 mg).

Previous studies have shown that *Opuntia* fruits contain ascorbic acid, which ranges from 20 to 40 mg/100 g of fresh weight. In a study conducted by Guerrero-Beltrán and Ochoa-Velazco [46], the vitamin C content was also assessed, with results ranging from 20 to 24.1 mg/100 g. In a study by Medina [56], green and orange prickly pears from Tenerife (Spain) only yielded 17.1 and 17.2 mg/100 mg, respectively. Vitamins are not only found in the pulp but also fat-soluble vitamins (alpha, beta, and delta tocopherols, beta carotene, and vitamin K1) found in prickly pear seed oils [53].

In a study conducted by Jorge and Troncoso [62] on red prickly pears collected in Huancavelica (Peru), the vitamin C content was 36.1 mg/100 g. Similarly, Monroy-Gutiérrez et al. [63] evaluated different varieties of prickly pears and observed that

Minerals	Red prickly pear	Orange prickly pear	Purple prickly pear
Calcium	12.80	35.8	13.2
Iron	0.40	0.20	11.5
Magnesium	16.1	11.8	11.5
Sodium	0.6	0.90	0.50
Phosphorus	32.8	8.5	4.9
Potassium	217.0	117.7	19.6

Taken from Guerrero-Beltrán and Ochoa-Velazco [46].

Table 10.
Chemical and nutritional composition of prickly pears of different colors (mg/100 g).

Variety	Evaluation days								
	0	3	6	9	12	15	18	21	24
Red prickly pear "Rojo Pelón"	37.5	28.1	33.9	25.7	25.8	29.3	29.8	42.5	38.9
Red prickly pear "Liso Forrajero"	19.7	16.0	19.4	21.6	16.9	24.7	—	—	—

Taken from Monroy-Gutiérrez et al. [63].

Table 11.
Vitamin C content in red prickly pear cultivars (mg/100 mg).

the vitamin C content in the cultivars "Rojo Pelón" and "Liso Forrajero" (both red prickly pears) ranged from 42.54 to 16.00 mg/100 mg, where the highest value was presented 21 days after evaluation in the "Rojo Pelón" variety (**Table 11**).

3.2 Bioactive phytochemicals

Different studies indicate that all parts of *O. ficus-indica* are rich in polyphenols, flavonoids, and phenolic acids. This includes pulp, seeds, peel, flowers, and cladodes, with the pulp having the highest amount of bioactive compounds [35, 53, 58, 60, 64]. According to a report by Gonçalves Albuquerque et al. [53], the method most widely used to assess phenolic and polyphenolic antioxidants is Folin-Ciocalteu, although results may present considerable differences between studies, ranging from 5.54 to 1000 mg GAE/100 g of pulp. About phenolic compounds, quercetin is the most abundant component in prickly pear pulp, followed by isorhametin-3-rutinoside [55].

When analyzing green and orange prickly pears from the island of Tenerife (Spain), Medina [56] found that the phenol content in pulp was 45.0 and 45.4 mg/100 mg, respectively. This study also determined that one portion of *O. ficus-indica*, regardless of color, represents only 68% of the recommended total phenol intake per day. Similarly, Monroy-Gutiérrez et al. [63] evaluated the content of phenols in the pulp of red prickly pears from the Zacatecas region (Mexico). The authors observed that the phenol content in both varieties decreased during the evaluation period (**Table 12**), with the higher phenol content found in the variety "Rojo Pelón" from the beginning of the study (30.32 mg GAE/100 g). According to these authors, this is a consequence of the environmental conditions (temperature, relative humidity, and light), as well as the crop nutrients and the pre and post-harvest handling.

Variety	Evaluation days								
	0	3	6	9	12	15	18	21	24
Red prickly pear "Rojo Pelón"	30.3	15.7	16.0	12.5	14.2	13.7	13.6	18.2	17.8
Red prickly pear "Liso Forrajero"	6.3	5.8	5.6	4.6	4.7	5.1	—	—	—

Taken from Monroy-Gutiérrez et al. [63]

Table 12.
Total phenol content in red prickly pear cultivars (mg GAE/100 g).

Variety	Pulp		Peel		Seeds	
	Red prickly pear	Green prickly pear	Red prickly pear	Green prickly pear	Red prickly pear	Green prickly pear
Total phenolic content (mg GAE/g)	3.62	5.00	4.32	4.34	4.30	2.93
Total flavonoids (mg catechin equivalents [CE]/g)	3.14	3.58	3.18	3.40	3.10	3.17
DPPH (mmol TE/g)	7.74	6.74	7.90	6.87	8.08	7.08
FRAP (mmol TE/g)	1.38	2.03	5.18	7.37	0.80	2.08
$ABTS^{\bullet+}$ (mmol TE/g)	11.87	16.33	16.42	17.65	14.19	17.18

Adapted from Valero-Galván et al. [60].

Table 13.
Phytochemical profile and antioxidant capacity of commercial varieties of prickly pears (red and green).

In the study conducted by Valero-Galván et al. [60], the content of total phenols in pulp was higher than in the peel and seeds while the content of flavonoids was higher in the peel than in the pulp or seeds (**Table 13**). Likewise, the antioxidant capacity determined by DPPH was similar among the three types of tissues. However, the antioxidant activity determined by the ABTS assay was higher in the seeds and peel than in the pulp. In contrast, the results obtained by the FRAP assay showed higher antioxidant activity in the peel than in the pulp and the seeds.

The antioxidant capacity of prickly pear may be due to one or more components, but a synergistic effect is also possible. A study by Gonçalves Albuquerque et al. [53] indicates that prickly pear extracts have higher antioxidant activity than other fruits such as pears, apples, grapes, oranges, grapefruits, and tomatoes. Furthermore, compared with other *Opuntia* species, prickly pear has the lowest content of polyphenols and flavonoids but the highest antioxidant activity. Flavonoids are especially efficient antioxidants owing to their ability to inhibit pro-oxidative processes on DNA, proteins, and lipids, and to prevent the generation of stable radicals [55, 61, 63].

Monroy-Gutiérrez et al. [63] also analyzed the antioxidant capacity over time in red prickly pears, with results similar to previous studies. As days go by, the antioxidant activity decreased in all varieties (**Table 14**). The authors pointed out that *Opuntia* fruits have the moderate antioxidant capacity, which could be directly associated with the fruit pigment content and the growing conditions. This could be observed in the study conducted by Ordoñez et al. [64] on prickly pears collected in Huánuco (Peru), which showed different antioxidant capacities between the yellow

Variety	Evaluation days								
	0	3	6	9	12	15	18	21	24
Red prickly pear "Rojo Pelón"	0.51	0.52	0.50	0.44	0.44	0.44	0.46	0.51	0.53
Red prickly pear "Liso Forrajero"	0.42	0.42	0.42	0.39	0.39	0.39	—	—	—

Taken from Monroy-Gutiérrez et al. [63].

Table 14.
Antioxidant capacity in red prickly pear cultivars expressed in vitamin C equivalents (mg/g).

and purple peel varieties. The peel of purple prickly pear (18.50 mg/mL) presented greater antioxidant activity against the radical DPPH compared with the remaining tissues of the yellow prickly pear (16.73–27.99 mg/mL). In contrast, Jorge and Troncoso [62] found that the antioxidant capacity was provided by vitamin C, its contribution is greater than 50% of the total antioxidant capacity of the prickly pear.

Prickly pears have pulps and peels of different colors. These can be pale green, yellow, orange, magenta, red, and red-purple, indicating that these varieties have different pigments [46]. Betalains are water-soluble pigments that give the red-purple (betacyanins) and yellow (betaxanthins) colors to the fruits of several cactus spices, such as *Opuntia* sp., *H. polyrhizus*, or *Myrtillocactus geometrizans* [65]. The concentration of these pigments is responsible for the color variation in prickly pears. In contrast, the pale green pigments are the result of chlorophylls. Betaxanthins include indicaxanthin, miraxanthin, portulaxanthin, and vulgaxanthin. In addition, plant betacyanins include betanin, isobetanin, neobetanin, and probetanin [46].

A study conducted by Guerrero-Beltrán and Ochoa-Velasco [46] compared three prickly pears of different colors. While their nutritional composition differed slightly from each other, the concentration of β-carotene in green prickly pear (0.5 mg/100 g) and orange prickly pear (2.3 mg/100 g) was highlighted, as well as the content of betanin in purple prickly pear (100 mg/100 g). Similarly, a study by Fernández-López et al. [55] assessed the content of betacyanins (15.2 mg betanin/100 g fresh fruit) and betaxanthins (25.4 mg indicaxanthin/100 g of fresh fruit) in red peeled prickly pears collected in Murcia (Spain). Valera-Galván [60] reported the betacyanin and betaxanthin content in green and red prickly pears. The pigment content was found to be higher in the fruit pulp, in contrast to the peel, with the concentration being greater in red prickly pears. However, there were no statistical differences between the antioxidant capacities of both varieties.

Monroy-Gutiérrez et al. [63] compared the content of betanins and indicaxanthins in two varieties of red prickly pear. The concentration of betanin (red-violet) was higher in the "Rojo Pelon" variety and increased during the evaluation days. The same was observed in said variety for indicaxanthin, starting at 9.52 mg/100 mg, and increasing to 19.91 mg/100 mg after 24 days of evaluation. Conversely, the concentration of both pigments in the "Liso Forrajero" variety decreased until day 15 (**Table 15**). According to Ortega-Hernández et al. [66], it is possible to increase the content of betalains, ascorbic acid, and phenolic compounds in prickly pears using UVB light and inflicting wounds in the plant tissue. This could explain the higher amount of betanin found in the "Rojo Pelón" variety. Caused by the red or purple coloration of these varieties, it was not possible to assess the chlorophyll and total carotenes content.

According to Cota-Sánchez [52], prickly pear seeds are rich in minerals and sulfur-containing amino acids. Additionally, the composition of fatty acids in *O. ficus-indica* seeds can improve food properties and be used as seasonings in

Pigment/variety	Evaluation days								
	0	3	6	9	12	15	18	21	24
Betanin									
Red prickly pear "Rojo Pelón"	12.6	14.6	17.9	15.6	16.7	17.4	16.7	19.1	22.2
Red prickly pear "Liso Forrajero"	7.9	5.5	7.7	11.8	6.9	7.4	—	—	—
Indicaxanthin									
Red prickly pear "Rojo Pelón"	9.5	13.6	13.3	14.1	18.1	21.7	11.6	11.2	19.9
Red prickly pear "Liso Forrajero"	8.8	9.2	8.5	9.8	4.9	7.5	—	—	—

Taken from Monroy-Gutiérrez et al. [63].

Table 15.
Betanin and indicaxanthin content in red prickly pear cultivars (mg/100g).

Fatty acid profile	Orange prickly pear	Red prickly pear
C14:0	—	1.09
C16:0	25.08	22.74
C18:0	7.14	5.22
C18:1 n-9	22.61	18.81
C18:2 n-6	28.77	28.21
C20:0	—	1.22
C18:3 n-3	18.20	17.12

Taken from Romero et al. [61].

Table 16.
Fatty acid profile in prickly pear seeds with orange and red peel.

culinary art. Corzo-Rios et al. [35] pointed out that the lipids present in *Opuntia* cladodes are palmitic, oleic, and linoleic. Romero et al. [61] was able to identify and quantify the fatty acid profile in orange and red prickly pears, where linoleic acid stood out (28.77 and 28.21 mg/100 g in orange and red prickly pears, respectively) (**Table 16**).

3.3 Commercial fruit application

The *O. ficus-indica* fruit is valued for its sweet and slightly sour taste. Different food products such as jams, jellies, nectars, dehydrated fruits, syrups, liqueurs, wines, and vinegars can be made from prickly pear [53]; however, this plant also has alternative uses in the Andean region. In Bolivia, prickly pear is used to treat heat stroke, sunburn, yellow fever, and gastritis. In Colombia, anti-inflammatory infusions are prepared from the leaves. Used in poultices, *O. ficus-indica* can relieve skin irritations or remove swellings. In Peru, fresh fruit is used to treat hair loss and diabetes [67].

Several biological activities have been reported for *Opuntia*; these encompass potential applications in health and nutrition. Prickly pear extracts are used for the treatment of diabetes, cholesterol, and immune diseases. Additionally, the extracted polysaccharides can protect the liver from organophosphate pesticide damage. The antioxidant capacity of betalains present in prickly pear can prevent ovarian and cervical cancer [35, 52]. These nutritional benefits increase the interest in making products from prickly pear.

The production of drinks or juices is the first step in the manufacture of other processed products, although heat treatment is usually a determining factor. The alteration of the sensory or nutritional characteristics usually occurs in products based on prickly pear or other fruits [46]. Nonetheless, Lagua Yanchaguano et al. [68] obtained a drink based on prickly pear and passion fruit (*Passiflora edulis*) collected in Ecuador with a high degree of acceptability. This was achieved by evaluating different concentrations of both fruits. According to this study, the most suitable concentrations for the drink were 15% *O. ficus-indica* pulp, 85% passion fruit pulp and 12% sucrose. The physicochemical properties (pH, °Brix, and density) of this drink remained stable for 72 h.

Prickly pear has been identified as a source of pectin and has great potential in the food industry as a gelling, thickening, and stabilizing agent [69]. Montilla et al. [70] quantified the pectins in the pulp of 3 different species of *Opuntia* from the semi-arid regions of Venezuela. The results were expressed as a percentage of an hydrogalacturonic acid (% AAG) and *O. ficus-indica* stood out with a value of 0.1531% ± 0.0087. Similarly, Chaparro et al. [69] evaluated the application of pectin extracted from prickly pear in a pineapple candy. The extraction yield was 9.14%, with a degree of esterification of 62%, indicating high methoxyl content and slow gelation. This suggests that it is suitable for the food industry and the production of preserves, such as jams and sweets. Nonetheless, the yield was low compared with commercial pectin sources such as orange or apple peel. The sensory quality characteristics of the pineapple candy with prickly pear pectin achieved an acceptable level of satisfaction.

Several studies have focused on obtaining pigments from the pulp and peel of prickly pear. This fruit has an attractive color that varies from pale green, green, yellow, orange, and red to violet tones; these are due to betacyanins (red-violet) and betaxanthins (yellow-orange) [46]. Pigments can be obtained by different methods; Coba Carrera et al. [71] got a higher pigment yield lyophilizing at 60°C. The colorant obtained met all the requirements (pH 6.1, 13.71°Brix, 1.35 of nD and 7.73% total solids) according to the specified standard. In a study by Otárola et al. [72], the microencapsulation of betalains from purple prickly pears (Santiago del Estero, Argentina) was evaluated by spray drying. Encapsulation was supplemented with maltodextrin and cladode mucilage to improve stability. Pigment retention was greater than 70% at 18°C, and relative humidity was below 57%; these being the most stable conditions. The use of cladode mucilage improved encapsulation efficiency, reducing moisture content and increasing dietary fiber content. Furthermore, the authors concluded that betalains from purple prickly pears have the potential to be an encapsulating agent for atomization in the food industry.

In a study conducted by Romero et al. [61], the effect of lyophilized pulps from *Eugenia uniflora* L. and *O. ficus-indica* fruits on the oxidative stability of meat patties was evaluated. In addition, the effect of lyophilized pulps on the cooking performance, color, texture parameters, and sensory acceptance was assessed. The authors determined that the lyophilized pulps limit the oxidation of lipids stored in a refrigerator to an acceptable level for up to 15 days, with *O. ficus-indica* (red peel variety) having the greatest antioxidant activity. Moreover, this variety had the highest preference among consumers for sensory parameters. The authors highlighted the effectiveness of the studied fruits to reduce lipid oxidation in meat pies.

4. Sanky

Among endemic Andean region, edible fruits are the sanky or sancayo (*C. brevistylus*) (**Figure 3**) a member of the succulent plant Cactaceae family [54, 73, 74].

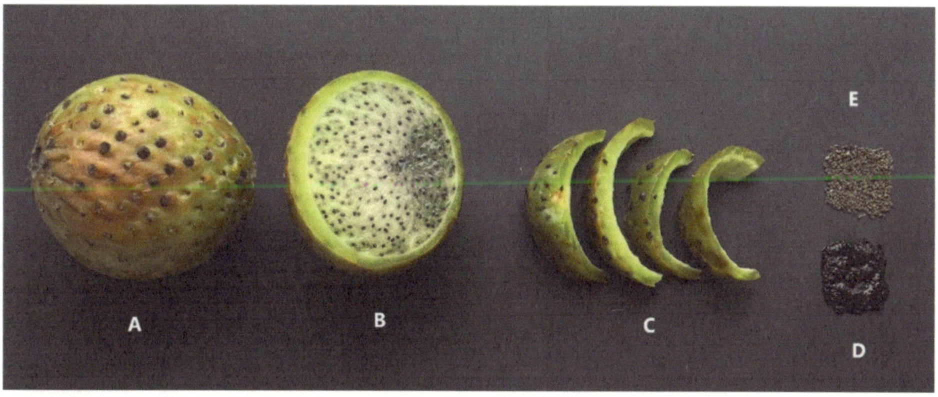

Figure 3.
Major parts of sanky (Corryocactus brevistylus subsp. puquiensis (Rauh & Backeberg) Ostolaza) fruits. (A) Fruit, (B) mesocarp, (C) peel, (D) mucilaginous coating around sanky seeds (wet) and (E) shade-dried sanky seeds.

C. brevistylus subsp. brevistylus (K. Schum. ex Vaupel) Britton & Rose is a native erect, branched cactus, up to 5 m tall. Branches with 7–8 ridges, spines up to 24 cm. Flowers yellow, fruit pear like, 9 cm long, with plenty of small spines [75]. Fruit is a large berry, between 7 and 10 cm long, round, olive green, with numerous brown seeds inside [73]. Grows on rocky slopes, shrubland, from 2000 to 3500 m above sea level and is endemic southern Peru and northern Chile [73, 75].

C. brevistylus subsp. *puquiensis* (Rauh & Backeberg) Ostolaza is an endemic arboreal cactus of Peru, of basal branching, more than 5 m tall, branches 20 cm in diameter, 7–8 tuberculate ribs near the apex, spines up to 20 cm, yellow flowers, 7 cm long and edible fruit. It differs from the previous subspecies by being taller and having smaller flowers [54]. This species is distributed in the provinces of Arequipa (Huanca, Majes, Caravelí, and Caylloma), Ayacucho (Lucanas) and grows between 2500 and 3000 m above the sea level [73].

Currently, consumption and use in the food products manufactured such as drinks based on sanky pulps have increased [76]. In addition, sanky pulp processing residues can be used as an additive (stabilizer) in the food and pharmaceutical industries [77]. On the other hand, the peels, and seeds of sanky have provided interesting antioxidant and nutritional properties [78]. Sanky peel powder-derived functional ingredients showed good potential as natural llama burger-making additives as well as after their incorporation improved the sensory and chromatic properties [79].

4.1 General chemical composition

Sanky fruits from *C. brevistylus* subsp. puquiensis (Rauh & Backeberg) Ostolaza have a weight between 341 to 413 g and a diameter in the range of 7.7–8.4 cm (**Figure 4A** and **B**). The pulp has a pH and soluble solids (°Brix) of 2.54 and 3.99, respectively. The pulp is white flesh with small black seeds and *sui generis* flavor (**Figure 4C**). In addition, it has a gelatinous appearance and an acid taste (unpublished data). Further studies on sanky pulp are still necessary, especially in the proximal chemical composition and physicochemical properties of the different parts of the fruit (pulp, seed, and peel). **Table 17** shows the nutritional composition of the peel and seed of sanky.

The carbohydrate (~46%) and fiber (~16%) content are the main components of the sanky peels, while seeds are an important source of dietary fiber (~29%) and

Figure 4.
(A) Sanky fruits of different weights, (B) diameter range of 7.7–8.4 cm, (C) cross-section in sanky fruits showing seeds and pulp, (D) sanky pulp used in jam and nectar, (E) fiber, mucilage and seeds, and (F) sanky peel used as a stabilizer.

	***Corryocactus brevistylus* subsp. puquiensis (Rauh & Backeberg) Ostolaza**		
Origin	Ayacucho (Peru)		
Crop locus	Saisa		
Altitude (m.a.s.l)	3075		
	Pulp	Peel	Seed
Moisture (g/100 g)	95.3	10.74	4.36
Protein (g/100 g)	0.2	9.19	15.56
Lipid (g/100 g)	0.1	2.68	26.10
Carbohydrate (g/100 g)	3.3	46.25	22.53
Fiber (g/100 g)	0.5	16.39	29.53
Ash (g/100 g)	0.5	14.75	2.22
Calcium (ppm)		329.56	207.81
Iron (ppm)		5.95	39.36
Zinc (ppm)		1.06	9.40
Ascorbic acid (mg/100 g)	57.1		

Taken from Muñoz et al. [78]; Nolazco and Guevara [80] and Areche et al. [81].

Table 17.
Levels of nutrients and minerals in sanky fruit.

lipids (~26%). The dietary fiber content of the peels compared to other cacti such as prickly pear (40.8%) and dragon fruit (23.75%) is lower [82, 83]. While the contribution of the seeds is superior to the prickly pear seeds (12.47%) [84]. Dietary fibers are defined as "macromolecules present in the diet that resist digestion by endogenous enzymes in the small intestine of humans" [85]. The main effects related to dietary fiber consumption are to improve intestinal function, increase microbial biomass, blood cholesterol decrease, lower risk of cardiovascular disease, type 2 diabetes, and colorectal cancer [86, 87]. On the other hand, the fiber-rich by-products have been used to fortify various foods (as corn and wheat tortillas, bakery products, snack foods, noodles, and cooked meat products) to increase their dietary fiber content and to drive the nutraceutical industry [88–92].

The protein content of peel and seed of sanky is ~9% and ~15% (dry basis), respectively. The amino acid composition of sanky fruit proteins has not been fully characterized. The protein level of sanky peels compared to other cacti was slightly superior to that of prickly pear (6.12%) and dragon fruit (6.0%), while the seeds of these species were 4.78% and 26.3%, respectively [84, 93, 94].

The ash content in both peels and seeds was 14.75% and 2.22%, respectively. In prickly pear, seeds were found around 1.27%, while the dragon fruit seeds had a content of 6.1% (*H. polyrhizus*) and 3.1% (*H. undatus*) [84, 94]. Sanky fruit is a good source of calcium. The peels contain more calcium than the seeds, while the latter are high in iron and zinc (**Table 1**). The calcium content in cactus fruits was reported to be 750 ppm in jiotilla (*Escontria chiotilla*), 50 ppm in dragon fruit (*H. undatus*), 440 ppm in berry cactus (*M. geometrizans*), and 560 ppm in prickly pear (*Opuntia* sp.). While the iron content was 32.6 ppm, 7.50 ppm, 380 ppm and 12.34 ppm, respectively [95].

	Sanky	Tuna or prickly pear	Dragon fruit	
			Hylocereus undatus	*Hylocereus polyrhizus*
Palmitic acid (C16:0)	14.09	12.23	14.95	18.39
Palmitoleic acid (16:1 ω-7)	0.37		0.79	1.04
Stearic acid (C18:0)	2.52	0.15	6.93	8.05
Oleic acid (18:1 ω-9)	23.04	25.52	18.67	23.61
Linoleic acid (C18:2 ω-6)	56.44	61.01	55.43	45.21
α-Linolenic acid (C18:3 ω-3)	0.56	0.40–0.69	0.21	0.15
Arachidic acid (C20:0)	1.12		0.95	1.04
SFA	17.72	12.38	22.83	27.48
MUFA	23.41	25.52	19.46	24.65
PUFA	57.00	61.51	55.64	45.36

Taken from Muñoz et al. [78]; Özcan and Al Juhaimi [84] and Liaotrakoon et al. [97].

Table 18.
Fatty acids composition of sanky seed oils and other Cactaceae fruits oils.

Overall, the lipid content of sanky peel is minimal (~2%). The oil content of the peels is comparable to that of orange peel (1.45%), prickly pear peel (1.55%), mango peel (2.12%), watermelon peel (2.61%), banana peel (4.51%) and dragon fruit peel (6.20%) [83, 93], while the sanky seeds are a source of lipids (~26%). Compared with prickly pear seeds (2.24–5.69%) it is superior, whereas with dragon fruit seeds (*H. undatus*) (27.5%) they are slightly similar [94, 96]. The sanky seed oil was highly unsaturated of which 23.41% are monounsaturated fatty acids (MUFA) and 57% are polyunsaturated fatty acids (PUFA) (**Table 18**). The linoleic acid (56.44%) was the major unsaturated fatty acid found in the sanky seed oil and similar to dragon fruit seed oil (*H. undatus*) [97], followed by oleic (23.04%) and palmitic (14.09%) acids. These values are comparable to prickly pear seed and dragon fruit seed oils. Palmitic, stearic and arachidic acids accounted for 14.09%, 2.52 and 1.12% of sanky seed oil, respectively. Linolenic acid (C18:3 ω-3) (0.56%) in sanky seed oil was similar to Cactaceae fruits oils [84, 97].

The proximal chemical composition of *C. brevistylus* subsp. brevistylus (K. Schum. ex Vaupel) Britton & Rose is currently limited. The ascorbic acid content in this subspecies (38.45 mg/100 g) was higher than that reported for another cactus fruits of 17.3 mg/100 g in jiotilla, 25.8 mg/100 g in dragon fruit, 32.5 mg/100 g in berry cactus, 14 mg/100 g in prickly pear and 17 mg/100 g in pitaya (*Stenocereus* sp.). While the "chilito" species (*Mammillaria uncinata*) has an ascorbic acid content of 1390 mg/100 g, much higher than other Cactaceae fruits, this content was very comparable with the content of ascorbic acid in the camu-camu fruit around 1451 mg/100 g [98].

4.2 Bioactive phytochemicals

There are few studies on bioactive compound profiles in sanky fruits. Phytochemical screening of the pulp and peel reported the presence of reducing sugars, lactones, triterpenes, anthocyanidins, mucilages and catechins [80].

4.2.1 Bioactive compound profile

Bioactive compounds have recently been identified in lyophilized sanky pulp powder (*C. brevistylus* subsp. brevistylus (K. Schum. ex Vaupel) Britton & Rose) [81]. Those compounds were extracted with ethanol (three times) followed by sonication. The bioactive compounds identified were organic acids, hydroxycinnamic acids, isoamericanol derivatives, flavonoids, and sterols, and the main classes found were organic acids (12), hydroxycinnamic acids (9), and flavonoids (6). The main phytochemicals were rutin and coumaroyl isocitric acid.

The organic acids identified were malic acid, isocitric acid, hydroxyglutaric acid, hydroxyglutaric acid isomer, homo isocitric acid, hydroxybenzoic acid, dehydroshikimic acid, ascorbic acid, benzoyl aspartic acid derivative, and azelaic acid [81]. These organic acids are widely distributed in nature and have therapeutic action. They are also used in the food industry as acidulants, preservatives, inactivating or inhibiting the growth of microorganisms and establishing a protective barrier at acid pH [99]. The hydroxycinnamic acids identified were caffeoyl-*O*-hexoside, caffeoyl isocitric acid (RT: 10.14 min; *m/z*: 353.0513), caffeoyl isocitric acid (RT: 10.42 min; *m/z*: 353.0513), coumaroyl isocitric acid (RT: 10.68 min; *m/z*: 337.0564), coumaroyl isocitric acid (RT: 11.16 min; *m/z*: 337.0563), feruloyl isocitric acid, methylcoumaroyl isocitric acid and methylferuloyl isocitric acid [81]. These compounds are of interest because of their biological properties that include antioxidant activity, anticancer activity, improves blood pressure and metabolic syndrome [100, 101]. The flavonoids identified were rutin, isorhamnetin-*O*-rutinose, taxifolin, quercetin, isorhamnetin and rhamnetin (**Table 19**). The reports highlighted that bioactive molecules of sanky exerts strong antioxidant-antiradical activity, gastroprotective effects and other bioactivities. Flavonoids are the most studied bioactive compounds [115] and are widely distributed in different parts of Cactaceae fruits (pulp, peel, seeds, and cladodes) [95, 116]. In addition, they provide a high antioxidant potential [117]. The quantification of the detailed phytochemical compounds of the different parts of the sanky fruit (pulp, seed, and peel) is still necessary.

4.2.2 Polyphenols and antioxidant activity

The total phenolic contents (TPC) of peels from sanky (*C. brevistylus* subsp. *puquiensis* (Rauh & Backeberg) Ostolaza) ranged from 14.2 to 43.9 mg gallic acid equivalent (GAE)/g dry weight [118], while the TPC of lyophilized sanky pulp powder from sanky (*C. brevistylus* subsp. brevistylus (K. Schum. ex Vaupel) Britton & Rose) was 24.24 mg GAE/g dry weight, and total flavonoid content (TFC) was 13.33 mg quercetin equivalents/g dry weight [81]. When compared to sanky pulp, white and red pitayas showed lower TPC (3.52–4.91 mg GAE/g) [119, 120]. While in different varieties of cactus pear fruit pulp the TPC varied between 1.68 and 22.08 mg GAE/kg [119].

The phenolic compounds are known for their antioxidant capacity [117]. Antioxidant activity measured in lyophilized sanky pulp powder extract by DPPH, ABTS and FRAP assays presented the following values: IC_{50} = 47.45 μg/mL, IC_{50} = 225.12 μg/mL, and 155.34 μmol Trolox/g dry weight, respectively. Antioxidant activity is probably due to the presence of the following phytochemicals: organic acids (ascorbic acid), hydroxycinnamic acids, isoamericanol derivatives, and flavonoids (rutin, taxifolin, quercetin, isorhamnetin and rhamnetin), these compounds found in food matrices have shown high antioxidant activity [12].

Bioactive compounds	Structures	Functional effects
Rutin MW = 610.517 g/mol MF=C27H30O16	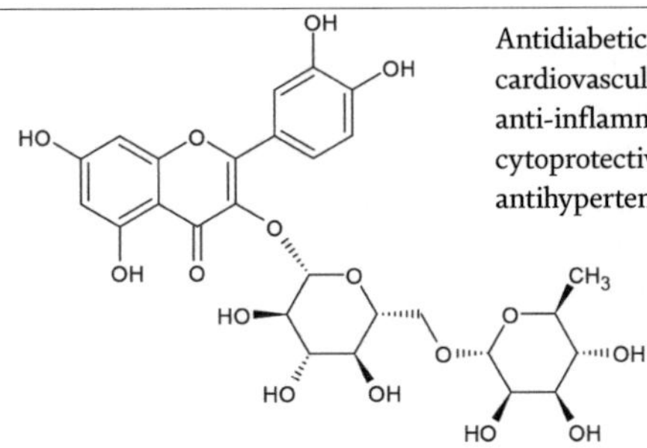	Antidiabetic, antioxidant, cardiovascular, anticancer, anti-inflammatory, antiviral, cytoprotective, hypolipidemic, and antihypertensive effects
Coumaroyl isocitric acid MW = 338.266 g/mol MF=C15H14O9	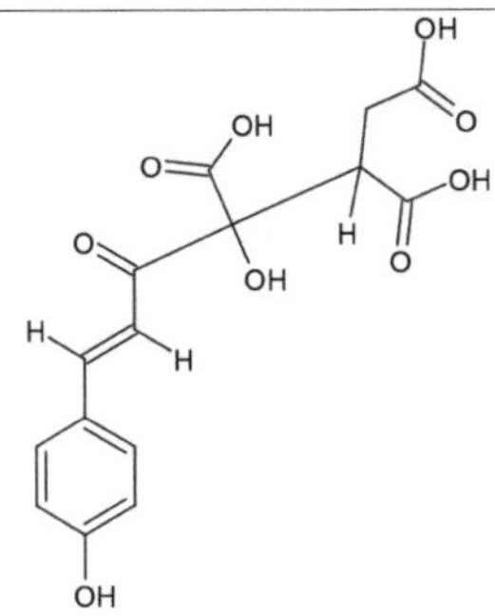	
Ascorbic acid MW = 176.124 g/mol MF=C6H8O6	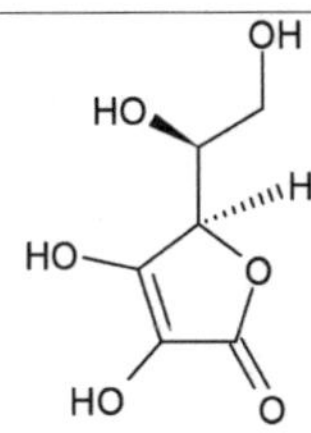	Considered one of the most effective natural antioxidants, promising anti-cancer agent
Taxifolin MW = 304.251 g/mol MF=C15H12O7	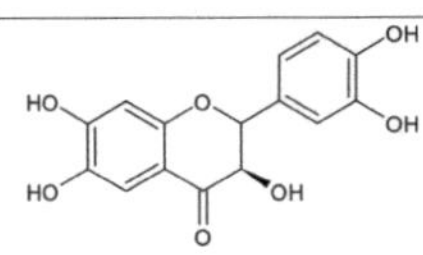	Inhibits the proliferation, migration, and invasion of breast cancer cells
Isorhamnetin-O-rutinose MW = 624.544 g/mol MF=C28H32O16	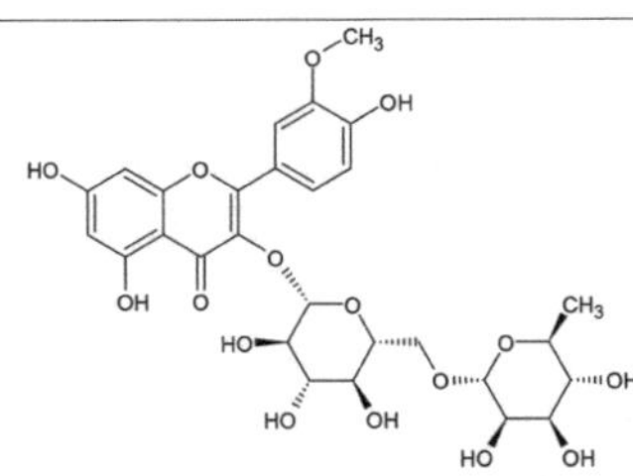	Insulin sensitizer, and hepatoprotective effect
Quercetin MW = 302.235 g/mol MF=C15H10O7	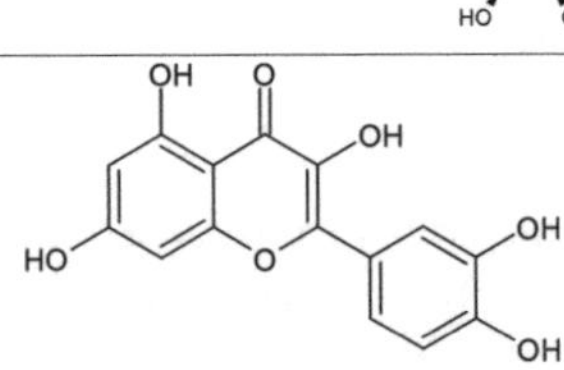	Antioxidant, anti-inflammatory, anticancer, cardioprotective, and neuroprotective effects

Bioactive compounds	Structures	Functional effects
Isorhamnetin MW = 316.262 g/mol MF=C16H12O7	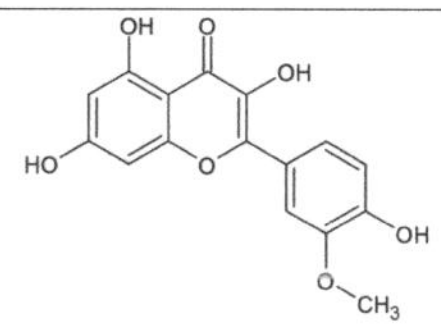	Regulation of apoptosis and anti-osteoporosis, anti-inflammatory, cerebrovascular protection, and antioxidant effects
Rhamnetin MW = 316.262 g/mol MF=C16H12O7	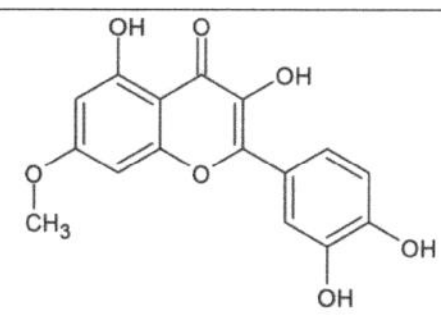	Suppresses the growth of human breast cancer cells

Taken from Ghorbani [102]; Patel and Patel [103]; Caparica et al. [104]; Sun, et al. [105]; Barba et al. [106]; Shenoy et al. [107]; Van Gorkom et al. [108]; Jia et al. [109]; Gevrenova et al. [110]; Li et al. [111]; Ay et al. [112]; Gong et al. [113]; and Lan et al. [114].

Table 19.
Some bioactive compounds in lyophilized sanky pulp powder.

4.3 Commercial applications of sanky fruits

Due to their technological characteristics, sanky fruits can be well used for industrial processes. Currently, within the cosmetic derivatives creams, shampoos, and soaps have been developed. For food applications, sanky jam (pulp and seed) (**Figure 4D** and **E**), sanky nectar (pulp) and sanky peel (**Figure 4F**) extract a stabilizer in meat products during refrigerated storage have been developed [121]. The demand for sanky by consumers in the domestic market continues to grow, especially in the form of pulp and fresh fruit.

The production of jams from sanky fruits is being developed recently. Some studies are focused on evaluating the sensory and physicochemical properties as well as the influence of the addition of pectin and carboxymethyl cellulose. The conditions of acceptability of the sanky jam presented a concentration of soluble solids (67.92°Bx); acidity (2.50%, expressed as citric acid); pH 3.17; and viscosity of 8878.40 cP [122]. In another study, the nutritional and sensory properties of a sanky jam sweetened with fructooligosaccharides (FOS) were evaluated. The final product reached a pH of 3.7, soluble solids of 66°Brix, titratable acidity of 3.41%, the crude fiber of 0.3%, ascorbic acid of 44 mg/100 g, total polyphenols of 381 mg GAE/100 g, and an antioxidant capacity of 65.92 mg Trolox/100 g [123]. Some unpublished data refers to the elaboration of sanky compote sweetened with panela and bee honey and others to the formulation of fruit mix compote (sanky, banana, and mango). Featherstone [124] mentions that jams are products of a combination of fruits (or fruit pulp, puree, juice, or concentrates) and sugar followed by heat treatment of them to produce a tasty product of sufficiently high sugar (>65%) content. Fruit jams are a source of energy and carbohydrate [125]. In addition to their nutritional composition, fruit jams are a source of bioactive components that have shown antioxidant activity. Fruit jams (such as blueberry, blackberry, blackcurrant, cranberry, and raspberry) have shown a polyphenol content between 170.32 and 473.91 mg GAE/100 g, the total flavonoids ranged between 2.61 and 11.43 mg quercetin equivalents (QE)/100 g and the antioxidant activity by ABTS assay as ranging from 6.10 to 36.56 μM Trolox/g [126].

Other sanky-based products are nectar and functional beverages. Neves et al. [127] define nectar as a category of packaged beverage that presents a juice content ranging from 25 to 99%. In addition, nectar can contain sweeteners, coloring, and preservatives. As part of the use of sanky fruits, **Figure 5** shows the block diagram of sanky nectar processing. The sanky fruit presents the following yields for the whole fruit. The pulp represents around ~57% (fresh weight), this fraction contains mucilage that could be used as a thickener in the preparation of various food products such as:

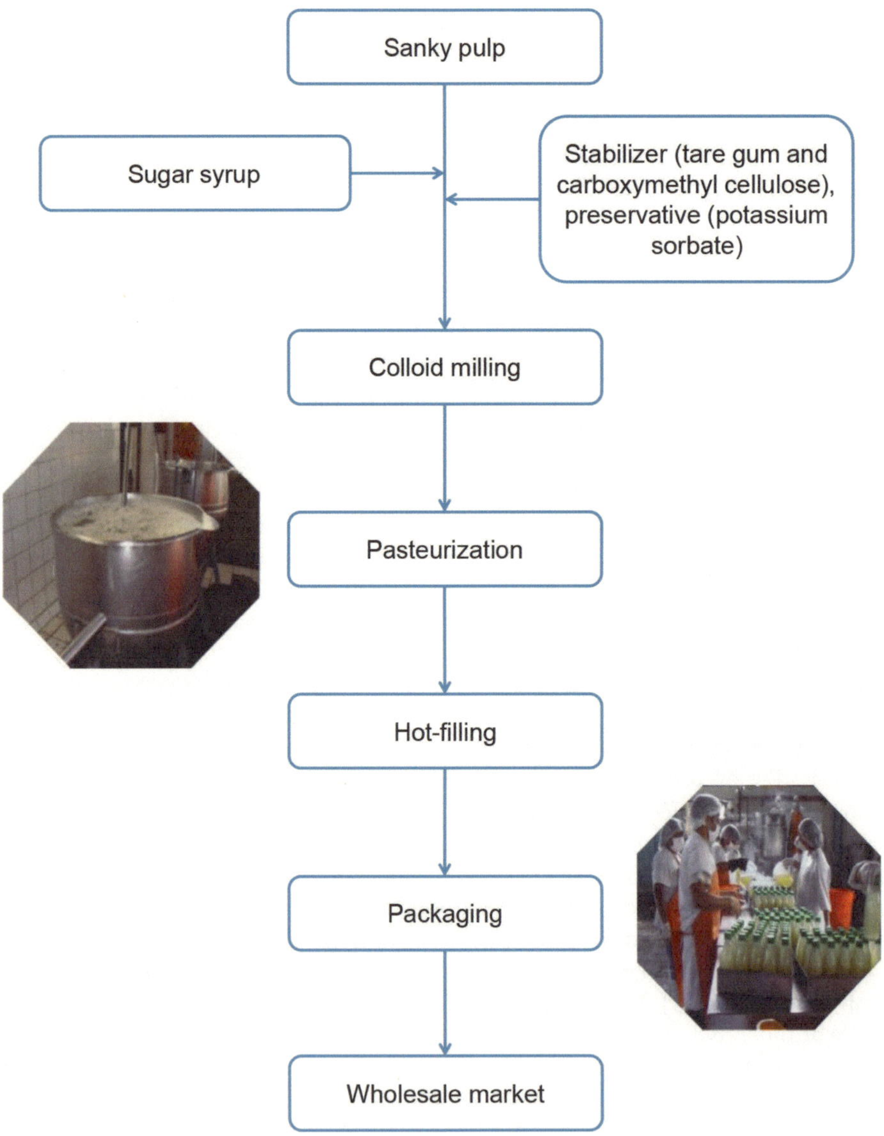

Figure 5.
Block diagram for obtaining sanky nectar (source: "Aprovechamiento industrial en los bosques naturales de sanky," n.d.).

baby food, compotes, pasta, etc. [77]. Mucilages can also be used as pharmaceutical excipients and wall materials for vegetable oils, essential oils, emulsions, etc. [128, 129]. The peels represent around ~35% (fresh weight). The peels are an important source of dietary fiber, in addition, they contain antioxidant phenolic compounds [14, 130]. While the seeds represent ~7%. Fruit seeds have a high oil content, are rich in monounsaturated and polyunsaturated fatty acids. They also contain other phytochemicals among them phytosterols, phospholipids, glycolipids, tocopherols, tocotrienols, carotenoids and polyphenols [131].

Sanky fruits are generally harvested using a long-handled fruit picker, followed by collection in crates. After harvesting, the sanky fruits are selected and classified (the fruits are selected manually, removing those that show signs of deterioration and/or breakage). Washing is carried out with water by immersion to remove foreign substances and particles, while disinfecting is done with sodium hypochlorite solution at 100 ppm. Peeling is carried out by cutting the fruits in half with stainless steel knifes, allowing the separation of pulp and peel. The sanky pulp is pasteurized at atmospheric pressure until boiling temperature for 1 second. Then, the pasteurized pulp is packed in 1 kg high-density polyethylene bags and then stored at a temperature of 5°C. Sanky pulp is used for the processing of mixed fruit drinks (noni, sanky and graviola) (http://vidanatural.pe/), according to the product information, this beverage stimulates the immune system, has antioxidant activity and reduces the levels of osteoporosis due to its calcium content. Fruit juices, beverages and nectars have shown antioxidant activity due to their high-value nutrient and bioactive components [132, 133].

Sanky fruit peels have been used to improve the chromatic and sensory characteristics of llama meat (*Lama glama*) during refrigerated storage, however, it did not inhibit microbial growth [121]. The effect of the incorporation of fruit peel powder on the quality and shelf-life characteristics of meat and derived products has been demonstrated. In addition, many extracts have shown an inhibitory effect on the growth of Gram-positive and negative strains [134, 135]. Bioactive compounds as polyphenols present in fruit peels provide an antioxidant effect and inhibit lipid oxidation. In many cases, the natural antioxidants present in these matrices provide better protection compared to synthetic antioxidants [135, 136].

5. Conclusions and future research

Edible cacti (Cactaceae) from the South American Andes contain a range of nutrients including macro- and micronutrients and bioactive compounds. The bioactive components present in pitahaya, tuna or prickly pear, been shown a wide range of biological activities. Even though there is abundant information about pitahaya and prickly pear, these fruits have been used as natural colorants, due to the presence of betalains, anthocyanins, carotenoids, and chlorophylls. In addition to having several pharmacological properties and great potential as functional foods.

The sanky belongs to the Cactaceae family. This fruit is currently being marketed as fresh fruit, pulp, and processed products. Sanky seeds are a source of proteins, lipids, fiber, and iron. Sanky peels are a source of proteins, lipids, fiber, and iron. While sanky peels are a source of carbohydrates and calcium and the pulp is rich in ascorbic acid. Sanky fruit contains several beneficial bioactive compounds in its parts (pulp, seed, and peel), including organic acids, hydroxycinnamic acids, isoamericanol derivatives, flavonoids, sterols, and fatty acids. Sanky extracts showed a biological activity as including antioxidant and gastroprotective. These effects could be due to the components such as hydroxycinnamic acids, flavonoids, and ascorbic acid or other bioactive compounds present in the fruit. The most abundant bioactive compounds in lyophilized sanky pulp powder are coumaroyl isocitric acid (hydroxycinnamic acid) and rutin (flavonoid). Some processed products based on sanky pulp are nectar and jam. Stabilizers for application in the food industry are obtained from by-products such as peels.

There are research opportunities for sanky fruit focused on human consumption and applications in the food industry. The chemical composition focused on the characterization of the macro- and micronutrients and bioactive compounds of the by-products (pulp, seed, and peel) of both species should be studied. Polysaccharide and sugar water-soluble characterization and evaluate the antioxidant activity *in vitro* in the mucilage remains to be studied. The impact of sanky fruit mucilage on human gut microbiota remains to be studied. Evaluate the sanky fruit mucilage as a new wall material for microencapsulation by spray drying of Sacha inchi oil.

Conflict of interest

The authors declare no conflict of interest.

Author details

Frank L. Romero-Orejon[1], Ana María Muñoz[1,2], Luciana de la Fuente-Carmelino[?], Diana Jimenez-Champi[1], Eliana Contreras-López[3], Ivan Best[1,2], Luís Aguilar[1,4] and Fernando Ramos-Escudero[1,4*]

1 Unidad de Investigación en Nutrición, Salud, Alimentos Funcionales y Nutraceúticos, Universidad San Ignacio de Loyola (UNUSAN-USIL), Lima, Peru

2 Instituto de Ciencias de los Alimentos y Nutrición, Universidad San Ignacio de Loyola (ICAN-USIL), Lima, Peru

3 Facultad de Farmacia y Bioquímica, Grupo de Investigación Revalorización de Fuentes Naturales y Alimentos Funcionales-REVALF, Universidad Nacional Mayor de San Marcos, Lima, Peru

4 Facultad de Ciencias de la Salud, Universidad San Ignacio de Loyola, Lima, Peru

*Address all correspondence to: diomedes.fernando@gmail.com; dramos@usil.edu.pe

IntechOpen

© 2022 The Author(s). Licensee IntechOpen. This chapter is distributed under the terms of the Creative Commons Attribution License (http://creativecommons.org/licenses/by/3.0), which permits unrestricted use, distribution, and reproduction in any medium, provided the original work is properly cited.

References

[1] Nazareno MA. Phytochemical of nutraceutical importance from cactus and their role in human health. In: Prakash D, Sharma G, editors. Phytochemicals and Nutraceutical Importance. Oxfordshire, UK: CAB International; 2014. pp. 103-115

[2] Ferreira MA, Bispo SV, Filho RRR, Urbano SA, Costa CTF. The use of cactus as forage for dairy cows in semi-arid regions of Brazil. In: Konvalina P, editor. Organic Farming and Food Production. London, UK: IntechOpen; 2012. pp. 169-189

[3] Shetty AA, Rana MK, Preetham SP. Cactus: A medicinal food. Journal of Food Science and Technology. 2012;**49**:530-536

[4] Isaac AA. Overview of cactus (*Opuntia ficus-indica* (L)): A myriad of alternatives. Ethno-Medicine. 2016;**2**:195-205

[5] Guerrero-Beltrán JA, Ochoa-Velasco CE. Figo da india-*Opuntia* spp. In: Rodrigues S, Silva EO, de Brito ES, editors. Exotic Fruits Reference Guide. London, UK: Elsevier Inc; 2018. pp. 187-201

[6] Mercado-Silva EM. Pitaya-*Hylocereus undatus* (Haw). In: Rodrigues S, Silva EO, de Brito ES, editors. Exotic Fruits Reference Guide. London, UK: Elsevier Inc; 2018. pp. 339-349

[7] Mondragón-Jacobo C, Méndez-Gallegos SJ. Nopalitos or vegetables cactus, production, and utilization. In: Inglese P, Mondragon C, Nefzaoui A, Sáenz C, editors. Crop Ecology, Cultivation and Uses of Cactus Pear. Rome, Italy: Food and Agriculture Organization of the United Nations (FAO) and International Center for Agricultural Research in the Dry Areas (ICARDA); 2017. pp. 93-104

[8] Sanz C. Processing and utilization of fruits, cladodes, and seed. In: Inglese P, Mondragon C, Nefzaoui A, Sáenz C, editors. Crop Ecology, Cultivation and Uses of Cactus Pear. Rome, Italy: Food and Agriculture Organization of the United Nations (FAO) and International Center for Agricultural Research in the Dry Areas (ICARDA); 2017. pp. 136-149

[9] Otálora MC, Barbosa HJ, Perilla JE, Osorio C, Nazareno MA. Encapsulated betalains (*Opuntia ficus-indica*) as natural colorants. Case study: Gummy candies. LWT-Food Science and Technology. 2019;**103**:222-227

[10] Gouws C, Mortazavi R, Mellor D, McKune A, Naumovski N. The effects of prickly pear fruit and cladode (*Opuntia* spp.) consumption on blood lipids: A systematic review. Complementary Therapies in Medicine. 2020;**50**:102384

[11] Agostini-Costa TS. Bioactive compounds and health benefits of Pereskioideae and Cactoideae: A review. Food Chemistry. 2020;**327**:126961

[12] Muñoz AM, Ramos-Escudero F. Phenolics compounds of the diet and his biomedicinal properties. Horizons Médicaux. 2007;**7**:23-31

[13] Campos D, Chirinos R, Gálvez Ranilla L, Pedreschi R. Bioactive potential of Andean fruits, seeds, and tubers. Advances in Food and Nutrition Research. 2018;**84**:287-343

[14] Domínguez-Rodríguez G, García MC, Plaza M, Marina ML. Revalorization of *Passiflora* species peels

as a sustainable source of antioxidant phenolic compounds. Science of the Total Environment. 2019;**696**:134030

[15] Quiroga Ledezma CC. Native food crops for present and future generations: Their role in nutrition and health. In: Betoret N, Betoret E, editors. Sustainability of the Food System: Sovereignty, Waste, and Nutrients Bioavailability. London, UK: Elsevier Inc; 2020. pp. 3-23

[16] Verona-Ruiz A, Urcia-Cerna J, Paucar-Menacho LM. Pitahaya (Hylocereus spp.): Culture, physicochemical characteristics, nutritional composition, and bioactive compounds. Scientia Agropecuaria. 2020;**11**(3):439-453. DOI: 10.17268/sci.agropecu.2020.03.16

[17] Ochoa Velasco CE, García Vidal V, Luna Guevara JJ, Luna Guevara ML, Hernández Carranza P, Guerrero Beltrán JÁ. Características antioxidantes, fisicoquímicas y microbiológicas de jugo fermentado y sin fermentar de tres variedades de pitahaya (Hylocereus spp). Scientia Agropecuaria. 2012;**3**:279-289

[18] Huachi L, Yugsi E, Paredes M, Coronel D, Verdugo K, Coba SP. Desarrollo de la pitahaya (Cereus SP.) en Ecuador. La Granja Rev Ciencias la Vida. 2015;**22**(2):50-58. DOI: 10.17163/lgr.n22.2015.05

[19] Arivalagan M, Karunakaran G, Roy TK, et al. Biochemical and nutritional characterization of dragon fruit (Hylocereus species). Food Chemistry. 2021;**353**:129426. DOI: 10.1016/j.foodchem.2021.129426

[20] Torre Grisales Y, Melo Sabogal DV, Torres-Valenzuela LS, Serna-Jiménez JA, Sanín VA. Evaluation of bioactive compounds with functional interest from yellow pitahaya (Selenicereus megalanthus haw). Revista Facultad Nacional de Agronomía Medellín. 2017;**70**(3):8311-8318. DOI: 10.15446/RFNA.V70N3.66330

[21] Mejia NM, Castro JP, Ocampo YC, Salas RD, Delporte CL, Franco LA. Evaluation of antioxidant potential and total phenolic content of exotic fruits grown in Colombia. Journal of Applied Pharmaceutical Science. 2020;**10**(9):50-58. DOI: 10.7324/JAPS.2020.10906

[22] Sotomayor A, Pitizaca S, Sánchez M, et al. Evaluación físico química de fruta de pitahaya (*Selenicereus megalanthus*) en diferentes estados de desarrollo. Enfoque UTE. 2019;**10**(1):89-96. DOI: 10.29019/enfoqueute.v10n1.386

[23] Obregón-La Rosa AJ, Augusto-Elías-Peñafiel CC, Contreras-López E, Arias-Arroyo GC, Bracamonte-Romero M. Características fisicoquímicas, nutricionales y morfológicas de frutas nativas. Revista de Investigaciones Altoandinas Journal of High Andean Research. 2021;**23**(1):17-25. DOI: 10.18271/ria.2021.202

[24] Cañar D, Caetano C, Bonilla-Morales M. Caracterización fisicoquímica y proximal del fruto de pitahaya amarilla [*Selenicereus megalanthus* (K. Schum)]. Agronomía. 2014;**22**(1):77-87

[25] Enriquez Paredes I, Ore AF. Elaboración de una bebida funcional a base de malta de *Amaranthus caudatus* L. y pulpa de Hylocereus triangularis. Ciencia Latina Revista Científica Multidisciplinar. 2021;**5**(3):3353-3366. DOI: 10.37811/cl_rcm.v5i3.536

[26] Vásquez-Castillo W, Aguilar K, Vilaplana R, Viteri P, Viera W, Valencia-Chamorro YS. Calidad del

fruto y pérdidas poscosecha de pitahaya amarilla (*Selenicereus megalanthus* Haw.) en Ecuador. Agronomia Colombiana. 2016;**34**(1):S1081-S1083. DOI: 10.15446/agron.colomb.v34n1supl.58279

[27] Chauca Aguilar MA, Chávez Quintana SG. Fenoles y capacidad antioxidante de Psidium guajava, Vaccinium myrtillus, Selenicereus megalanthus y Physalis peruviana de diferentes procedencias [Phenols Antioxid Capacit Psidium guajava, Vaccinium myrtillus, Selenicereus Megal Physalis peruviana.]. Bioagro. 2020;**32**(3):225-230. Available from: http://search.ebscohost.com/login.aspx?direct=true&db=fsr&AN=147260939&lang=es&site=ehost-live

[28] Suárez Román RS, Caetano CM, Ramírez H, Morales Osorio JG. Multiplicación de *Selenicereus megalanthus* (pitahaya amarilla) e *Hylocereus polyrhizus* (pitahaya roja) vía organogénesis somática. Acta Agronomica. 2014;**63**(3):272-281. DOI: 10.15446/acag.v63n3.40980

[29] Coral Caycho ER, Calixto Cotos MR, Soberón Lozano MM. Actividad inhibitoria in vitro de los extractos acuosos de los frutos de Hylocereus megalanthus Y Passiflora tripartita var. mollisima sobre las enzimas α-amilasa y α-glucosidasa. Rev la Soc Química del Perú. 2020;**86**(2):93-104. DOI: 10.37761/rsqp.v86i2.279

[30] Daza-Ramírez LD, Murillo-Perea E, Pardo DA. Evaluation of the antioxidant capacity of fruits grown in the department of Tolima and agro-industrial wastes. Revista Tumbaga. 2015;**2**(10):3-14

[31] Bhat R, Paliyath G. Fruits of tropical climates: Dietary importance and health benefits. Encyclopedia of Food and Health. 2016;**1**:144-149. DOI: 10.1016/B978-0-12-384947-2.00338-X

[32] Mercado-Silva EM. Pitaya—*Hylocereus Undatus* (Haw). In: Rodrigues S, Silva EO, Brito ES, editors. Exotic Fruits Reference Guide. London, United Kingdom: Elsevier Inc; 2018. pp. 339-349. DOI: 10.1016/b978-0-12-803138-4.00045-9

[33] Ortiz T, Assari L. Pitaya fruit quality (Hylocereus undatus [Haworth] britton & Rose) according to physiological maturity. A review. Revista colombiana de Ciencias Hortícolas. 2020;**14**(1):63-75

[34] Freitas BAG, Ribeiro JS, Viana EBM, De Souza CCE, Zanuto ME. Aspectos químicos, fitoquímicos e funcionais das pitaias Hylocereus undatus, Hylocereus monacanthus e Hylocereus megalanthus: Uma revisão/chemical, phytochemical and functional aspects of pitayas Hylocereus undatus, Hylocereus monacanthus and Hylocereus. Brazilian Journal of Health Review. 2021;**4**(5):19986-20024. DOI: 10.34119/bjhrv4n5-123

[35] Corzo-Rios LJ, Bautista Ramírez ME, Gómez y Gómez YM, Torres-Bustillos LG. Frutas de cactáceas: Compuestos bioactivos y sus propiedades nutracéuticas. In: Ramírez Ortiz ME, editor. Propiedades Funcionales de Hoy. Barcelona, España: OmniaScience; 2017. pp. 35-66. DOI: 10.926/oms.60

[36] Alves T d B, Afonso MRA, Costa JMC d. Efeitos da adição de agentes carreadores sobre o pó da polpa de pitaia vermelha (*H. polyrhizus*) liofilizada. *Res*. Social Development. 2020;**9**(8):e950986105. DOI: 10.33448/rsd-v9i8.6105

[37] Rocha LdeJFG, Godoy RLdO, Cunha CPd. Estudo De Alguns Compostos

Bioativos Das Pitayas De Polpas Branca E Vermelha (Cereus Undatus, Sinonímia: Hylocereus Guatemalensis, H.Undatus) / Study of Some Bioactive Compounds of White and Red Pulp Pitayas (Cereus Undatus, Synonymy: Hylocereus Guatemale). Brazilian Journal of Development. 2020;**6**(9):66217-66223. DOI: 10.34117/bjdv6n9-160

[38] Villalobos-Gutiérrez MG, Schweiggert RM, Carle R, Esquivel P. Chemical characterization of central American pitaya (Hylocereus sp.) seeds and seed oil. CyTA—Journal of Food. 2012;**10**(1):78-83. DOI: 10.1080/19476337.2011.580063

[39] Holanda MO, Lira SM, da Silva JYG, et al. Intake of pitaya (Hylocereus polyrhizus (F.A.C. Weber) Britton & Rose) beneficially affects the cholesterolemic profile of dyslipidemic C57BL/6 mice. Food Bioscience. 2021;**42**:101181. DOI: 10.1016/J.FBIO.2021.101181

[40] Som AM, Ahmat N, Abdul Hamid HA, Azizuddin NM. A comparative study on foliage and peels of Hylocereus undatus (white dragon fruit) regarding their antioxidant activity and phenolic content. Heliyon. 2019;**5**(2):e01244. DOI: 10.1016/J.HELIYON.2019.E01244

[41] Esquivel P, Araya YQ. Características del fruto de la pitahaya (Hylocereus sp.) y su potencial de uso en la industria alimentaria Pitahaya (Hylocereus sp.): Fruit characteristics and its potential use in the food industry. Revista Venezolana de Ciencia y Tecnología de Alimentos. 2012;**3**(1):113-129. Available from: http://www.rvcta.org

[42] Tongkham N, Juntasalay B, Lasunon P, Sengkhamparn N. Dragon fruit peel pectin: Microwave-assisted extraction and fuzzy assessment. Agriculture and Natural Resources. 2017;**51**(4):262-267. DOI: 10.1016/J.ANRES.2017.04.004

[43] Jiang H, Zhang W, Li X, Shu C, Jiang W, Cao J. Nutrition, phytochemical profile, bioactivities and applications in food industry of pitaya (Hylocereus spp.) peels: A comprehensive review. Trends in Food Science and Technology. 2021;**116**:199-217. DOI: 10.1016/J.TIFS.2021.06.040

[44] Castro Carranza JD, Vera Rodríguez LR, Palacios Cedeño AC, Dueñas Rivadeneira AA. Bebida funcional a base de pitahaya (*Hylocereus undatus*) y extractos de hierba luisa (*Cymbopogon citratus*) y albahaca (*Ocimum tenuiflorum*). Revista Técnica de la Facultad de Ingeniería Universidad del Zulia. 2020;**1**(2):90-95. DOI: 10.22209/rt.ve2020n2a13

[45] Erazo A, Suárez A. Caracterización socio productivo de la Opuntia ficus indica (Tuna) en las comunidades del cantón Guano. Bosques Latid cero. 2021;**11**(1):155-165

[46] Guerrero-Beltrán JÁ, Ochoa-Velasco CE. Figo da India—*Opuntia* spp. In: Rodrigues S, Silva EO, Brito ES, editors. Exotic Fruits Reference Guide. London, United Kingdom: Elsevier Inc; 2018. pp. 187-201. DOI: 10.1016/B978-0-12-803138-4.00024-1

[47] Assunção Alves M, Mendonça de Souza AC, Gamarra-Rojas G, Barbosa GN. Fruto de palma [*Opuntia fícus-indica* (L) Miller, Cactaceae]: Morfologia, compocicão química, fisiologia, índices de colheita e fisiologia pós-colheita. Revista Iberoamericana de Tecnología Postcosecha. 2008;**9**(1):16-25

[48] Hernández-Carranza P, Rivadeneyra-MataM,Ramos-CassellisME,

et al. Characterization of red prickly pear peel (*Opuntia ficus-indica* L.) and its mucilage obtained by traditional and novel methodologies. Journal of Food Measurement and Characterization. 2019;**13**(2):1111-1119. DOI: 10.1007/s11694-018-00026-y

[49] Fideicomiso de Riesgo Compartido. La TUNA, una fruta muy mexicana. Blog Fideicomiso de Riesgo Compartido. 2017. Available from: https://www.gob.mx/firco/articulos/la-tuna-una-fruta-muy-mexicana?idiom=es [Accessed: November 4, 2021]

[50] León Carrasco JC. Perú todavía no aprovecha el potencial de negocio de las tunas. Agencia Agraria de Noticias 2017. Available from: https://agraria.pe/noticias/peru-todavia-no-aprovecha-el-potencial-de-negocio-de-las-13798 [Accessed: November 5, 2021]

[51] Arancibia D, Talamilla M, Prat L. La Tuna—Manejo Del Cultivo y Usos. 2016. Available from: http://bibliotecadigital.fia.cl/bitstream/handle/20.500.11944/147594/FichatecnicaTuna.pdf?sequence=15&isAllowed=y

[52] Cota-Sánchez JH. Nutritional Composition of the Prickly Pear (*Opuntia Ficus-Indica*) Fruit. In: Simmonds MSJ, Preedy VR, editors. Nutritional Composition of Fruit Cultivars. London, United Kingdom: Elsevier Inc; 2016. pp. 691-712. DOI: 10.1016/B978-0-12-408117-8.00028-3

[53] Gonçalves Albuquerque T, Pereira P, Silva MA, Vicente F, Ramalho R, Costa HS. Prickly pear. In: Jaiswal AK, editor. Nutritional Composition and Antioxidant Properties of Fruit and Vegetables. London, United Kingdom: Elsevier Inc; 2020. pp. 709-728. DOI: 10.1016/b978-0-12-812780-3.00044-1

[54] Ostolaza C. Todos Los Cactus Del Perú. Lima, Perú: Editorial Franco EIRL.; 2014. p. 538

[55] Fernández-López JA, Almela L, Obón JM, Castellar R. Determination of antioxidant constituents in Cactus pear fruits. Plant Foods for Human Nutrition. 2010;**65**(3):253-259. DOI: 10.1007/s11130-010-0189-x

[56] Medina EMD, Rodríguez EMR, Romero CD. Chemical characterization of Opuntia dillenii and Opuntia ficus indica fruits. Food Chemistry. 2007;**103**(1):38-45. DOI: 10.1016/j.foodchem.2006.06.064

[57] Rössel-Kipping ED, Ortiz-Laurel H, Moreno-Ovalle FM, López-Martínez LA, Amante-Orozco A. Physical and chemical attributes of prickly pear cactus (*Opuntia ficus-indica*) varieties Copena, Pelon Blanco and Pelon Rojo. Agro Product. 2021;**14**(4):127-134. DOI: 10.32854/AGROP.V14I4.1843

[58] Aruwa CE, Amoo SO, Kudanga T. Opuntia (Cactaceae) plant compounds, biological activities and prospects—A comprehensive review. Food Research International. 2018;**112**(February):328-344. DOI: 10.1016/j.foodres.2018.06.047

[59] Anculle Arenas A, Castro Cepero V, Julca OA. Caracterización de fincas productoras de tuna (*Opuntia ficus indica*) para la producción de cochinilla del carmín (*Dactylopius coccus*) en La Joya (Arequipa, Perú). Aporte Santiaguino. 2017;**10**(2):245. DOI: 10.32911/as.2017.v10.n2.167

[60] Valero-GalvánJ,González-FernándezR, Sigala-Hernández A, et al. Sensory attributes, physicochemical and antioxidant characteristics, and protein profile of wild prickly pear fruits (*O. macrocentra* Engelm., *O. phaeacantha*

Engelm., and *O. engelmannii* Salm-Dyck ex Engelmann.) and commercial prickly pear fruits (*O. ficus*). Food Research International. 2021;**140**(November). DOI: 10.1016/j.foodres.2020.109909

[61] Romero MC, Fogar RA, Fernández CL, Doval MM, Romero AM, Judis MA. Effects of freeze-dried pulp of *Eugenia uniflora* L. and *Opuntia ficus-indica* fruits on quality attributes of beef patties enriched with n-3 fatty acids. Journal of Food Science and Technology. 2021;**58**(5):1918-1926. DOI: 10.1007/s13197-020-04703-z

[62] Jorge P, Troncoso L. Capacidad antioxidante del fruto de la Opuntia apurimacensis (ayrampo) y de la *Opuntia ficus-indica* (tuna). Anales de la Facultad de Medicina. 2016;**77**(2):105. DOI: 10.15381/ANALES.V77I2.11812

[63] Monroy-Gutiérrez T, Martínez-Damián MT, Barrientos-Priego AF, et al. Compuestos bioactivos y capacidad antioxidante en frutos de xocotuna, tuna y xoconostle (Opuntia spp.). Chilean Journal of Agricultural & Animal Sciences. 2017;**33**(3):263-272. DOI: 10.4067/S0719-38902017005000704

[64] Ordoñez ES, Leon-Arevalo A, Rivera-Rojas H, Vargas E. Quantification of total polyphenols and antioxidant capacity in skins and seeds from cacao (*Theobroma cacao* L.), tuna (*Opuntia ficus indica* Mill), grape (*Vitis Vinífera*) and uvilla (*Pourouma cecropiifolia*). Scientia Agropecuaria. 2019;**10**(2):175-183. DOI: 10.17268/SCI.AGROPECU.2019.02.02

[65] Villaño D, García-Viguera C, Mena P. Colors: Health effects. Encyclopedia of Food and Health. 2016;**1**:265-272. DOI: 10.1016/B978-0-12-384947-2.00190-2

[66] Ortega-Hernández E, Welti-Chanes J, Jacobo-Velázquez DA. Effects of UVB light, wounding stress, and storage time on the accumulation of Betalains, phenolic compounds, and ascorbic acid in red prickly pear (Opuntia ficus-indica cv. Rojo vigor). Food and Bioprocess Technology. 2018;**11**(12):2265-2274. DOI: 10.1007/S11947-018-2183-5

[67] Paniagua-Zambrana NY, Bussmann RW, Romero C. *Opuntia ficus-indica* (L.) Mill. Cactaceae. In: Ethnobotany of the Andes. Cham: Springer; 2020. pp. 1315-1320. DOI: 10.1007/978-3-030-28933-1_210

[68] Lagua Yanchaguano HI, Garzón Carrera JG, Domínguez Narváez V, Alta TA. Elaboración de una bebida nutritiva a base de pulpa de opuntia ficus indica (nopal) enriquecida y saborizada con jugó de passiflora edulis, (maracuyá). Ciencia Digital. 2020;**4**(4.1):6-17. DOI: 10.33262/CIENCIADIGITAL.V4I4.1.1447

[69] Chaparro SP, Márquez RA, Sánchez JP, Vargas ML, Gil JH. Extracción de pectina del fruto del higo (*Opuntia ficus indica*) y su aplicación en un dulce de piña. Revista U.D.C.A Actualidad & Divulgación Científica. 2015;**18**(2):435-443. DOI: 10.31910/rudca.v18.n2.2015.171

[70] Montilla A, Tovar I, Pachero H. Cuantificación de pectinas en la pulpa del fruto de tres especies de tuna (Opuntioideae, Cactaceae). La Técnica: Revista de las Agrociencias. 2020;23:25-32. DOI: 10.33936/la_tecnica.v0i23.1875

[71] Coba Carrera RL, Mariuxi Flores L, Elizabeth Valverde P, Salazar Llangari KG. Obtención del colorante natural de tuna (Opuntia ficus-indica). Ciencia Digital. 2019;**3**(3.2):232-240. DOI: 10.33262/CIENCIADIGITAL.V3I3.2.729

[72] Otálora MC, Carriazo JG, Iturriaga L, Nazareno MA, Osorio C. Microencapsulation of betalains obtained

from cactus fruit (*Opuntia ficus-indica*) by spray drying using cactus cladode mucilage and maltodextrin as encapsulating agents. Food Chemistry. 2015;**187**:174-181. DOI: 10.1016/j.foodchem.2015.04.090

[73] Pauca Tanco A, Quipuscoa Silvestre V. Catálogo de las cactáceas del departamento de Arequipa, Perú. Arnaldoa. 2017;**24**:447-496

[74] Echevarría J, Paniagua-Zambrana NY, Bussmann RW. *Corryocactus brevistylus* (K. Schum. ex Vaupel) Britton & Rose Cactaceae. In: Paniagua-Zambrana NY, Bussmann RW, editors. Ethnobotany of the Andes. Ethnobotany of Mountain Regions. Cham, Basel, Switzerland: Springer; 2020. pp. 611-613

[75] Heim E. Flora de Arequipa, Peru: A Field Guide for Nature Lovers. Norderstedt, Germany: BoD-Books on Demand; 2014. p. 128

[76] Arévalo Minchola A, García Vásquez J, Armas García V, Rodríguez Ávila JO, Iparraguirre Cortavitarte R, Bartolo Rodríguez L, et al. Uso de biopectinasa y filtración al vacío para la clarificación de una mezcla diluida de pulpa de sancayo (*Corryocactus brevistylus*) y tuna (*Opuntia ficus-indica*) a diferentes temperaturas. Agoindustrial Science. 2012;**1**:84-90

[77] Alvarez-Yanamango E, Vietti F, Huayta F. Use of waste from the processing of Sanky pulp (*Corryocactus brevistylus*) to obtain a food additive. Proceedings of the International Multi-Conference for Engineering, Education, and Technology, Lima, Peru. 2018. DOI: 10.18687/LACCEI2018.1.1.412

[78] Muñoz AM, Contreras-López E, Fernández-Jerí Y, Arosena Chao A, Chavez Cerna R, Alvarado Yarasca A, et al. Evaluación de la actividad antioxidante y características nutricionales de cascara y semilla de sanky. Congreso Iberoamericano de Química, Lima, Perú, 2018

[79] Contreras-López E, Salvá Ruiz BK. Sensory characterization of llama burger with sanky peel. Revista de Investigaciones Altoandinas. 2018;**20**:155-168

[80] Nolazco CD, Guevara PA. Estudio de las principales características fisicoquímicas y comportamiento del sanqui (*Corryocactus brevistylus* subsp. *puquiensis* (Rauh & Backeberg) Ostolaza) en almacenamiento. Anales Científicos UNALM. 2009;**70**:1-11

[81] Areche C, Hernandez M, Cano T, Ticona J, Cortes C, Simirgiotis M, et al. *Corryocactus brevistylus* (K. Schum. ex Vaupel) Britton & Rose (Cactaceae): Antioxidant, gastroprotective effects, and metabolomic profiling by ultrahigh pressure liquid chromatography and electrospray high resolution orbitrap tandem mass spectrometry. Frontiers in Pharmacology. 2020;**11**:417

[82] El Kossori RL, Villaume C, El Boustani E, Sauvaire Y, Méjean L. Composition of pulp, skin and seeds of prickly pears fruit (*Opuntia ficus indica* sp.). Plant Foods for Human Nutrition. 1998;**52**:263-270

[83] Chia SL, Chong GH. Effect of drum drying on physico-chemical characteristics of dragon fruit peel (*Hylocereus polyrhizus*). International Journal of Food Engineering. 2015;**11**:285-293

[84] Özcan MM, Al Juhaimi FY. Nutritive value and chemical composition of prickly pear seeds (*Opuntia ficus indica* L.) growing in Turkey. International Journal of Food Sciences and Nutrition. 2011;**62**:533-536

[85] Oomah BD, Patras A, Rawson A, Sing N, Compos-Vega R. Chemistry of pulses. In: Tiwari BK, Gowen A, McKenna B, editors. Pulse Foods: Processing, Quality and Nutraceutical Applications. London, UK: Elsevier Inc; 2011. pp. 9-55

[86] Park Y. Dietary fiber and health: Cardiovascular disease and beyond. In: Watson RR, Preedy VR, editors. Fruits, Vegetables, and Herbs: Bioactive Foods in Health Promotion. London, UK: Elsevier Inc; 2016. pp. 423-449

[87] Dreher ML. Whole fruits and fruit fiber emerging health effects. Nutrients. 1833;**2018**:10

[88] Elleuch M, Bedigian D, Roiseux O, Besbes S, Blecker C, Attia H. Dietary fibre and fibre-rich by-products of food processing: Characterization, technological functionality and commercial applications: A review. Food Chemistry. 2011;**124**:411-421

[89] Serna-Saldivar SO. Nutrition and fortification of corn and wheat tortillas. In: Rooney LW, Serna-Saldivar SO, editors. Tortillas: Wheat Flour and Corn Products. Saint Paul, MN: AACC International, Inc; 2015. pp. 29-63

[90] Wang Q, Shi A, Liu H, Liu L, Zhang Y, Li N, et al. Peanut by-products utilization technology. In: Wang Q, editor. Peanuts: Processing Technology and Product Development. Beijing, China: China Science Publishing & Media Ltd; 2016. pp. 221-325

[91] Mu T, Sun H, Zhang M, Wang C. Sweet potato dietary fiber. In: Mu T, Sun H, Zhang M, Wang C, editors. Sweet Potato Processing Technology. Beijing, China: China Science Publishing & Media Ltd; 2017. pp. 121-181

[92] Díaz-Vela J, Totosaus A, Pérez-Chabela ML. Integration of agroindustrial co-products as functional food ingredients: Cactus pear (*Opuntia ficus indica*) flour and pineapple (*Ananas comosus*) peel flour as fiber source in cooked sausages inoculated with lactic acid bacteria. Journal of Food Processing & Preservation. 2015;**39**:2630-2638

[93] Garcia-Amezquita LE, Tejada-Ortigoza V, Heredia-Olea E, Serna-Saldívar SO, Welti-Chanes J. Differences in the dietary fiber content of fruits and their by-products quantified by conventional and integrated AOAC official methodologies. Journal of Food Composition and Analysis. 2018;**67**:77-85

[94] Chemah TC, Aminah A, Noriham A, Wan Aida WM. Determination of pitaya seeds as a natural antioxidant and source of essential fatty acids. International Food Research Journal. 2010;**17**:1003-1010

[95] Ramírez-Rodríguez Y, Martínez-Huélamo M, Pedraza-Chaverri J, Ramírez V, Martínez-Tagüeña N, Trujillo J. Ethnobotanical, nutritional and medicinal properties of Mexican drylands cactaceae fruits: Recent findings and research opportunities. Food Chemistry. 2020;**312**:126073

[96] Labuschagne MT, Hugo A. Oil content and fatty acid composition of cactus pear seed compared with cotton and grape seed. Journal of Food Biochemistry. 2010;**34**:93-100

[97] Liaotrakoon W, De Clercq N, Van Hoed V, Dewettinck K. Dragon fruit (*Hylocereus* spp.) seed oils: Their characterization and stability under storage conditions. Journal of the American Oil Chemists' Society. 2013;**90**:207-215

[98] Imán CI, Bravo Zamudio L, Sotero Solís V, Oliva Cruz C. Contenido de vitamina C en frutos de camu camu

Myrciaria dubia (H.B.K) Mc Vaugh, en cuatro estados de maduración, procedentes de la colección de Germoplasma del INIA Loreto, Perú. Scientia Agropecuaria. 2011;**2**:123-130

[99] Coban HB. Organic acids as antimicrobial food agents: Applications and microbial productions. Bioprocess and Biosystems Engineering. 2020;**43**:569-591

[100] Teixeira J, Gaspar A, Garrido EM, Garrido J, Borges F. Hydroxycinnamic acid antioxidants: An electrochemical overview. BioMed Research International. 2013; ID 251754

[101] Coman V, Vodnar DC. Hydroxycinnamic acids and human health: Recent advances. Journal of the Science of Food and Agriculture. 2020;**100**:483-499

[102] Ghorbani A. Mechanisms of antidiabetic effects of flavonoid rutin. Biomedicine & Pharmacotherapy. 2017;**96**:305-312

[103] Patel K, Patel DK. The beneficial role of rutin, a naturally occurring flavonoid in health promotion and disease prevention: A systematic review and update. In: Watson RR, Preedy VR, editors. Bioactive Food as Dietary Interventions for Arthritis and Related Inflammatory Diseases. London, UK: Elsevier Inc; 2019. pp. 457-579

[104] Caparica R, Júlio A, Araújo MEM, Baby AR, Fonte P, Costa JG, et al. Anticancer activity of rutin and its combination with ionic liquids on renal cells. Biomolecules. 2020;**10**:233

[105] Sun C, Wang L, Sun J, Wang Z, Tang Z. Hypoglycemic and hypolipidemic effects of rutin on hyperglycemic rats. Journal of Traditional Chinese Medicine. 2020;**40**:640-645

[106] Barba FJ, Esteve M, Frígola A. Bioactive components from leaf vegetable products. In: Rahman A, editor. Studies in Natural Products Chemistry. Oxford, UK: Elsevier BV; 2014. pp. 321-346

[107] Shenoy N, Creagan E, Witzig T, Levine M. Ascorbic acid in cancer treatment: Let the phoenix fly. Cancer Cell. 2018;**34**:700-706

[108] Van Gorkom GNY, Lookermans EL, Van Elssen CHM, Bos GMJ. The effect of vitamin C (ascorbic acid) in the treatment of patients with cancer: A systematic review. Nutrients. 2019;**11**:977

[109] Jia Y, Ma Y, Cheng G, Zhang Y, Cai S. Comparative study of dietary flavonoids with different structures as α-glucosidase inhibitors and insulin sensitizers. Journal of Agricultural and Food Chemistry. 2019;**67**:10521-10533

[110] Gevrenova R, Kondeva-Burdina M, Denkov N, Zheleva-Dimitrova D. Flavonoid profiles of three *Bupleurum* species and *in vitro* hepatoprotective of activity *Bupleurum flavum* Forsk. Pharmacognosy Magazine. 2015;**11**:14-23

[111] Li J, Hu L, Zhou T, Gong X, Jiang R, Li H, et al. Taxifolin inhibits breast cancer cells proliferation, migration, and invasion by promoting mesenchymal to epithelial transition via β-catenin signaling. Life Sciences. 2019;**232**:116617

[112] Ay M, Charli A, Jin H, Anantharam V, Kanthasamy A, Kanthasamy AG. Quercetin. In: Gupta RC, editor. Nutraceuticals: Efficacy, Safety and Toxicity. London, UK: Elsevier Inc; 2016. pp. 447-452

[113] Gong G, Guan Y-Y, Zhang Z-L, Rahman K, Wang S-J, Zhou S, et al. Isorhamnetin: A review of pharmacological effects. Biomedicine & Pharmacotherapy. 2020;**128**:110301

[114] Lan L, Wang Y, Pan Z, Wang B, Yue Z, Jiang Z, et al. Rhamnetin induces apoptosis in human breast cancer cells via the miR-34a/Notch-1 signaling pathway. Oncology Letters. 2019;**17**:676-682

[115] Ruiz-Cruz S, Chaparro-Hernández S, Hernández-Ruiz KL, Cira-Chávez LA, Estrada-Alvarado MI, Gassos Ortega LE, et al. Flavonoids: Important biocompounds in food. In: Justino G, editor. Flavonoids—from Biosynthesis to Human Health. Vol. 2017. London, UK: IntechOpen; 2017. pp. 353-369

[116] De Santiago E, Pereira-Caro G, Moreno-Rojas JM, Cid C, De Peña MP. Digestibility of (poly)phenols and antioxidant activity in raw and cooked Cactus cladodes (*Opuntia ficus-indica*). Journal of Agricultural and Food Chemistry. 2018;**66**:5832-5844

[117] De Wit M, du Toit A, Osthof G, Hugo A. Cactus pear antioxidants: A comparison between fruit pulp, fruit peel, fruit seeds and cladodes of eight different cactus pear cultivars (*Opuntia ficusindica* and *Opuntia robusta*). Journal of Food Measurement and Characterization. 2019;**13**:2347-2356

[118] Rojas T, Fuentes Campos ME, Contreras-López E, Gómez S, Muñoz AM. Extracción asistida por ultrasonido de compuestos fenólicos de la cáscara de sanky (*Corryocactus brevistylus*). Revista de la Sociedad Química del Perú. 2019;**85**:258-267

[119] Kim H, Choi H-K, Moon JY, Kim YS, Mosaddik A, Cho SK. Comparative antioxidant and antiproliferative activities of red and white pitayas and their correlation with flavonoid and polyphenol content. Journal of Food Science. 2011;**76**:C38-C45

[120] Huyut Z, Beydemir Ş, Gülçin G. Antioxidant and antiradical properties of selected flavonoids and phenolic compounds. Biochemistry Research International. 2017; ID 7616791

[121] Contreras-López E, Muñoz AM, Salvá-Ruiz BK. Evaluation of sanky peel extract in the stability of llama meat. Revista de Investigaciones Altoandinas. 2020;**22**:123-134

[122] Carrión-Galindo MG. Influencia de la adición de gelificantes sobre la aceptabilidad sensorial y las características fisicoquímicas en la elaboración de una mermelada de sancayo (*Corryocactus brevistylus*) [bachelor thesis]. Tacna, Perú: Universidad Nacional Jorge Basadre Grohmann; 2018

[123] Muñoz AM, Contreras-López E, Vega R, Benavides E, Bautista N, Peña C, et al. Aporte nutricional y evaluación sensorial de mermelada de sanky (*Corryocactus brevistylus* K. Schum) elaborada con FOS. Archivos Latinoamericanos de Nutrición. 2019;**69**:PO444

[124] Featherstone SA. Complete Course in Canning and Related Processes. 14th ed. Cambridge, UK: Elsevier Ltd; 2016. pp. 313-349

[125] Mohd Naeem MN, Mohd Fairulnizal MN, Norhayati MK, Zaiton A, Norliza AH, Wan Syuriahti WZ, et al. The nutritional composition of fruit jams in the Malaysian market. Journal of the Saudi Society. 2017;**16**:89-96

[126] Shinwari KJ, Rao PS. Stability of bioactive compounds in fruit jam and jelly during processing and storage: A review. Trends in Food Science and Technology. 2018;**75**:181-193

[127] Neves MF, Trombin VG, Lopes FF, Kalaki R, Milan P. The Orange Juice Business: A Brazilian Perspective.

Wageningen, Netherlands: Wageningen Academic Publishers; 2011. pp. 117-117

[128] Choudhary PD, Pawar HA. Recently investigated natural gums and mucilages as pharmaceutical excipients: An overview. Journal of Pharmaceutics. 2014;**2014**:204849

[129] Cortés-Camargo S, Cruz-Olivares J, Barragán-Huerta BE, Dublán-García O, Román-Guerrero A, Perez-Alonso C. Microencapsulation by spray drying of lemon essential oil: Evaluation of mixtures of mesquite gum–nopal mucilage as new wall materials. Journal of Microencapsulation. 2017;**34**:395-407

[130] Wanlapa S, Wachirasiri K, Sithisam-ang D, Suwannatup T. Potential of selected tropical fruit peels as dietary fiber in functional foods. International Journal of Food Properties. 2015;**18**:1306-1316

[131] Alves E, Simoes A, Dominguez MR. Fruit seeds and their oils as promising sources of value-added lipids from agro-industrial byproducts: Oil content, lipid composition, lipid analysis, biological activity, and potential biotechnological applications. Critical Reviews in Food Science and Nutrition. 2021;**61**(8):1305-1339. DOI: 10.1080/10408398.2020.1757617

[132] Bartoszek M, Polak JA. Comparison of antioxidative capacities of fruit juices, drinks and nectars, as determined by EPR and UV–vis spectroscopies. Spectrochimica Acta. Part A, Molecular and Biomolecular Spectroscopy. 2016;**153**:546-549

[133] Maia GA, Ribeiroda LM, Prado SGM, Fonseca AVV, de Sousa PHM, Figueiredo RW. Development of mixed beverages based on tropical fruits. In: Grumezescu AM, Holban AM, editors. Non-alcoholic Beverages. Duxford, UK: Elsevier Inc; 2019. pp. 129-162

[134] Palmeri R, Parafati L, Restuccia C, Fallico B. Application of prickly pear fruit extract to improve domestic shelf life, quality and microbial safety of sliced beef. Food and Chemical Toxicology. 2018;**118**:355-360

[135] Sharma P, Yadav S. Effect of incorporation of pomegranate peel and bagasse powder and their extracts on quality characteristics of chicken meat patties. Food Science of Animal Resources. 2020;**40**:388-400

[136] Lourenço SC, Moldão-Martins M, Alves VD. Antioxidants of natural plant origins: From sources to food industry applications. Molecules. 2019;**24**:4132

Chapter 14

Antibacterial Activity of Plant Polyphenols

Galina Satchanska

Abstract

This chapter focuses on methods of polyphenol isolation and on the antibacterial activity of different polyphenols found in herbs, spices, fruits and vegetables. Polyphenols are secondary metabolites which protect plants from different pathogens, such as viruses, bacteria, fungi, insects, and herbivores. Currently, about 9000 polyphenols found in more than 480 plants are known. Their amount fluctuates across different species and varieties. This chapter describes conventional and novel methods for extraction, the influence of the type of solvents, solvent concentration and temperature on the yield. The highest yield is obtained at 70% of methanol and ethanol, and at 90% of acetone. Extraction at 80°C leads to higher amounts of polyphenols than extraction at 100°C. Polyphenols are usually metabolized in the human liver but can also remain unaffected as they pass through the gastrointestinal tract. The main location for their uptake is the colon. They exhibit a wide range of antibacterial activity against *Pseudomonas aeruginosa*, *Staphylococcus aureus*, *Streptococcus epidermidis*, *Klebsiella pneumoniae, E. coli*, *Listeria monocytogenes*, *Acinetobacter* sp., *Proteus* sp., *Micrococcus* sp., and *Bacillus* sp. All these plants, rich in antimicrobial polyphenols, represent a promising and powerful source of highly effective novel antibacterial substances in the current era of ubiquitous antibiotic resistance.

Keywords: plant polyphenols, isolation, antibacterial activity

1. Introduction

The widespread antibiotic resistance in the last 20 years has become one of the biggest worldwide threats to mankind. Plants are valuable reservoir of novel antimicrobials and their secondary metabolites as polyphenols demonstrate strong antimicrobial activity at extremely low concentrations. Precursor of polyphenols is phenol which consist of one aromatic ring and a hydroxyl group. Polyphenols as more complex substances are polyaromatic and contain a few hydroxyl groups. They are divided into four main groups: Flavonoids, Lignans, Phenolic acids, and Stilbenes. Among them, flavonoids are the most numerous. All polyphenols play an important role in the defense of plants against bacteria, viruses, fungi, insects and herbivores. Polyphenol synthesis derives from two aromatic amino acids – tyrosine and phenylalanine. As secondary plant metabolites, their amount is estimated at only around 10% of plant metabolites [1].

Among the 300,000 plant species that exist in the world only 15% have been investigated for their pharmacological potential, the rest of them are a potential source of novel natural antimicrobial products [2]. According to the WHO, the global market of plant products is estimated at the huge amount of US $83 billion and currently continues to grow. Usually, the daily polyphenol human intake varies between 20 and 500 mg, taken *via* onions, tomatoes, red wine and many other foods and beverages [3]. Once inside the body, phenol is retained there bound to other molecules, most often to proteins. Absorbed or unabsorbed, while passing through the gastrointestinal tract polyphenols strongly influence human microbiota. They inhibit gastrointestinal pathogenic bacteria and enrich the beneficial intestinal bacteria. In this way, they significantly improve human health [4].

Currently, more than 9000 polyphenols have been identified. Some major representatives of polyphenols are shown in **Figure 1**.

Salicylic acid (2), an ortho-hydroxylated benzoic acid, is a beta hydroxy acid that occurs as a natural compound in plants. More than 2000 years ago, Hippocrates (c. 300 BC) cured rheumatism and inflammation with white willow (*Salix alba*) leaves and bark extracts, both containing the precursor of salicylic acid – salicin. Cardamom seeds, a typical aroma additive to tea and Arabian coffee in Asia and the Middle East, also contains salicin. It is highly active as an anti-inflammatory agent and a proven antibacterial agent. Salicylic acid is detectable in most of human organs and tissues being most abundant in saliva. It persist in all eukaryotic organisms. Foods processed from cereals are rich salicylic acid easily recognizable by its phenol-like smelling.

Gallic acid (3), also named gallate falls in the class Gallic acids [5]. It comprise of 3, 4, 5-trihydroxybenzoic acid. Gallic acid can be found in various foods, such as apple, ginger, yellow pepper, hazelnuts, and oak bark. Gallic acid is recognized as strong antioxidant.

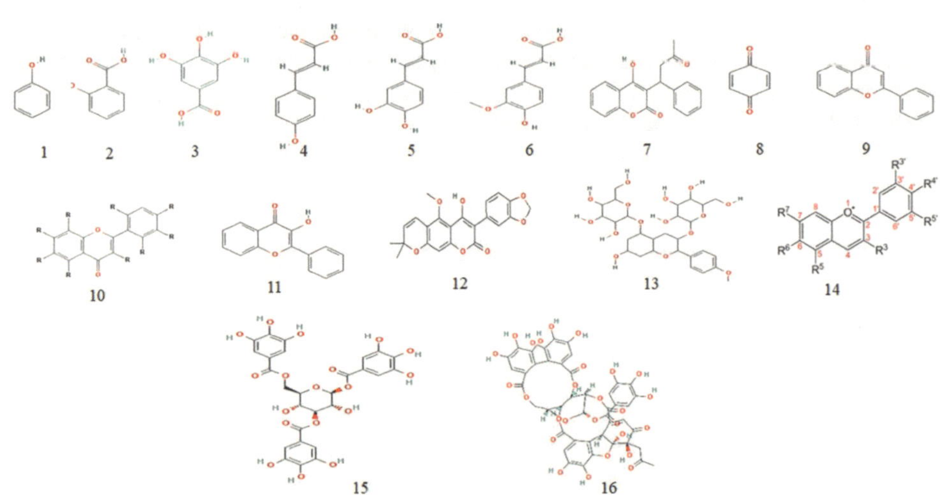

Figure 1.
Structure and molecular formula of main polyphenols: (1) phenol (C_6H_5OH); (2) 2-Hydroxybenzoic acid (salicylic acid) (HOC_6H_4COOH); (3) 3,4,5-Trihydroxybenzoic acid (Gallic acid) ($C_6H_2(OH)_3COOH$); (4) 4-Hydroxycinnamic (p-Coumaric acid) ($C_9H_8O_3$); (5) 3,4-Dihydroxycinnamic acid (Caffeic acid) ($C_9H_8O_4$); (6) 4-Hydroxy-3-methoxycinnamic acid (Ferulic acid) ($C_{10}H_{10}O)_4$; (7) 4-hydroxy-3-(3-oxo-1-phenylbutyl)chromen-2-one, Warfarine (Coumarins) ($C_{19}H_{16}O_4$); (8) 1,4- benzoquinone (p-benzoquinone) (Quinone)($C_6H_4O_2$); (9) 2,3-Dihydroflavone (flavone) ($C_{15}H_{12}O_2$); (10) 2-phenyl-4H-chromen-4-one (flavone) ($C_{15}H_{10}O_2$); (11) 3-Hydroxy-2-phenyl-4H-chromen-4-one (Flavonol) ($C_{15}H_{10}O_3$); (12) 7-(1,3-benzodioxol-5-yl)-6-hydroxy-5-methoxy-2,2-dimethylpyrano[3,2-g]chromen-8-one (Robustin) (Isoflavonoid) ($C_{22}H_{18}O_7$); (13) 2-(4-Hydroxyphenyl)chromenylium-3,5,7-triol (Pelargonidine) (Anthocianidine) ($C_{15}H_{11}O_5^+$); (14) Anthocianins; (15) 1,3,6-tri-O-galloyl-beta-D-glucose (Gallotannin) ($C_{27}H_{24}O_{18}$); and (16) Elagitannin ($C_{44}H_{32}O_{27}$) (source of figures and short description of substances below [5].

Hydroxycinnamic acids (4), known also as coumaric acid, contains cinnamic acid, where the benzene ring is hydroxylated at C-4. Inside the cell, hydroxycinnamic acids are located in the cytoplasm and mitochondria. Similarly to salicylic acid, *trans*-4-coumaric acid is present in all eukaryotes. Plants like green pepper, apricot, and blueberry are excellent source of this acid. Like gallic acid, hydroxycinnamic acids can be detected in human feces, urine, and blood.

Caffeic acid (5) is a hydroxycinnamic acid derivative. It exhibits antioxidant, anti-inflammatory, and antineoplastic activities and protect DNA from free radicals damage. Bountiful in skin and prostate gland, caffeic acid shows suppressive effect on prostate cancer proliferation. Along with its beneficial effect, the substance is classified as a possible carcinogen (IARC – International Agency for Research on Cancer (WHO) classification of cancerogenic xenobiotics – Group 2B) and toxic compound [5]. Many plants are rich in caffeic acid: apricot, prunes, salvia, spearmint, thyme, aronia, sunflower seeds, barley and rye. Oddly, coffee contains modest concentrations of caffeic acid in contrast to argan oil.

Ferulic acid (6), a tyrosin-similar compound comprises of the water soluble *trans*-cinnamic acid. Main location of the substance is the plant cell is the cell wall. Its name originates from the giant fennel (*Ferula communis*). Rich in ferulic acid are root vegetables and sweet poprcorn. Being a constituent of the plant biopolymer lignocellulose the ferulic acid is involved in accumulation of this most abundant biowaste on Earth. Pronounced apoptosis inhibitor and cardioprotector, ferulate also helps the skin aging retardation inhibiting melanin formation. Used widely as food preservative [5] it can be successfully excreted *via* human epidermis.

Warfarin (7) falls in the class of 4-hydroxycoumarins and is one of the best synthetic oral anticoagulants. It constrains the synthesis of blood clotting factors which depends on Vitamin K. The key role of Vitamin K is the synthesis of one unusual for proteins amino acid –*gamma*-carboxyglutamic acid, an important for the biological activity of clotting proteins component. Warfarin is applied in the treatment of various types of embolism like cerebral or lung embolism. Vit. K exists in two forms: Vit. K1 (phylloquinone) which is synthesized by plants and can be found mainly in green leafy vegetables, and Vit. K2 (menaquinon) which is synthesized by the probiotic lactic acid bacteria in the human intestine and is also abundant in the fermented dairy products.

1,4-*Benzoquinone* (8) is a member of *p*-benzoquinones and a metabolite of benzene. 1,4-Benzoquinone possess two C=O groups attached at the 1- and 4-positions at of the aromatic ring. Inside the cell, mitochondria and the cytoplasm are the cell structures docking quinone which enforce specific enzymatic reactions. Quinone is capable to turn into orotic acid and when accumulated in the human blood orotic acid leads to aciduria resulting in quick liberation of ammonia. Often 1,4-Benzoquinone is transformed to glycerol 3-phosphate, a key substance of Glycolysis. Quinone is also responsible for the Vit. B_{12} (ciancobalamin) catabolism. Vit. B_{12} is synthesized by probiotic bacteria in the human colon and its deficiency cause anemia. Anise is one of the richest sources of 1,4-Benzoquinone.

Flavone (9) is a lipid molecule and member of the class of flavanones. In the cell it is harbored in the cytoplasmic membrane and among the human tissues the most abundant is placenta. Amid fruits pomegranate is excessive in this substance while out of spices rosemary is the wealthiest one.

3-*Hydroxyflavone* (10) belongs to flavonols [5]. Similarly to flavanone, it is spotted in the cell membrane and is a water insoluble compound. 3-Hydroxyflavone is precursor of tambulin known an anti-aging and anti-Parkinsonian medicine. Foods abundant in this polyphenol are brassicas and papaya.

Vegetables	Fruits	Grains	Beans	Herbs&Spices	Beverages
Artichoke	Apples	Oat	Black beans	Basilicum	Black tea
Asparagus	Apricots	Rye	Soy meat	Black tea	Coffee
Broccoli	Black chockeberry	Whole grains	Soy milk	Celery	Dark chocolate
Capers	Black currant	Wheat	Sprout	Cinnamon	Ginger
Carrots	Black elderberry		White beans	Cummin	Green tea
Cayenne pepper	Black grapes			Curry	Olive oil
Garlic	Blackberry			Ginger	Rapeseed oil
Olives	Blueberry			Green tea	Red wine
Potatoes	Cherry sour			Majoran	Vinegar
Red lettuce	Cherry sweet			Oregano	
Onion	Grapefruit			Parsley	
Spinach	Nectarines			Peppermint	
	Peaches			Rosemary	
	Pears			Sage	
	Pomegranate			Spearmint	
	Plum			Star anise	
	Raspberry				
	Strawberry				

Table 1.
Foods of daily diet supplying polyphenols (Adapted after Perez-Jumenez et al. [7] and Mustafa et al. [8]).

Another derivative of flavone is *Flavonol* (11). Due to its rich yellow color it was used for centuries in wool and silk dyeing. Flavonol is capable to bind essential and heavy metal ions.

Generally, flavonoids are main contributors to the flavor of fruits and the bitterness of citruses. Naringin, tangeritin, quercetin and neohesperidin impart the bitter taste in citruses, while the bitterness of wine is due to catechins and epicatechins [6].

Pelargonidin (13) can be found in almost all berries – blueberries, blackberries, cranberries, raspberries, strawberries, and aronia. Large amounts of pelargonidin are typically found also in plums and pomegranates, a polyphenol responsible for the color of radishes.

Gallotannins (15) is a class of tannins obtained by condensation of the carboxy group of gallic acid and the hydroxy group of glucose. Rich in gallotanins are pomegranates, strawberries and gallnuts. In **Table 1** are presented the foods supplying polyphenols.

2. Isolation of polyphenols

Isolation of polyphenols is a challenging procedure due to the instability and complex structure of these compounds. Most often polyphenols are harbored in plant leaves and gymnosperm, and within the cell in the cell wall and vacuoles associated with the nuclei. The covalent bonding of polyphenols with the plant structures is a limiting factor for their liberation [9]. Additionally, other factors influence the recovery of phenolic compounds from plant samples: location in plant tissues, extraction method, sample size, storage conditions and possible subsequent chemical conversions. A wide spectrum of plant secondary metabolites, including polyphenols can be obtained using water or organic solvents

2.1 Conventional extraction using solvents

These methods are widely used. Washed or dried plant material is finely ground and subjected to solvent extraction. The most commonly used solvents are water, hexane, ether, chloroform, acetone, benzene, ethanol and methanol. All these solvents are effective in taking out bioactive compounds i.e. polyphenols from the cell.

EtOH dissolves alkaloids, glycosides and dyes but does not dissolve gum, waxes or fats. It easily penetrates the cell membrane and fast and effortlessly extracts the cell metabolites. A disadvantage of ethanol is its volatility and flammability.

Acetone dissolves large amount of substances both hydrophilic and lipophilic. It is excellent for tannin extraction. Its advantage is not only the prevention of microorganism growth but also the easy evaporation and low cost extraction. Like ethanol, it is flammable and volatile.

Chloroform, *ether* and *bischloromethanol*. Chloroform is suitable for tannin extraction, ether for coumarins and tannins, and bischloromethanol – for terpenoids but not for phenolic compounds. All three solvents evaporate easily and are low-cost. Their disadvantages besides volatility and flammability are explosiveness and toxicity.

According to the paper of Alothman et al. [10], who studied pineapple, banana and guava, the percentage of different solvents strongly influence the polyphenol yield. Investigating polyphenol extraction with methanol, ethanol and acetone at concentration 90, 70, and 50%, authors found the highest yield at 70% of ethanol and methanol, and at 90% of acetone. Among the three fruits, guava showed to be the most abundant in polyphenols.

The main steps of polyphenol extraction are: (1) Sample grinding, (2) Extraction, (3) Filtration, (4) Concentration, and (5) Drying.

The conventional methods of extraction are maceration, infusion, percolation, Soxhlet extraction, and water-alcoholic extraction *via* fermentation. The most preferable of these methods is the Soxhlet extraction. It operates as follows: the finely ground plant sample is placed into a filter bag, the solvent is heated using a heating device and its vapors condense in a condenser. The condensed solvent drips on the plant material (bag) and extracts the polyphenols from it. After the extraction, the solvent is discarded from the polyphenol extract using vacuum evaporation. The advantages of the Soxhlet method are the high amount of extract yielded with a small amount of solvent, the low cost and the ease of conducting the process.

2.2 Novel methods of extraction

Ultrasonic extraction. During the process, the plant cell wall is broken by waves with a frequency of 10 kHz to 10 MHz. The polyphenol yield is improved up to 35% [9]. The best yield can be obtained at a 4:1 solvent/solute ratio for 200 ms extraction time at 400 W ultrasonic power. The disadvantage of this method is the degradation of anthocyanins due to the formation of OH radicals by the sonolysis of water and the high amplitude of treatment.

Microwave extraction. The solvent and plant sample are treated with microwaves and thus the phenolic compounds are released in the solvent. The advantages of the method are the low cost, smaller volume of solvent, temperature control and higher antioxidant activity of the product [11]. As the power increases, the amount of phenolic acids increases. In contrast, the extended treatment at 250 W leads to a decrease in total flavonoids.

Pressurized liquid extraction. Using this method, the temperature increases from 50 to 200°C and the pressure from 3.5 to 200 MPa. The advantage of the method is the better penetration of the solvent in the plant sample due to the increased pressure at high temperature.

Pulse electric field extraction. This method is used for pre-treatment of the plant material resulting in enhanced amount of polyphenols. Even under low electric field, the permeability of the plant cell is increased. Anthocyanins yield from grape pulp with grape skin is improved up to 28% when using the following conditions: 50 pulses and 1 Hz at 10 kV [12].

The best way to determine polyphenols remains HPLC-DAD (high-performance liquid chromatography with diode array detection). Many phenolic compounds, such as epicatechin, vanillic acid, quercetin, kaempferol, epigallocatechin, rutin, and myricetin were analyzed using this method. Structure elucidation of polyphenols can be implemented using gas chromatography-Gassmass spectrometry [13].

3. Antimicrobial activity of polyphenols from herbs and spices

Longevity of the community inhabiting Mediterranean area is due to their polyphenols rich diet. Most of Mediterranean herbs containing polyphenols were described to possess antibacterial activity against both Gram (+) and Gram (−) bacteria. Shehadi et al. [14] reported growth inhibition of *Bacillus subtilis* by *Rosemary officinalis*, *Eugenia caryofillata Menta piperita*, and *Prunus avium*. According the authors the major inhibition was observed by the extract of wild cherry (4 mg/ml) followed by cloves (1.6 mg/ml). As described, the phenolic extracts were recovered at 80°C. An important finding of the study was the extracts' strength depends on the temperature at which they were obtained. Extracts revealed at 100°C possess lower bactericidal activity (2.4 mg/ml – wild cherry and 0.6 mg/ml – cloves) compared to those eluted at 80°C.

Alamri and Moustafa [15] reported action against different bacteria of *Ricinnus communis* L.) and *Allium ampeloprasum* var. *porrum.* Both extracts were able to inhibit the growth of Gram (+) pathogens *Staphylococcus aureus*, *Streptococcus epidermidis*, and of Gram (−) *Pseudomonas aeruginosa, E. coli*, *Klebsiella pneumoniae*, *Proteus* sp., and the pronounced agent of nosocomial infections – *Acinetobacter.* Authors discussed the abundant polyphenolic content of both extracts including at least six polyphenols and the higher activity of *R. communis* (27 mm) compared to *Allium ampeloprasum* (23 mm). Strongest inhibition was observed mainly against the pathogenic Gram (−) *P. aeruginosa.*

Data about the influence of different solvents for polyphenol extraction on the antibacterial activity were published by Harfouch et al. [16], Rizwana et al. [17]. Studying the effectiveness of methanolic and ethanolic extract of *Martricaria aurea* L. the authors found that the methanolic one is more powerfull against the Gram(+) chemolitic *Streptococcus pyogenes* (23 mm) and skin pathogen *S. aureus* (21 mm) compared to ethanolic extract. The inhibition is performed *via* bacterial cell wall damage.

Besides the herbs, various medicinal plants demonstrate high polyphenol concentrations [18]. *Hypericum perforatum* L., *Origanum vulgare* L., and *Melissa officinalis*

L. along with four other medicinal plants were investigated for phenolic presence. Recently, the antibacterial activity of methanolic extracts of both flowers and leaves was proved, including wild hypericum collected from Kashmir, Himalaya.

Bactericidal effect of *Geranium macrorrhizum* was described by Ivancheva et al. [19]. This plant is known with its high polyphenol concentrations consisting mainly of flavonoids and tannins and in this particular study water alcoholic extracts of the plant were investigated. The tested extracts inhibited the growth of several pathogenic bacteria *S. aureus*., *E. coli*, *K. pneumonia,* and *P. aeruginosa*.

Another plant used as medicinal plant, spice and herb – the sanogenous parsley also showed antibacterial properties. Several classes of polyphenols persist in parsley mostly flavonoids like kaempferol, apigenin and luteolin. Average content of flavonoids is approximately 100 mg/100 g fresh weight [20]. According to Tomov et al. [21] the green minced parsley leaves demonstrated weak antibacterial effect against *E. coli* and *B. subtilis* (*see* **Table 2**).

The antimicrobial activity of lavandula against *Staphylococcus epidermidis* was described in detail by Zou et al. [22]. The antimicrobial activity of the lavandula phenolic extract was reported by a Moroccan research team [23]. The authors described that Lavandula inhibits the growth of clinical *Listeria monocytogenes* and *S. aureus* isolates from a Moroccan hospital. Georgiev et al. [24] also reported on the antioxidant activity of *Lavandula vera*.

Mihajlova et al. [25] studied the phenolic profile and the antibacterial activity of mallow (*Malvia silvestris*).

Green tea is also excessive in polyphenols and demonstratie robust antimicrobial action [26]. Green tea polyphenols consist mainly of flavonoids. Catechins are in the highest concentration of 30–40%. Four main catechins were isolated from tea: epicatechin (EC), epicatechin-3-gallate (ECG), epigallocatechin (EGC), and epigallocatechin-3-gallate (EGCG), as reported by Raygaert [27]. In green tea, EGCG is the most abundant, representing approximately 59% of the total catechins. It is important to note that during the initial steaming process of tea production the enzyme polyphenol oxidase is destroyed and thus the polyphenol content is protected. Catechins of green tea damage the bacterial cell membrane, inhibit the fatty acid synthesis of

No.	Vegetable/plant vegetative organ	Inhibition zone d on *B. subtilis* NIBMCC 8752	Inhibition zone d on *E. coli* NIBMCC 8751
1.	Parsley (leaves)	2	0
2.	Tomato (seeds)	5	0
3.	Cayenne pepper (tissue discs)	24	25
4.	Cayenne pepper (seeds)	7	11
5.	Onion brown skin (mature bulbs)	27	3
6.	Onion red skin (mature bulbs)	25	3
7.	Onion young (fresh bulbs)	0	0
8.	Garlic (mature bulbs)	7	30
9.	Garlic young (fresh bulbs)	2	0

Table 2.
Antibacterial activity of polyphenol containing vegetables (inhibition zones d in mm) against B. subtilis *and* E. coli *type strains, personal results.*

bacteria and DNA-gyrase during bacterial replication. In the same paper the inhibitory effect of green tea catechins on the binding of *Helicobacter pylori* to the Toll-like receptor-4 (TLR-4) on gastric epithelial cells was described.

4. Antimicrobial activity of polyphenols from fruits

Polyphenol abundant fruits also exhibit antibacterial action. In pomegranate juice (*Pommes granatum* L.) were obtained caffeic acid, gallic acid and epigallocatechin. Latest substance can be found as prior component also in the green tea. Divyarhree and Kunniah [28] studied its hydrochloric extract on the oral cavity inhabiting bacteria colonizing the dental plaque. Authors' research showed the extract inhibited *Porphyromonas gingvinalis*, *Prevotella intermedia,* and *Aggregatibacter actinomycetemcomitans*, all responsible for assaultive periodontitis.

Noticeable Mediterranean fruits wealthy in polyphenols are olives (*Olea europea*). The general phenolic component responsible for their health beneficial effect is hydroxytyrosol [29]. Besides its antimicrobial activity, hydroxytyrosol is a superior antioxidant and radical scavenger, which induces apoptosis and arrests the cell cycle in cancer cells. Usually hydroxytyrosol is renally evacuated. Other phenolic compounds in olives are tyrosol, glycoside oleuropein, oleocanthal, and oleacein. Hydroxytyrosol and oleuropein demonstrated antimicrobial activity [30] against ATTC bacterial strains and clinical bacterial isolates.

In vitro antibacterial (MIC and MBC) effect of *Sida alba*, a polyphenol-containing and typical for India and Arabian peninsula plant was obtained by Konate et al. [31].

The antimicrobial activity of fruit extracts was reported by Marinova et al. [32], who examined more than 20 fruits for their polyphenolic content. The analysis included *Pyrus communis*, *Malus pumila*, *Prunus domestica*, *Prunus persica*, *P. avium*, *Prunus cerasus vulgaris*, *Rubus idaeus*, *Fragraria vesca*, *Vinis vinifera*, *Cornus mas*, *Rubus fruticosus*, *Viccinum mirtilus,* and *Ficus carica*. The authors found the highest polyphenol content in *Viccinum mirtilus* (European black berry) – 670 mg GAE/100 g fresh mass followed by C. mas (dogwood) – 429, and the lowest content was demonstrated by *P. persica* (peach) – 50 GAE/100 g.

Tannins, a common polyphenolic substances in all types of red wines were reported as natural antibacterial substances as well [33].

Polyphenols from tobacco leaves extracted with 80% ethanol manifest antibacterial activity against *Escherichia coli*, *S. aureus* and *B. subtilis with inhibition zones ranging* 13, 17 and to 20 mm [34].

Polyphenols play synergistic effect when applied in combinations with antibiotics [35, 36]. Their mode of action is straight inhibition of the pathogenic microorganisms' virulence factors.

5. Antimicrobial activity of polyphenols from vegetables

Some authors report antibacterial activity of tomato. Tomato ranks second in world consumption among all vegetables [37]. Our previous research [21] showed no significant difference in the effect of raw or cooked tomato products against bacteria. The antibacterial effect was not a strong one (up to 7 mm zone). Seeds of two out of six tomato varieties slightly inhibited the growth of *B. subtilis* and showed no antibacterial activity against

E. coli. With regular consumption of tomato this activity plays a preventive role against bacterial infections. The inhibition zones were 4–7 mm on agar plates. These results coincided with the results obtained by Unnisa et al. [38], who reported low antibacterial activity of tomato fruit against *E. coli*. The elucidation of the low antibacterial activity of tomatoes is related with the low polyphenol concentration in this vegetable. According to Marti et al. [37] the following amounts of polyphenols (mg/100 g fresh weight) were obtained from tomatoes: naringine chalcone (0.9–18.2), rutin (0.5–4.5), quercetin (0.7–4.4), chlorogenic acid (1.4–3.3), naringenin (0–1.3), kaempferol-3 rutinoside (0–0.8), *p*-coumaric acid (0.2–0.5), ferulic acid (0.2–0.5) and kaempferol (0–0.2).

Onions are vegetables with strong antimicrobial activity. The antibacterial activity of garlic (*Allium sativa*) was studied by Chen et al. [39], Tomov et al. [21] and Obied et al. [40]. As reported by Tomov et al. [21], *Allium sativa* possesses pronounced antibacterial activity against Gram (−) *E. coli* (30 mm), and lower activity towards Gram (+) *B. subtilis* (7 mm). However, *Allium cepa*, also rich in polyphenols, showed lower antibacterial activity in comparison to garlic [21] against the tested strains.

Ramos [41] described *A. cepa* extracts to be more effective against Gram (+) microorganisms, while Gram (−) bacteria were reported to be less susceptible. He discussed the water extracts from yellow onion skin and found that even onion skin is active against Gram (−) bacteria.

An interesting finding is that the synthesis of antibacterial substances in *Allium sativum* and *A. cepa* occurs intensively in mature onion and garlic, not in the green leafy ones. Moreover, the synthesis of the antibacterial compounds continues when they are stored at room temperature (22°C) but stops at refrigerator (5–8°C).

Anthocyanins and flavonols are two flavonoids found out in *A. cepa*. Anthocyanins give the red color of some varieties. Flavonols as quercetin are responsible for the orange-brown onion skin. More than 25 different flavonols are currently recovered from the onion. One of them – quercetin was ubiquitous in all onion varieties. About 80% of the total flavonols in *A. cepa* are represented by quercetin 4′-glucoside and quercetin 3,4′-glucoside.

Cayenne pepper is remarkable with its lofty phenol content [42]. The authors supplied data that the ripening and cooking processes lead to an increase in the polyphenol concentration in 16 out of 18 studied cultivars. Chili peppers lead the ranking of antimicrobial activity, as shown by Omolo et al. [43]. Our experiments [21] on cayenne pepper fruits and seeds showed growth inhibition of *E. coli* and *B. subtilis.* Smashed pepper tissues showed no antibacterial effect, while the pepper discs demonstrated pronounced antibacterial activity against both *E. coli* (25 mm) and *B. subtilis* (24 mm). The results of Mariângela et al. [44], Koffi-Nevri et al. [45], and Nascimento et al. [46] describe similar activity of the capsicum fruit against both Gram (+) and Gram (−) bacteria. An interesting finding was that similarly to tomatoes seeds, pepper seeds exhibited inhibition on the growth of *E. coli* (11 mm) and *B. subtilis* (7 mm). Data about the antibacterial activity of honey derived polyphenols is discussed by Cianciosi et al. [47], Uthurry et al. [48] and Sateriale et al. [49]. Useful information about the richest dietary sources of polyphenols are available at the Phenol Explorer Database, Rothwell et al. [50] and Perez-Jimenez et al. [7].

6. Conclusions

Selected herbs, spices, fruits and vegetables contain high polyphenol concentrations. They show pronounced antibacterial activity acting against a plethora of pathogenic Gram(−) and Gram(+) bacteria as *P. aeruginosa*, *S. aureus*, *Streptococcus epidermidis, K. pneumoniae*, *E. coli*, *L. monocytogenes*, *Acinetobacter* sp., *Proteus* sp., *Micrococcus* sp., and *Bacillus* sp. Extraction of polyphenols is challenging and depends on the method, solvent, solvent percentage, temperature of isolation and plant sample. Plant polyphenolic extracts obtained at 80°C possess higher antibacterial activity compared to those extracted at 100°C. Numerous herbs, spices, fruits, and vegetables rich in polyphenols are valuable sources of novel highly effective antimicrobial substances.

Acknowledgements

This work was funded by The Central Fund for Strategic Development, New Bulgarian University, Sofia, Bulgaria - Grant № 1593/21.12 2021.

Conflict of interest

The author declares no conflict of interest.

Author details

Galina Satchanska
Department of Natural Sciences, New Bulgarian University, Sofia, Bulgaria

*Address all correspondence to: gsatchanska@nbu.bg

IntechOpen

© 2022 The Author(s). Licensee IntechOpen. This chapter is distributed under the terms of the Creative Commons Attribution License (http://creativecommons.org/licenses/by/3.0), which permits unrestricted use, distribution, and reproduction in any medium, provided the original work is properly cited. (cc) BY

References

[1] Othman L, Sleiman A, Abdel-Massih R. Antimicrobial activity of polyphenols and alkaloids in Middle Eastern plants. Frontiers in Microbiology. 2019;**10**:911. DOI: 10.3389/fmicb.2019.00911

[2] De Luca V, Salim V, Atsumi S, Yu F. Mining the biodiversity of plants: A revolution in the making – A review. Science. 2012;**336**:1658-1661

[3] Wang TY, Qing LK. Bioactive flavonoids in medicinal plants: Structure, activity and biological fate. A review. Asian Journal of Pharmaceutical Sciences. 2018;**13**:12-23. DOI: 10.1016/j.ajps.2017.08.004

[4] Abbas M, Saeed F, Anjum F, Afzaal M, Tufail T, Bashir M, et al. Natural polyphenols: An overview. International Journal of Food Properties. 2017;**20**(8):1689-1699

[5] PubChem. National Library of Medicine, National Center for Biotechnology Information. Available from: https://pubchem.ncbi.nlm.nih.gov/

[6] Aschenbacher P, Zanger U. Metabolism of drugs and other xenobiotics. Weinheim, Germany: Willey-VCH Verlag und Co; 2012. p. 565

[7] Perez-Jimenez J, Neveu V, Vos F, Scalbert A. Identification of the 100 richest dietary sources of polyphenols: An application of the phenol-explorer database. European Journal of Clinical Nutrition. 2010;**64**:112-120

[8] Mustafa S, Oyouni A, Meshari M, Aljohani M, Ahmad MA. Polyphenols more than an antioxidant: Role and scope. Journal of Pure and Applied Microbiology. 2020;**14**:47-61

[9] Suwal S, Marciniak A. Technologies for the extraction, separation and purification of polyphenols – A review. Nepal Journal of Biotechnology. 2018;**6**:74-91. DOI: 10.3126/njb.v6i1.22341

[10] Alothman M, Rajeev B, Karim AA. Antioxidant capacity and phenolic content of selected tropical fruitsfrom Malaysia, extracted with different solvents. Food Chemistry. 2009;**115**:785-788

[11] Hayat K, Zhang X, Farooq U, Abbas S, Xia S, Jia C, et al. Effect of microwave treatment on phenolic content and antioxidant activity of citrus mandarin pomace. Food Chemistry. 2010;**123**:423-429. DOI: 10.1016/j.foodchem.2010.04.060

[12] López N, Puértolas E, Condón S, Álvarez I, Raso J. Effects of pulsed electric fields on the extraction of phenolic compounds during the fermentation of must of Tempranillo grapes. Innovative Food Science & Emerging Technologies. 2008;**9**:477-482. DOI: 10.1016/j.ifset.2007.11.001

[13] Tasioula-Margari M, Okogeri O. Isolation and Characterization of Virgin Olive Oil Phenolic Compounds by HPLC/UV and GC-MS. Hoboken, NJ: Wiley; 2006. DOI: 10.1111/j.1365-2621.2001.tb04597.x

[14] Shehadi M, Awada F, Oleik R, Chockr A, Hamze K, Hamdan H, et al. Comparative analysis of the anti-bacterial activity of four plant extracts. International Journal of Current Research and Academic Review. 2014;**2**:83-94

[15] Alamri S, Moustafa M. Antimicrobial properties of three medicinal plants from Saudi Arabia against some clinical

isolates of bacteria. Saudi Medical Journal. 2012;**33**:272-277

[16] Harfouch R, Mohammad R, Suliman H. Antibacterial activity of Syrian propolis extract against several strains *in vitro*. World Journal of Pharmacy and Pharmaceutical Sciences. 2017;**6**:42-46. DOI: 10.20959/wjpps20172-8162

[17] Rizwana H, Alwhibi M, Soliman D. Antimicrobial activity and chemical composition of flowers of *Matricaria aurea* a native herb of Saudi Arabia. International Journal of Pharmacology. 2016;**12**:576-586. DOI: 10.3923/ijp.2016.576.586

[18] Ivanova D, Gerova D, Chervenkov T, Yankova T. Polyphenols and antioxidant capacity of Bulgarian medicinal plant. Journal of Ethnopharmacology. 2005;**96**:145-150. DOI: 10.1016/j.jep.2004.08.033

[19] Ivancheva S, Manolova N, Serkedjieva J, Dimov V, Ivanovska N. Polyphenols from Bulgarian medicinal plants with anti-infectious activity. Basic Life Sciences. 1992;**59**:717-728. DOI: 10.1007/978-1-4615-3476-1_43

[20] Papuc C, Gheorghe V, Corina C, Predescu N, Nicorescu V, Stefan G. Plant polyphenols as antioxidant and antibacterial agents for shelf-life extension of meat and meat products: Classification, structures, sources, and action mechanisms. Comprehensive Reviews in Food Science and Food Safety. 2017;**16**:1243-1268. DOI: 10.1111/1541-4337.12298

[21] Tomov A, Doycheva A, Mladenova P, Satchanska G. Antibacterial activity of organically grown vegetables. Acta Microbiologica Bulgarica. 2015;**31**:54-60

[22] Zou M, Tao W, Ye X, Liu D. Evaluation of antimicrobial and antibiofilm properties of proanthocyanidins from Chinese bayberry (*Myrica rubra* Sieb. et Zucc.) leaves against *Staphylococcus epidermidis*. Food Science & Nutrition. 2019;**8**:139-149. DOI: 10.1002/fsn3.1283

[23] Ez Zoubi Y, Bousta D, Farah A. A Phytopharmacological review of a Mediterranean plant: *Lavandula stoechas* L. Clinical Phytoscience/Clinical Phytoscience. 2020;**6**:1-9. DOI: 10.1186/s40816-019-0142-y

[24] Georgiev M, Abrashev R, Krumova E, Demirevska K, Ilieva M, Angelova M. Rosmarinic acid and antioxidant enzyme activities in *Lavandula vera* MM cell suspension cultures: A comparative study. Applied Biochemistry and Biotechnology. 2009;**159**:415-425

[25] Mihaylova D, Popova A, Denkova R, Alexieva I, Krastanov A. *In vitro* antioxidant and antimicrobial activity of extracts of Bulgarian *Malva silvestris* L. Annuaire de l'Université de Sofia "St Kliment Ohridski". Faculte de Biologie. 2015;**100**:41-48

[26] Gopal J, Muthu M, Paul D, Kim DH, Chun S. Bactericidal activity of green tea extracts: The importance of catechin containing nano particles. Scientific Reports. 2016;**28**:19710. DOI: 10.1038/srep19710

[27] Raygaert W. The antimicrobial possibilities of green tea. Frontiers in Microbiology. 2014;**5**:434. DOI: 10.3389/fmicb.2014.00434

[28] Divyashree P, Kunnaiah R. *Punica granatum*: A review on its potential role in treating periodontal disease. Journal of Indian Society of Periodontology. 2014;**18**:428-432

[29] Gorzynik-Debicka M, Przychodzen P, Cappello F,

Kuban-Jankowska A, Gammazza A, Knap N, et al. Potential health benefits of olive oil and plant polyphenols – A Review. Journal of Molecular Sciences. 2018;**19**:686. DOI: 10.3390/ijms19030686

[30] Tuck K, Hayball P. Major phenolic compounds in olive oil: Metabolism and health effects. The Journal of Nutritional Biochemistry. 2002;**13**:636-644. DOI: 10.1016/S0955-2863(02)00229-2

[31] Konaté K, Hilou A, Mavoungou J, Lepengué A, Souza A, Barro N, et al. Antimicrobial activity of polyphenol-rich fractions from *Sida alba* L. (*Malvaceae*) against co-trimoxazol-resistant bacteria strains. Annals of Clinical Microbiology and Antimicrobials. 2012;**11**:1-5

[32] Marinova D, Ribarova F, Atanasova M. Total phenolics and total flavonoids in Bulgarian fruits. Journal of Chemical Technology and Metallurgy. 2005;**40**:255-260

[33] Vilhelmova-Ilieva N, Galabov AS. Tannins as natural helpers in reinforcement of human health. Acta Microbiologica Bulgarica. 2017;**33**:105-110

[34] Docheva H, Popova V, Ivanova T, Nikolova V, Hristeva T, Nikolov N. Polyphenol content and antioxidant activity of aqueous/methanol extracts from different tobacco species (Nicotiana). Bulgarian Chemical Communications. 2018;**50**:553-559

[35] Daglia M. Polyphenols as antimicrobial agents. Current Opinion in Biotechnology. 2012;**23**:174-181

[36] Mojzer E, Hrnčič K, Škerget M, Knez J, Bren U. Polyphenols: Extraction methods, antioxidative action, bioavailability and anticarcinogenic effects. Molecules. 2016;**21**:2-38

[37] Martí R, Roselló S, Cebolla-Cornejo J. Tomato as a source of carotenoids and polyphenols targeted to cancer prevention – A review. Cancers. 2016;**8**:58. DOI: 10.3390/cancers8060058

[38] Unnisa N, Tabassum H, Naiman A, PoniaK. Evaluation of antibacterial activity of five selected fruits on bacterial wounds isolates. International Journal of Pharma and Bio Sciences. 2012;**3**:531-554

[39] Chen S, Shen S, Cheng S, Li P, Du J, Chang Y, et al. Evaluation of garlic cultivars for polyphenolic content and antioxidant properties. PLoS One. 2013;**8**:e79730. DOI: 10.1371/journal.pone.0079730

[40] Obied H. Biography of biophenols: Past, present and future. Functional Foods in Health and Disease. 2013;**3**:230-241. DOI: 10.31989/ffhd.v3i6.51

[41] Ramos F, Takaishi Y, Shirotori M, Kawaguchi Y, Tsuchiya K, Shibata H. Antibacterial and antioxidant activities of quercetin oxidation products from yellow onion (*Allium cepa*) skin. Journal of Agricultural and Food Chemistry. 2006;**54**:3551-3557

[42] Hamed M, Kalita D, Bartolo M, Jayanty S. Capsaicinoids, polyphenols and antioxidant activities of *Capsicum annuum*: Comparative study of the effect of ripening stage and cooking methods. Antioxidants. 2019;**8**:364. DOI: 10.3390/antiox8090364

[43] Omolo M, Wong ZZ, Mergen A, Hastings J, Le N, Reiland H, et al. Antimicrobial properties of chili peppers. Journal of Infectious Diseases and Therapy. 2014;**22**:1-8. DOI: 10.4172/2332-0877.1000145

[44] Mariângela D, Carvalho A, Ribeiro S, Da Cunha M, Beltramini L, Rodrigues R, et al. Characterization, immunolocalization

and antifungal activity of a lipid transfer protein from chili pepper (*Capsicum annuum*) seeds with novel α-amylase inhibitory properties. Physiologia Plantarum. 2011;**142**:233-246. DOI: 10.1111/j.1399-3054.2011.01464.x

[45] Koffi-Nevry R, Clément K, Nanga Z. Antibacterial activity of two bell pepper extracts: *Capsicum annuum* L. and *Capsicum frutescens*. International Journal of Food Properties. 2010;**15**:961-971. DOI: 10.1080/10942912.2010.509896

[46] Nascimento P, Nascimento T, Ramos N, Silva G, Gomes J, Falcão R, et al. Quantification, antioxidant and antimicrobial activity of phenolics isolated from different extracts of *Capsicum frutescens* (*Pimenta Malagueta*). Molecules. 2014;**19**:5434-5447. DOI: 10.3390/molecules19045434

[47] Cianciosi D, Forbes-Hernández T, Afrin S, Gasparrini M, Reboredo-Rodriguez P, Manna P, et al. Phenolic compounds in honey and their associated health benefits: A review. Molecules. 2018;**23**:2322. DOI: 10.3390/molecules23092322

[48] Uthurry C, Hevia D, Gomez-Cordoves C. Role of honey polyphenols in health. ApiProduct and ApiMedical Science. 2011;**3**:141-159

[49] Sateriale D, Facchian S, Colicchio R, Pagliuca C, Varricchio E, Paolucc M, et al. Pagliarulo C. *In vitro* synergy of polyphenolic extracts from honey, myrtle and pomegranate against oral pathogens, *S. mutans* and *R. dentocariosa*. Frontiers in Microbiology. 2020. DOI: 10.3389/fmicb.2020.01465

[50] Rothwell JA, Pérez-Jiménez J, Neveu V, Medina-Ramon A, M'Hiri N, Garcia Lobato P, et al. Phenol-explorer 3.0: A major update of the phenol-explorer database to incorporate data on the effects of food processing on polyphenol content. Database (Oxford). 2013;**2013**:bat070. DOI: 10.1093/database/bat070 [Published online: 6 Oct 2013]